Lecture Notes on Data Engineering and Communications Technologies

Volume 3

Series editor

Fatos Xhafa, Technical University of Catalonia, Barcelona, Spain
e-mail: fatos@cs.upc.edu

The aim of the book series is to present cutting edge engineering approaches to data technologies and communications. It publishes latest advances on the engineering task of building and deploying distributed, scalable and reliable data infrastructures and communication systems.
The series has a prominent applied focus on data technologies and communications with aim to promote the bridging from fundamental research on data science and networking to data engineering and communications that lead to industry products, business knowledge and standardisation.

More information about this series at http://www.springer.com/series/15362

Gregorio Martinez Perez · Krishn K. Mishra
Shailesh Tiwari · Munesh C. Trivedi
Editors

Networking Communication and Data Knowledge Engineering

Volume 1

Springer

Editors
Gregorio Martinez Perez
University of Murcia
Murcia
Spain

Krishn K. Mishra
Department of Computer Science and Engineering
Motilal Nehru National Institute of Technology Allahabad
Allahabad, Uttar Pradesh
India

Shailesh Tiwari
Department of Computer Science and Engineering
ABES Engineering College
Ghaziabad, Uttar Pradesh
India

Munesh C. Trivedi
Department of Computer Science and Engineering
ABES Engineering College
Ghaziabad, Uttar Pradesh
India

ISSN 2367-4512 ISSN 2367-4520 (electronic)
Lecture Notes on Data Engineering and Communications Technologies
ISBN 978-981-10-4584-4 ISBN 978-981-10-4585-1 (eBook)
https://doi.org/10.1007/978-981-10-4585-1

Library of Congress Control Number: 2017938124

Printed on acid-free paper

This Springer imprint is published by Springer Nature
The registered company is Springer Nature Singapore Pte Ltd.
The registered company address is: 152 Beach Road, #21-01/04 Gateway East, Singapore 189721, Singapore

Preface

The International Conference on Recent Advancement in Computer, Communication and Computational Sciences (ICRACCCS 2016) has been held at Udaipur, India, during 25–26 November 2016. The ICRACCCS 2016 has been organized and supported by the **Janardan Rai Nagar Rajasthan Vidyapeeth University, Udaipur, India.**

The ICRACCCS 2016 is an international forum for researchers, developers, and end-users to explore cutting-edge ideas and results for the problems involved in the general areas of communication, computational sciences and technology to disseminate and share novel research solutions to real-life problems that fulfill the needs of heterogeneous applications and environments, as well to identify new issues and directions for future research and development. ICRACCCS also provides an international communication platform for educational technology and scientific research for the universities and engineering field experts, and professionals.

Nowadays globalization of academic and applied research is growing with great pace. Computer, communication, and computational sciences are hot areas with a lot of thrust. Keeping this ideology in preference, **Janardan Rai Nagar Rajasthan Vidyapeeth University, Udaipur, India,** has come up with international event. ICRACCCS 2016 has a foreseen objective of enhancing the research activities at a large scale. Technical Program Committee and Advisory Board of ICRACCCS include eminent academicians, researchers and practitioners from abroad as well as from all over the nation.

Udaipur, formerly the capital of the Mewar Kingdom, is a city in the western Indian state of Rajasthan. Founded by Maharana Udai Singh II in 1559, it is set around a series of artificial lakes and is known for its lavish royal residences. City Palace, overlooking Lake Pichola, is a monumental complex of 11 palaces, courtyards and gardens, famed for its intricate peacock mosaics. Udaipur city is also referred to as the **"Venice of the East"**, the "Most Romantic City of India" and the **"Kashmir of Rajasthan"**. Udaipur the **"City of Lakes"** is one among the most romantic and most beautiful cities of India. The city of Dawn, Udaipur is a lovely land around the azure water lakes, hemmed in by the lush hills of the Aravalis.

Janardan Rai Nagar Rajasthan Vidyapeeth University, which was recognized in 1987 and established on 21st August, 1937 by Manishi Pandit Janardan Rai Nagar, an eminent educationalist, social worker and freedom fighter, with his team of dedicated worker. The University is now all set to take Higher Education to the masses that are still not getting benefits of various researches done for the socioeconomic and cultural values. The institution is not only spreading its wings of education in the country itself but has also entered into the area of international studies through academic exchange of its students and faculty members to Slippery Rock University a prestigious University of USA. This is a step forward to the academic excellence and towards providing opportunity to the students and teachers of Rajasthan Vidyapeeth.

ICRACCCS 2016 received around 300 submissions from around 662 authors of 15 different countries such as USA, Algeria, China, Saudi Arabia, and many more. Each submission has been gone through the plagiarism check. On the basis of plagiarism report, each submission was rigorously reviewed by at least two reviewers with an average of 2.07 per reviewer. Even some submissions have more than two reviews. On the basis of these reviews, 48 high quality papers were selected for publication in this proceedings volume, with an acceptance rate of 17%.

We are thankful to the speakers: Prof. Mohan Kohle, University of Agder, Norway, Dr. B.K. Panigrahi, IIT Delhi, Mr. Subhash Jagota, CEO, Global Business Solutions Pvt. Ltd., India, delegates and the authors for their participation and their interest in ICRACCCS as a platform to share their ideas and innovation. We are also thankful to the Prof. Dr. Xhafa, Fatos, Series Editor, LNDECT, Springer and Mr. Aninda Bose, Senior Editor, Hard Sciences, Springer for providing continuous guidance and support. Also, we extend our heartfelt gratitude and thanks to the reviewers and Technical Program Committee Members for showing their concern and efforts in the review process. We are indeed thankful to everyone directly or indirectly associated with the conference organizing team, leading it towards the success.

We hope you enjoy the conference proceedings and wish you all the best.

Udaipur, India

Organizing Committee
ICRACCCS 2016

Technical Program Committee

Prof. Ajay Gupta, Western Michigan University, USA
Prof. Babita Gupta, California State University, USA
Prof. Amit K.R. Chowdhury, University of California, USA
Prof. David M. Harvey, GERI, UK
Prof. Ajith Abraham, Director, MIR Labs.
Prof. Madjid Merabti, Liverpool John Moores University, UK
Dr. Nesimi Ertugrual, University of Adelaide, Australia
Prof. Ian L. Freeston, University of Sheffield, UK
Prof. Witold Kinsner, University of Manitova, Canada
Prof. Anup Kumar, M.I.N.D.S., University of Louisville
Prof. Sanjiv Kumar Bhatia, University of Missouri, St. Louis
Prof. Prabhat Kumar Mahanti, University of New Brunswick, Canada
Prof. Ashok De, Director NIT Patna
Prof. Kuldip Singh, IIT Roorkee
Prof. A.K. Tiwari, IIT, BHU, Varanasi
Dr. Vivek Singh, BHU, India
Prof. Abdul Quaiyum Ansari, Jamia Millia Islamia, New Delhi, India
Prof. Aditya Trivedi, ABV-IIITM Gwalior, India
Prof. Ajay Kakkar, Thapar University, Patiala, India
Prof. Bharat Bhaskar, IIM Lucknow, India
Prof. Edward David Moreno, Federal University of Sergipe, Brazil
Prof. Evangelos Kranakis, Carleton University, Canada
Prof. Filipe Miguel Lopes Meneses, University of Minho, Portugal
Prof. Giovanni Manassero Junior, Universidade de São Paulo, Brazil
Prof. Gregorio Martinez, University of Murcia, Spain
Prof. Pabitra Mitra, Indian Institute of Technology Kharagpur, India
Prof. Joberto Martins, Salvador University-UNIFACS
Prof. K. Mustafa, Jamia Millia Islamia, New Delhi, India
Prof. M.M. Sufyan Beg, Jamia Millia Islamia, New Delhi, India
Prof. Jitendra Agrawal, Rajiv Gandhi Proudyogiki Vishwavidyalaya, Bhopal, MP, India

Prof. Rajesh Baliram Ingle, PICT, University of Pune, India
Prof. Romulo Alexander Ellery de Alencar, University of Fortaliza, Brazil
Prof. Youssef Fakhri, Université Ibn Tofail, Faculté des Sciences
Dr. Abanish Singh, Bioinformatics Scientist, USA
Dr. Abbas Cheddad, (UCMM), Umeå Universitet, Umeå, Sweden
Dr. Abraham T. Mathew, NIT, Calicut, Kerala, India
Dr. Adam Scmidit, Poznan University of Technology, Poland
Dr. Agostinho L S Castro, Federal University of Para, Brazil
Prof. Goo-Rak Kwon Chosun University, Republic of Korea
Dr. Adam Scmidit, Poznan University of Technology, Poland
Prof. Nishant Doshi, S V National Institute of Technology, Surat, India
Prof. Gautam Sanyal, NIT Durgapur, India
Dr. Agostinho L.S. Castro, Federal University of Para, Brazil
Dr. Alberto Yúfera, Seville Microelectronics Institute, IMSE-CNM
Dr. Alok Chakrabarty, IIIT Bhubaneswar, India
Dr. Anastasios Tefas, Aristotle University of Thessaloniki, Greece
Dr. Anirban Sarkar, NIT-Durgapur, India
Dr. Anjali Sardana, IIIT Roorkee, Uttarakhand, India
Dr. Ariffin Abdul Mutalib, Universiti Utara Malaysia, Malaysia
Dr. Ashok Kumar Das, IIIT Hyderabad, India
Dr. Ashutosh Saxena, Infosys Technologies Ltd., India
Dr. Balasubramanian Raman, IIT Roorkee, India
Dr. Benahmed Khelifa, Liverpool John Moores University, UK
Dr. Björn Schuller, Technical University of Munich, Germany
Dr. Carole Bassil, Lebanese University, Lebanon
Dr. M.A. Chao, Hong Kong Polytechnic University, Hong Kong
Dr. Chi-Un Lei, University of Hong Kong, Hong Kong
Dr. Ching-Hao Mao, Institute for Information Industry, Taiwan
Dr. Chung-Hua Chu, National Taichung Institute of Technology, Taiwan
Dr. Chunye Gong, National University of Defense Technology, Taiwan
Dr. Cristina Olaverri Monreal, Instituto de Telecomunicacoes, Portugal
Dr. Chittaranjan Hota, BITS Hyderabad, India
Dr. D. Juan Carlos González Moreno, University of Vigo, Spain
Dr. Danda B. Rawat, Old Dominion University, USA
Dr. Davide Ariu, University of Cagliari, Italy
Dr. Dimiter G. Velev, University of National and World Economy, Europe
Dr. D. S. Yadav, Director, GEC, Banda, India
Dr. Darius M. Dziuda, Central Connecticut State University, USA
Dr. Dimitrios Koukopoulos, University of Western Greece, Greece
Dr. Durga Prasad Mohapatra, NIT-Rourkela, India
Dr. Eric Renault, Institut Telecom, France
Dr. Felipe RudgeBarbosa, University of Campinas, Brasil
Dr. Fermín Galán Márquez, Telefónica I+D, Spain
Dr. Fernando Zacarias Flores, Autonomous University of Puebla, Mexico
Dr. Fuu-Cheng Jiang, Tunghai University, Taiwan

Prof. Aniello Castiglione, University of Salerno, Italy
Dr. Geng Yang, NUPT, Nanjing, PR of China
Dr. Gadadhar Sahoo, BIT-Mesra, India
Prof. Ashokk Das, IIIT, Hyderabad, India
Dr. Gang Wang, Hefei University of Technology, China
Dr. Gerard Damm, Alcatel-Lucent, USA
Prof. Liang Gu, Yale University, New Haven, CT, USA
Prof. K.K. Pattanaik, ABV-IIITM, Gwalior, India
Dr. Germano Lambert-Torres, Itajuba Federal University, Brazil
Dr. Guang Jin, Intelligent Automation, Inc., USA
Dr. Hardi Hungar, Carl von Ossietzky University Oldenburg, Germany
Dr. Hongbo Zhou, Southern Illinois University Carbondale, USA
Dr. Huei-Ru Tseng, Industrial Technology Research Institute, Taiwan
Dr. Hussein Attia, University of Waterloo, Canada
Prof. Hong-Jie Dai, Taipei Medical University, Taiwan
Prof. Edward David, UFS—Federal University of Sergipe, Brazil
Dr. Ivan Saraiva Silva, Federal University of Piauí, Brazil
Dr. Luigi Cerulo, University of Sannio, Italy
Dr. J. Emerson Raja, Engineering and Technology of Multimedia University, Malaysia
Dr. J. Satheesh Kumar, Bharathiar University, Coimbatore, India
Dr. Jacobijn Sandberg, University of Amsterdam, Netherland
Dr. Jagannath V. Aghav, College of Engineering Pune, India
Dr. Mathieu Jaume, LIP6 UPMC, France
Dr. Jen-Jee Chen, National University of Tainan, Taiwan
Dr. Jitender Kumar Chhabra, NIT-Kurukshetra, India
Dr. John Karamitsos, Tokk Communications, Canada
Dr. Jose M. Alcaraz Calero, University of the West of Scotland, UK
Dr. K.K. Shukla, IT-BHU, India
Dr. K.R. Pardusani, Maulana Azad NIT, Bhopal, India
Dr. Kapil Kumar Gupta, Accenture, Australia
Dr. Kuan-Wei Lee, I-Shou University, Taiwan
Dr. Lalit Awasthi, NIT Hamirpur, India
Dr. Maninder Singh, Thapar University, Patiala, India
Dr. Mehul S. Raval, DA-IICT, Gujarat, India
Dr. Michael McGuire, University of Victoria, Canada
Dr. Mohamed Naouai, University Tunis El Manar and University of Strasbourg, Tunisia
Dr. Nasimuddin, Institute for Infocomm Research, Singapore
Dr. Olga C. Santos, aDeNu Research Group, UNED, Spain
Dr. Pramod Kumar Singh, ABV-IIITM Gwalior, India
Dr. Prasanta K. Jana, IIT, Dhanbad, India
Dr. Preetam Ghosh Virginia Commonwealth University, USA
Dr. Rabeb Mizouni, (KUSTAR), Abu Dhabi, UAE
Dr. Rahul Khanna, Intel Corporation, USA

Dr. Rajeev Srivastava, CSE, ITBHU, India
Dr. Rajesh Kumar, MNIT, Jaipur, India
Dr. Rajesh Bodade, MCT, Mhow, India
Dr. Rajesh Kumar, MNIT, Jaipur, India
Dr. Ranjit Roy, SVNIT, Surat, Gujarat, India
Dr. Robert Koch, Bundeswehr University München, Germany
Dr. Ricardo J. Rodriguez, Nova Southeastern University, USA
Dr. Ruggero Donida Labati, Università degli Studi di Milano, Italy
Dr. Rustem Popa, University "Dunarea de Jos" in Galati, Romania
Dr. Shailesh Ramchandra Sathe, VNIT Nagpur, India
Dr. Sanjiv K. Bhatia, University of Missouri—St. Louis, USA
Dr. Sanjeev Gupta, DA-IICT, Gujarat, India
Dr. S. Selvakumar, National Institute of Technology, Tamil Nadu, India
Dr. Saurabh Chaudhury, NIT Silchar, Assam, India
Dr. Shijo. M. Joseph, Kannur University, Kerala
Dr. Sim Hiew Moi, University Technology of Malaysia
Dr. Syed Mohammed Shamsul Islam, The University of Western Australia
Dr. Trapti Jain, IIT Mandi, India
Dr. Tilak Thakur, PEC, Chandighar, India
Dr. Vikram Goyal, IIIT Delhi, India
Dr. Vinaya Mahesh Sawant, D. J. Sanghvi College of Engineering, India
Dr. Vanitha Rani Rentapalli, VITS Andhra Pradesh, India
Dr. Victor Govindaswamy, Texas A&M University-Texarkana, USA
Dr. Victor Hinostroza, Universidad Autónoma de Ciudad Juárez, Mexico
Dr. Vidyasagar Potdar, Curtin University of Technology, Australia
Dr. Vijaykumar Chakka, DAIICT, Gandhinagar, India
Dr. Yong Wang, School of IS & E, Central South University, China
Dr. Yu Yuan, Samsung Information Systems America—San Jose, CA, USA
Eng. Angelos Lazaris, University of Southern California, USA
Mr. Hrvoje Belani, University of Zagreb, Croatia
Mr. Huan Song, SuperMicro Computer, Inc., San Jose, USA
Mr. K.K. Patnaik, IIITM, Gwalior, India
Dr. S.S. Sarangdevot, Vice Chancellor, JRN Rajasthan Vidyapeeth University, Udaipur
Dr. N.N. Jani, KSV University Gandhinagar, India
Dr. Ashok K. Patel, North Gujarat University, Patan, Gujarat, India
Dr. Awadhesh Gupta, IMS, Ghaziabad, India
Dr. Dilip Sharma, GLA University, Mathura, India
Dr. Li Jiyun, Donghua Univesity, Shanghai, China
Dr. Lingfeng Wang, University of Toledo, USA
Dr. Valentina E. Balas, Aurel Vlaicu University of Arad, Romania
Dr. Vinay Rishiwal, MJP Rohilkhand University, Bareilly, India
Dr. Vishal Bhatnagar, Ambedkar Institute of Technology, New Delhi, India
Dr. Tarun Shrimali, Sunrise Group of Institutions, Udaipur, India
Dr. Atul Patel, CU Shah University, Vadhwan, Gujarat, India

Dr. P.V. Virparia, Sardar Patel University, VV Nagar, India
Dr. D.B. Choksi, Sardar Patel University, VV Nagar, India
Dr. Ashish N. Jani, KSV University Gandhi Nagar, India
Dr. Sanjay M. Shah, KSV University Gandhi Nagar, India
Dr. Vijay M. Chavda, KSV University Gandhi Nagar, India
Dr. B.S. Agarwal, KIT Kalol, India
Dr. Apurv Desai, South Gujarat University, Surat, India
Dr. Chitra Dhawale, Nagpur, India
Dr. Bikas Kumar, Pune, India
Dr. Nidhi Divecha, Gandhi Nagar, India
Dr. Jay Kumar Patel, Gandhi Nagar, India
Dr. Jatin Shah, Gandhi Nagar, India
Dr. Kamaljit I. Lakhtaria, Auro University, Surat, India
Dr. B.S. Deovra, B.N. College, Udaipur, India
Dr. Ashok Jain, Maharaja College of Engg, Udaipur, India
Dr. Bharat Singh, JRN Rajasthan Vidyapeeth University, Udaipur, India
Dr. S.K. Sharma, Pacific University Udaipur, India
Dr. Naresh Trivedi, Ideal Institute of Technology, Ghaziabad, India
Dr. Akheela Khanum, Integral University Lucknow, India
Dr. R.S. Bajpai, Ram Swaroop Memorial University, Lucknow, India
Dr. Manish Shrimali, JRN Rajasthan Vidyapeeth University, Udaipur, India
Dr. Ravi Gulati, South Gujarat University, Surat, India
Dr. Atul Gosai, Saurashtra Univesrity, Rajkot, India
Dr. Digvijai sinh Rathore, BBA Open University Ahmadabad, India
Dr. Vishal Goar, Govt Engg College, Bikaner, India
Dr. Neeraj Bhargava, MDS University Ajmer, India
Dr. Ritu Bhargava, Govt Womens Engg College, Ajmer, India
Dr. Rajender Singh Chhillar, MDU Rohtak, India
Dr. Dhaval R. Kathiriya, Saurashtra Univesrity, Rajkot, India
Dr. Vineet Sharma, KIET Ghaziabad, India
Dr. A.P. Shukla, KIET Ghaziabad, India
Dr. R.K. Manocha, Ghaziabad, India
Dr. Nandita Mishra, IMS Ghaziabad, India
Dr. Manisha Agarwal, IMS Ghaziabad, India
Dr. Deepika Garg, IGNOU New Delhi, India
Dr. Goutam Chakraborty, Iwate Prefectural University, Iwate Ken, Takizawa, Japan
Dr. Amit Manocha Maharaja Agrasen University, HP, India
Prof. Enrique Chirivella-Perez, University of the West of Scotland, UK
Prof. Pablo Salva Garcia, University of the West of Scotland, UK
Prof. Ricardo Marco Alaez, University of the West of Scotland, UK
Prof. Nitin Rakesh, Amity University, Noida, India
Prof. Mamta Mittal, G. B. Pant Govt. Engg. College, Delhi, India
Dr. Shashank Srivastava, MNNIT Allahabad, India
Prof. Lalit Goyal, JMI, Delhi, India
Dr. Sanjay Maurya, GLA University, Mathura, India

Prof. Alexandros Iosifidis, Tampere University of Technology, Finland
Prof. Shanthi Makka, JRE Engg. College, Greater Noida, India
Dr. Deepak Gupta, Amity University, Noida, India
Dr. Manu Vardhan, NIT Raipur, India
Dr. Sarsij Tripathi, NIT Raipur, India
Prof. Wg Edison, HeFei University of Technology, China
Dr. Atul Bansal, GLA University, Mathura, India
Dr. Alimul Haque, V.K.S. University, Bihar, India
Prof. Simhiew Moi, Universiti Teknologi Malaysia
Prof. Rustem Popa, "Dunarea de Jos" University of Galati, Romania
Prof. Vinod Kumar, IIT Roorkee, India
Prof. Christos Bouras, University of Patras and RACTI, Greece
Prof. Devesh Jinwala, SVNIT Surat, India
Prof. Germano Lambert Torres, PS Solutions, Brazil
Prof. Byoungho Kim, Broadcom Corp., USA
Prof. Aditya Khamparia, LPU, Punjab, India

About the Book

Data science, data engineering and knowledge engineering requires networking and communication as a backbone and have a wide scope of implementation in engineering sciences. Keeping this ideology in preference, this book includes the insights that reflect the advances in these fields from upcoming researchers and leading academicians across the globe. It contains high-quality peer-reviewed papers of 'International Conference on Recent Advancement in Computer, Communication and Computational Sciences (ICRACCCS 2016)', held at Janardan Rai Nagar Rajasthan Vidyapeeth University, Udaipur, India, during 25–26 November 2016. The volume covers a variety of topics such as *Advanced Communication Networks, Artificial Intelligence and Evolutionary Algorithms, Advanced Software Engineering and Cloud Computing, Image Processing and Computer Vision, and Security*. The book will help the perspective readers from computer industry and academia to drive the advances of next-generation communication and computational technology and shape them into real life applications.

Contents

About the Editors

Gregorio Martinez Perez is Full Professor in the Department of Information and Communication Engineering of the University of Murcia, Murcia, Spain. His research interests include security and management of distributed communication networks. He received Ph.D. degree in Computer Science from the University of Murcia.

Krishn K. Mishra is currently working as a Visiting Faculty, Department of Mathematics and Computer Science, University of Missouri, St. Louis, USA. He is an alumnus of Motilal Nehru National Institute of Technology Allahabad, India which is also his base working institute. His primary area of research includes evolutionary algorithms, optimization techniques and design, and analysis of algorithms. He has also published more than 50 publications in International Journals and in Proceedings of Internal Conferences of repute. He is serving as a program committee member of several conferences and also editing Scopus and SCI-indexed journals. He has 15 years of teaching and research experience during which he made all his efforts to bridge the gaps between teaching and research.

Shailesh Tiwari currently works as Professor in Computer Science and Engineering Department, ABES Engineering College, Ghaziabad, India. He is also administrative head of the department. He is an alumnus of Motilal Nehru National Institute of Technology Allahabad, India. He has more than 15 years of experience in teaching, research and academic administration. His primary areas of research are software testing, implementation of optimization algorithms and machine learning techniques in software engineering. He has published more than 40 publications in International Journals and in Proceedings of International Conferences of repute. He is also serving as a program committee member of several conferences and also editing Scopus and E-SCI-indexed journals. He has organized several international conferences under the banner of IEEE and Springer. He is a Senior Member of IEEE, member of IEEE Computer Society and Executive Committee member of IEEE Uttar Pradesh section. He is a member of reviewer and editorial board of several international journals and conferences.

Munesh C. Trivedi currently works as Professor in Computer Science and Engineering Department, ABES Engineering College, Ghaziabad, India. He has rich experience in teaching the undergraduate and postgraduate classes. He has published 20 textbooks and 80 research publications in different International Journals and Proceedings of International Conferences of repute. He has received Young Scientist Visiting Fellowship and numerous awards from different national as well as international forums. He has organized several international conferences technically sponsored by IEEE, ACM and Springer. He has delivered numerous invited and plenary conference talks throughout the country and chaired technical sessions in international and national conferences in India. He is on the review panel of IEEE Computer Society, International Journal of Network Security, Pattern Recognition Letter and Computer & Education (Elsevier's Journal). He is Executive Committee Member of IEEE UP Section, IEEE India Council and also IEEE Asia Pacific Region 10. He is an active member of IEEE Computer Society, International Association of Computer Science and Information Technology, Computer Society of India, International Association of Engineers, and life member of ISTE.

Part I
Advanced Communication Networks

Detection and Prevention of DDoS Attacks in Wireless Sensor Networks

Shivam Dhuria and Monika Sachdeva

Abstract Wireless Sensor Networks are emerging at a great pace due to their cost effective solutions for the sensitive and remote applications like military, medical and environmental applications (Chatterjee and Pandey in Int J Sci Eng Res 5, 2014) [1]. But due to limited range, memory, processing and power supply, gathering of important remote data from wireless sensors is really challenging. The use of ad hoc network and radio waves for data transmission has also increased the chance for attackers to attack on such networks. Various schemes have been proposed in the past to fight against the attacks in WSN (Sahu and Pandey in Mod Educ Comput Sci 1:65–71, 2014) [2], (Paul et al. in Wireless Sensor Network Security: A Survey. Auerbach Publications, Florida, 2006) [3]. In this paper two methods have been introduced, one is light weight two way authentication method that will prevent majority of attacks in WSN and other is traffic analysis based data filtering method that will detect and prevent DDoS attacks in WSN. The results have been verified using the Network Simulator 2 (NS2) on several performance metrics i.e. throughput, delay, lost packets, energy consumption and PDR.

Keywords Data filtration · Authentication · Network performance Simulator · Data rate · Sensor · Data packets · Data traffic · Base station Node

1 Introduction

WSN is a network of spatially distributed autonomous sensors who cooperatively pass their data to the main location called base station. In WSN the nodes passes their data in number of hops to the base station to reduce their power consumption

S. Dhuria (✉)
CSE, SBSSTC, Ferozepur, India
e-mail: sdhuria13@gmail.com

M. Sachdeva
CSE, IKGPTU, Main Campus, Kapurthala, India

G.M. Perez et al. (eds.), *Networking Communication and Data Knowledge Engineering*, Lecture Notes on Data Engineering and Communications Technologies 3, https://doi.org/10.1007/978-981-10-4585-1_1

and increase the life span of their batteries because the power required to send data between two nodes is directly proportional to the distance between them. As most of the WSN are deployed in remote areas where humans can't reach, and once deployed their batteries are very hard to replace, so it becomes a very challenging task to produce a battery powered network that is scalable and also has long life span. Another challenge of WSN is its wireless nature of communication. As the important sensed data propagate through wireless channel in form of radio frequencies, it becomes vital for the network administrator to secure the network information from the attackers and intruders. Implementation of various security algorithms like Authentication, encryption, jamming detection etc. keeping in mind the limited power supply is one of the important concern in WSN.

WSN operate in an environment where there is lesser or no user intervention and also where there is a risk of node capturing, so it becomes important for the sensor node to detect any attack and take some corrective measures on its own [4] in order to prevent the attacker from draining all of its battery. Generally attackers in WSN can be described in two ways [5] i.e. (a) Outsider Attackers who have little or no secret information about the network. (b) Insider Attackers who have all the secret information of network and are legitimate part of the network. Outsider attackers attack the WSN by (a) Passive attack via connecting to appropriate frequency and getting all important information without causing any harm to data. (b) Active attacks via jamming the network [6], collision attacks, replay attack, authentication attacks etc. where as Insider attackers can attack the network more severely as being a legitimate user of the network it can directly communicate with any reachable node of the network and can also transmit wrong information in the network resulting in severe attacks like black hole and sink hole attacks.

Denial of service is a type of attack in which attacker attacks by overwhelming the legitimate node with continuous authentication packets or continuous stream of data so making its services unavailable to other legitimate nodes. All the legitimate node resources become busy in replying to authentication packets or receiving data packets from attacker thus leading to drainage of its battery soon [7]. DDoS is same as DoS but with more than one attackers involved resulting in more severe attack. In the research paper we have tackled both insider and outsider DDoS attacks by following authentication and data filtration methods.

2 Related Work

(1) Sahu and Pandey [8] have followed a probabilistic approach to mitigate the DDoS attack by calculating the receiving rate of intermediate nodes. It slashes the sending rate if receiving rate is found abnormal. Using PPFS mechanism, packet flow rate is reduced which gradually reduces the flooding of packets from attackers. The proposed mechanism however does not eliminate the attack

completely but is the first step to handle the attack, buffer mechanism to discard the packets from attackers after certain limit has been proposed for future work.

(2) Kaushal and Sahni [9] have followed a scheme to detect DDoS attacks in early stages in order to prevent the resources from getting wasted. They divide the network into grids and deploy the examiner nodes, if any node sending data at faster rate, then its PDR will be compared with the neighbor node by the examiner node as examiner node has all the information of nodes present in its grid. If PDR is abnormal then that node will be marked as malicious and the network will stop communicating with that node.

(3) Buch and Jinwala [5] have described the various types of DOS attacks and also various defense mechanisms to tackle them in WSN at different network layers. They concluded that majority of the attacks in WSN can be prevented by authentication and anti replay mechanism, other methods also exists to detect and recover from attacks but they can be defeated by some counter mechanism, so the reason to find some concrete solutions to overcome from DOS attacks in WSN.

(4) Fiore et al. [10] have explained that the network anomalies evolve in the network dynamically due to various attacks, so they purpose a restricted Boltzmann machine that will observe the different network behaviors in different situations and thus isolate the attacks like DoS, DdoS.

(5) Palmieri et al. [11] have isolate the various attacks like malicious/hostile activities, denial of service attacks, and network intrusions based upon the machine behavior learning process. The main significance is that the model can be deployed in different environments. The whole process consists of building the baseline traffic model over time and then comparing that baseline graph with the current traffic and isolating the malicious behavior in the system.

3 Proposed Model

To prevent the majority of attacks in WSN, prevention of both outsider attacks and insider attacks are important. Authentication is the best method to prevent the outsider attacks. Route request flooding is also one of the frequent attacks in WSN because attackers attack the legitimate nodes by bombarding the authentication packets in the network, in that case traffic analysis becomes important to prevent this attack. Insider attacks are also frequent in WSN because of their deployment in remote areas and lesser maintenance. Insider attacks can also be prevented by data flow analysis.

3.1 Two Way Authentication

In the research project a method is introduced under which a node will only be able to communicate with other node in the network if they know about the shared secret of network. Authentication scheme is based upon some mathematical formula which is only known to legitimate nodes in the network. For authentication, each node generate some random seed and generate query and answer hash key based upon that seed and send it to other nodes before they start actual communication. When other nodes verify this secret information successfully, they also authenticate themselves vice versa.

```
1) Node X sends REQ (11) + RREQ to Node y.
2) Node Y generates a random seed and send [REP(22) + SEED
(Y) + HASH(KEY(SEED(Y)))] to Node X.
3) Node X authenticate Node Y by generating the hash key at its
own end and matching the same with the hash came from Node Y.
4) If Auth == Success then Node X sends [REP(22) + SEED(X) +
HASH(KEY(SEED(X)))] to Node Y.
5) Node Y authenticate Node X same as STEP(3).
6) If Auth == Success then Node Y update its routing table and
start receiving data from Node X.
7) If TIMER(NODE X) expires, Repeat STEP(2) to STEP(6).
```

FUNCTION KEY(SEED)

```
1) KEY = SIN(SEED) * COS(SEED) * LOG10(SEED)
2) KEY = -KEY IF KEY < 0
3) KEY = ROUND(KEY*1000)
4) RETURN KEY
```

3.2 Data Filtration Method

For traffic analysis, each node in the network checks the input data coming from other nodes of the network. The individual data volumes are then scanned against threshold value to find the traffic abnormalities. As every node in WSN has data transfer rate in a limited range, if any node is found with data transmission range exceeding that particular range then that node will be considered as attacker node and all the communication from that particular node gets blocked by all other nodes in the network.

The methodology of above work is divided into three parts:

I Algorithm Steps
II Implementation using NS2
III Results and Discussions.

3.3 Algorithm Steps

```
1) Node X send some data to Node Y.
2) If Node X == Authenticated Node and Node Y == destination
node, then Node Y check the data rate of the source node.
3) If data rate(s) > Threshold Limit, then communication from
that data source gets blocked.
4) Node Y informs the neighbor nodes of the source node about
this attack.
5) If Node X ! = Authenticated Node, Node Y checks its
requesting rate or data rate, if it is exceeding the threshold
limit, then Node X gets blocked in this case.
6) Else Data from Node X is accepted and forwarded to desti-
nation or the base station.
```

4 Implementation Using NS-2

Above algorithms has been implemented using ns2 simulator [12] with following implementation details (Table 1):

Following are the node communication details used under the simulation (Table 2).

In the above table, N represents the normal nodes, I-Attackers represent the internal attackers in the network and E-Attackers represent the external attackers in the network. Normal nodes send their data to the gateway nodes 10, 20 and 30 respectively. In this DDoS attack simulation both external attackers and internal

Table 1 NS2 implementation details

NS2 version	2.35	Transmission radius	250 m
Energy model	Energy model	Initial energy	1000 J
Ideal power	0.2	Receiving power	0.3
Transmission power	0.4	Sleep power	0.01
Transition power	0.1	Number of nodes	33
Simulation time	100 s	Simulation modes	Normal, Attack, Control
Gateway nodes	10, 20, 30 (Blue)	Attacker nodes	7, 9, 18, 25, 28, 31, 32, 33
Normal data rate/attack rate	1/40 kbps	Time of attack	From 30 to 70 s
Routing protocol	AODV [13,14]		

Table 2 NS2 node communication details

Sender node	Receiver node	Data rate (Kbps)
1, 2, 3, 4, 5, 6, 8 (N) (Green)	10 (Blue)	1
11, 12, 13, 14, 15, 16, 17, 19 (N) (Green)	20 (Blue)	1
21, 22, 23, 24, 26, 27, 29 (N) (Green)	30 (Blue)	1
7, 9 (I-Attackers) (Red)	10 (Blue)	40
18 (I-Attackers) (Red)	20 (Blue)	40
25, 28 (I-Attackers) (Red)	30 (Blue)	40
31, 32, 33 (E-Attackers) (Yellow)	6, 19, 30 (respectively)	40

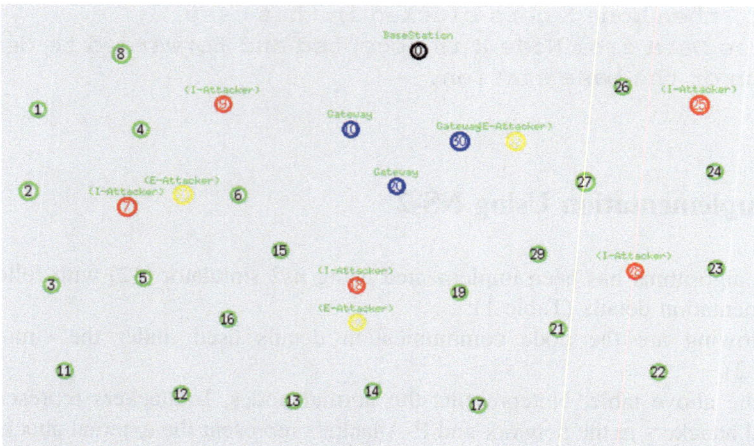

Fig. 1 Network animator screen-shot

attackers try to bombard the packets either to the normal nodes or to the gateway nodes between 30 to 70 s. Nodes which are attacked by the attackers get busy in receiving the packets from the attacker and the normal functioning of the network get degraded.

Under the simulation three modes of operation have been maintained to check the network performance.

(1) Normal Mode: When their is no attack and all the nodes work normally in the network.
(2) Attack Mode: When attackers attack the nodes and network is unable to handle the situation.
(3) Control Mode: When attackers attack the nodes and network has the ability to detect the attack and block the attacker nodes (Fig. 1).

5 Result Analysis

When simulation is performed under control mode and attack is done between 30 to 70 s, attackers are detected one by one and get eliminated from the system. All the simulation results under different modes of simulation operations (Normal mode, Attack mode and the Control mode) have been combined for each performance metrics. The deviations in the curves when comparing under different modes shows the effectiveness of the data filtering approach.

(1) Average Throughput Graph (Fig. 2):
(2) Average PDR Graph (Fig. 3)
(3) Lost Packets Graph (Fig. 4)
(4) Average Delay Graph (Fig. 5):
(5) Average Energy Consumption (Fig. 6)

The following table provide the values of various performance metrics under different modes at time t = 80 s i.e. just 20 s before the end of simulation (Table 3).

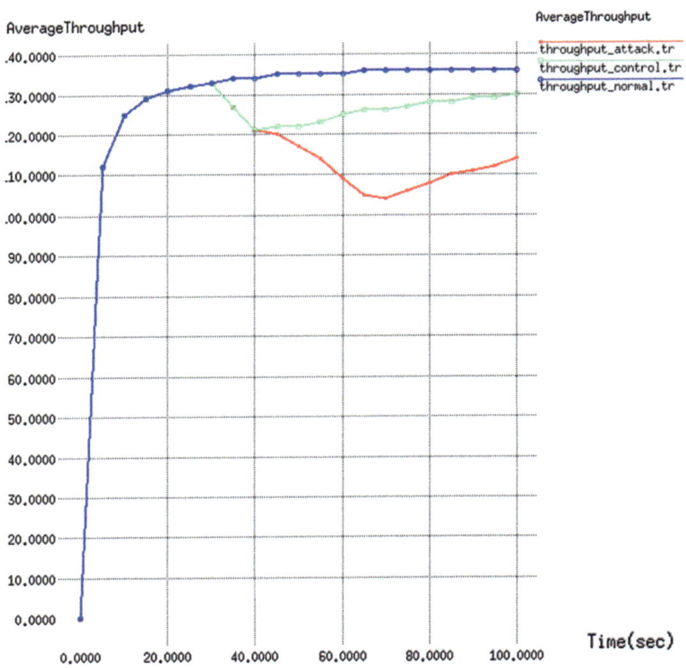

Fig. 2 Average throughput graph combined for 3 modes of operations

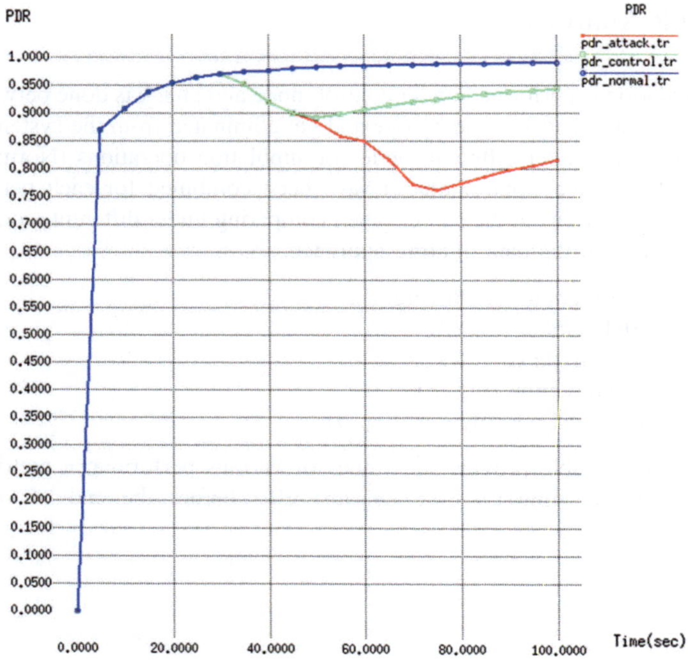

Fig. 3 Average PDR graph combined for 3 modes of operations

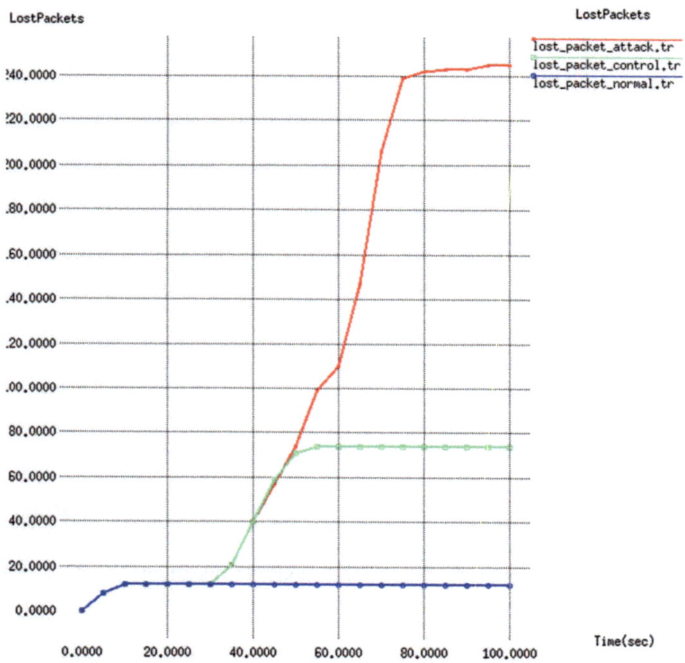

Fig. 4 Lost packets graph combined for 3 modes of operations

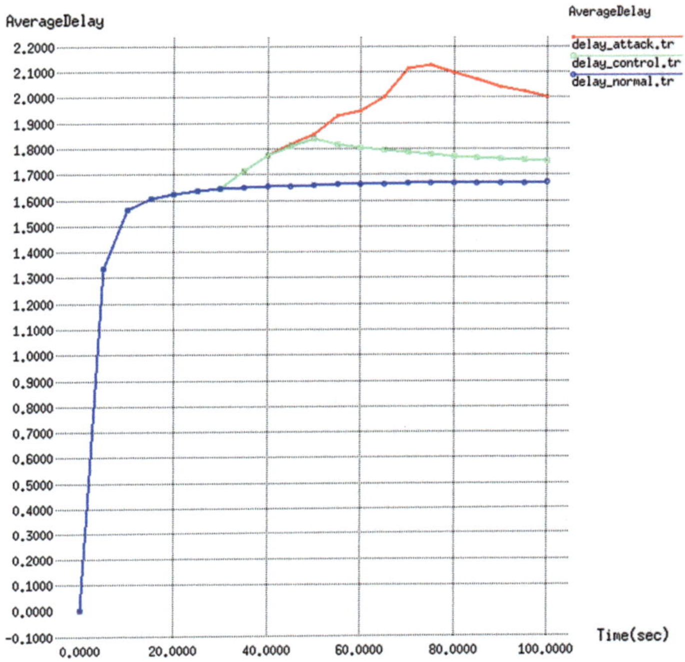

Fig. 5 Average delay graph combined for 3 modes of operations

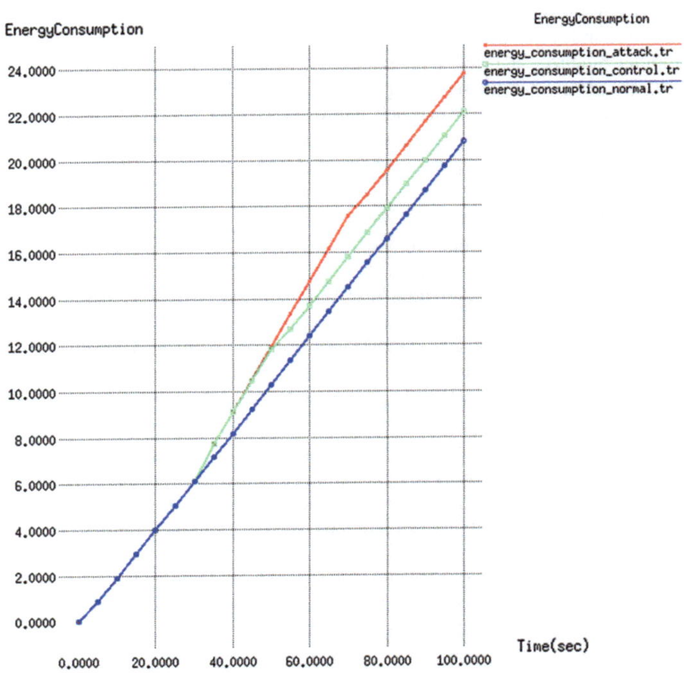

Fig. 6 Average energy consumption graph combined for 3 modes of operations

Table 3 Values found for various performance metrics at time t = 80 s

Simulator mode	Normal mode	Attack mode	Control mode
Avg. throughput	135 bytes/s	107 bytes/s	128 bytes/s
Avg. PDR	0.99	0.77	0.93
Lost packets	15	242	75
Avg. delay	1.68 s	2.1 s	1.77 s
Avg. energy	16.4 J	19.8 J	17.9 J

6 Conclusion

The authentication and data filtering approach detects and prevents the majority of DDoS attacks in WSN. As compared to the existing methods this approach is very simple and can be deployed at each node in the network. A little computation of tracking the data rates from the neighbor nodes prevents the whole drainage of battery source that can be caused by the DDoS attacks.

References

1. Ananya Chatterjee, Manjusha Pandey “Practical Applications of Wireless Sensor Network Based on Military, Environmental, Health and Home Applications: A Survey”, International Journal of Scientific & Engineering Research, Volume 5, Issue 1, January-2014.
2. S.S. Sahu, M. Pandey “Distributed Denial of Service Attacks: A Review”, Modern Education and Computer Science, 2014, 1, 65–71.
3. John Paul, Walters, Zhengqiang Liang, Weisong Shi, Vipin Chaudhary “Wireless Sensor Network Security: A Survey”, Auerbach Publications, CRC Press, 2006.
4. Marco Tiloca, Domenico De Guglielmo, Gianluca Dini and Giuseppe Anastasi “SAD-SJ: a Self-Adaptive Decentralized solution against Selective Jamming attack in Wireless Sensor Networks”, Department of Information Engineering, University of Pisa
5. Dhara Buch, D.C. Jinwala “Denial of Sevice Attacks in Wireless Sensor Networks”, Institute of Technology, Nirma Universiry Ahmedabad – 382 481, 09–11, December 2010.
6. Mahreen Shaikh, Abid. H Syed “A Survey on Jamming Attacks, Detection and Defending Strategies in Wireless Sensor Networks”, IJRET, Vol: 03 Issue: 03 | Mar 2014
7. Francesco Palmieri, Sergio Ricciardi, Ugo Fiore, Massimo Ficco, Aniello Castiglione “Energy-oriented denial of service attacks: an emerging menace for large cloud infrastructures”, The Journal of Supercomputing, May 2015, Volume 71, Issue 5, pp 1620–1641
8. S.S. Sahu, M. Pandey “A Probabilistic Packet Filtering-Based Approach for Distributed Denial of Service Attack in Wireless Sensor Network”, Springer India 2015.
9. Kanchan Kaushal, Varsha Sahni “Early Detection of DDoS Attack in WSN”, International Journal of Computer Applications (0975-8887) Volume 134 – No. 13, January 2016.
10. Ugo Fiore, Francesco Palmieri, Aniello Castiglione, Alfredo De Santisc “Network anomaly detection with the restricted Boltzmann machine”, Neurocomputing Volume 122, 25 December 2013, Pages 13–23
11. Francesco Palmieri, Ugo Fiore, Aniello Castiglione “A distributed approach to network anomaly detection based on independent component analysis”, Concurrency and Computation: Practice and Experience, 1532–0634, 2014

12 Genita Gautam, Biswaraj Sen "Design and Simulation of Wireless Sensor Network in NS2", International Journal of Computer Applications (0975 – 8887) Volume 113 – No. 16, March 2015.
13 P.S. Karadge, Dr. S. V. Sankpal "A Performance Comparison of Energy Efficient AODV Protocols in Mobile Ad hoc Networks", IJARCCE, Vol 2, Issue 1, January 2013
14 Muthana Najim Abdulleh, Salman Yussof, Hothefa Shaker Jassim "Comparative Study of Proactive, Reactive and Geographical MANET Routing Protocols", Communications and Network, 2015, 7, 125–137.

Wired LAN and Wireless LAN Attack Detection Using Signature Based and Machine Learning Tools

Jaspreet Kaur

Abstract There are various attack which is possible in the network, it may be from externally or internally. But internal attacks are more dangerous than external. So, my mainly concern upon Wireless LAN and Wired LAN attacks which occurs internally. There are various Signature based tools, IDS/IPS (Intrusion detection or prevention system) available now-a-days for detecting these types of attacks but these are not sufficient due to high false alarm rate. So, I detect these types of attacks with three ways: through Wireshark, with signature based tools (Snort and Kismet) and with machine learning tools (WEKA). In wired LAN attack, my mainly concern on PING scan or PING flood, NMAP scan (portsweep) and ARP spoofing attacks. In wireless LAN attacks, I take care of Deauthentication attack, Disassociation attack and Access point (AP) spoofing attack. Signature based tools detect these types of the attacks based on the stored signature and timing threshold. But machine learning tools take several different feature to detect these types of attacks with more accuracy and low false positive rate.

Keywords Wireless LAN · Wired LAN · Snort · WEKA
Kismet · Wireshark · Spoofing attack

1 Introduction

In the today generation attacks are very dangerous whether its happen due to the internal situations or to the external situations. But internal attacks are very difficult to detect and prevent due to the unawareness of these attacks. These attacks are done either from wired environment or from the wireless environment. In the wireless security (WLAN), my mainly concern on the 3 protections of any packet transmitted in the air: confidentiality, integrity, availability. Confidentiality and

J. Kaur (✉)
Department of Information Technology, Indira Gandhi Delhi Technical University for Women, Delhi 110006, India
e-mail: jaspreetkaur9817@gmail.com

G.M. Perez et al. (eds.), *Networking Communication and Data Knowledge Engineering*, Lecture Notes on Data Engineering and Communications Technologies 3, https://doi.org/10.1007/978-981-10-4585-1_2

integrity are mainly managed by various protocols such as: Wired Equivalent Privacy (WEP), WI-FI Protected Access (WPA) etc. But WLAN is still vulnerable from availability attacks such as DOS attacks. My mainly concern on the MAC layer DOS attacks such as Deauthentication attack, Disassociation attack and AP spoofing attack [1]. In the case of PING scan or PING flood, NMAP scan (portsweep) and ARP spoofing attacks which are to be occurred either through wirily or wirelessly are also very dangerous because PING scan and NMAP scan are the first attack which are done by the attacker for checking the vulnerable systems as PING scan tells the attacker system is up and how distant the system is as TTL, NMAP is used to find out the port and operating system vulnerability. In ARP table changes made the information leakage easily captured by the attacker as man in the middle attack. Firstly I used Wireshark for manually analysis of these attacks, then after used the signature based tools such as Snort (open source) widely used for PING scan, NMAP scan and ARP spoofing attack detection and Kismet as IDS for the Deauthentication attack, Disassociation Attack and Access point spoofing attack detection. Finally I used machine learning tool, WEKA with classification techniques such as Naive Bayes and J48 tree for detecting these types of attacks using appropriate parameters.

This paper is continue with Related Work, Proposed Approach, Setup Environment of Lab, Results and Observations and finally completed with Conclusions and Future Work.

2 Related Work

There are various papers on this topic summarized as follows: In this paper authors discussed MAC layer management frame attacks. Management frames are neither authenticated nor encrypted that's why these can be easily spoofed by the attacker and perform DOS attacks. They gave various solutions as signal print scheme, MAC Spoofing Detection. But solutions are not worked well [2]. In the 2nd paper author mainly consider on the detection of probe request frame DOS attack in 802.11 network. They used back propagation algorithm to find spoofed frames. But what if training data is corrupted [3]. Infrastructure networks based on an AP as a central node through which every communication is started, thus an AP can easily become a weak point for the entire network. They described software platform to detection of the WLAN attacks. But they said that it is not a long term solution [4]. They proved that management frames attacks can be executed by any malicious station, without being neither associated nor authenticated to the access point. APs main vulnerability is unacked frame retransmission. According to them the effective solution should reside at the firmware level [5]. Mainly three algorithm are proposed for detecting and preventing mac layer dos attack as Intrusion detector and manager, Letter envelop protocol with traffic pattern filtering, Medium access protocol spoofed detection and prevention. These 3 algorithms are implemented together at the AP providing the reliable solution and WLAN security. But it has little computational overhead [6]. In this they used IDS/IPS approach. IDS program sniff WI-FI data and did analysis. Based on the throughput of the network and Deauthentication frame in the network or at the particular client side its decide DOS attack has to be performed or not [7].

An another author suggest a tool named as the IJAM tool for performing WLAN attack. The author take some assumption such as: The attacker needs high transmit power etc. [8]. In this survey paper author suggest the various vulnerability occur within the WLAN such as Eaves dropping, Message modification, management frame attacks. Then discussed various available solution such as: Pseudo random number based authentication, Letter Envelop Protocol etc. But still DOS attack is possible [9]. In this I studied out various types of layer 2 attacks detection and their countermeasures. They mainly focused on the ARP spoofing and mac attacks and tell that Dynamic ARP inspection prevents current ARP attacks [10]. With the extension of ARP attacks they also consider Deauth attack and rouge Access point attack using some parameters [11]. In this paper they propose a layered architecture called as WISA guard. They uses the OS Fingerprinting, AP Fingerprinting, RSS Fingerprinting technique for attack detection [12]. In next paper, they uses a IDPS method in which Snort uses as IDS and for the prevention technique uses Aireplay-ng tool as sending the deauth packets to the attacker. For attacking purpose they uses the ICMP flood attack [13]. In one paper, they use the machine learning tool called as WEKA which includes the variety of data mining algorithms. But they mainly used the J48 tree and Naive Bayes algorithm for detection of ping sweeps and port sweeps attack [14].

3 Proposed Approach

My mainly concern on the detection of Wired LAN and Wireless LAN attacks using three approaches such as:

1. Manually through Wireshark tool.
2. With the help of signature based tool such as Snort and Kismet.
3. With the help of machine learning based tool such as WEKA using different algorithms. The proposed approach is shown at Fig. 1.

In Fig. 1 myself represent these 3 proposed approaches as (1), (2), (3). In the Wireshark I take the packet dump and manually inspect the various features. In signature based tools like Snort and Kismet, my mainly focus on the signature of the attacks and threshold limit of time such as 5 packets/min for Deauth or Disas packets in the alert rule option. In the machine learning tool WEKA, my mainly concern on the 2 algorithms of classification such as Naive Bayes and J48 tree for attacks detection. Here I take the various parameters for detection of these kinds of attacks as discussed later in this paper.

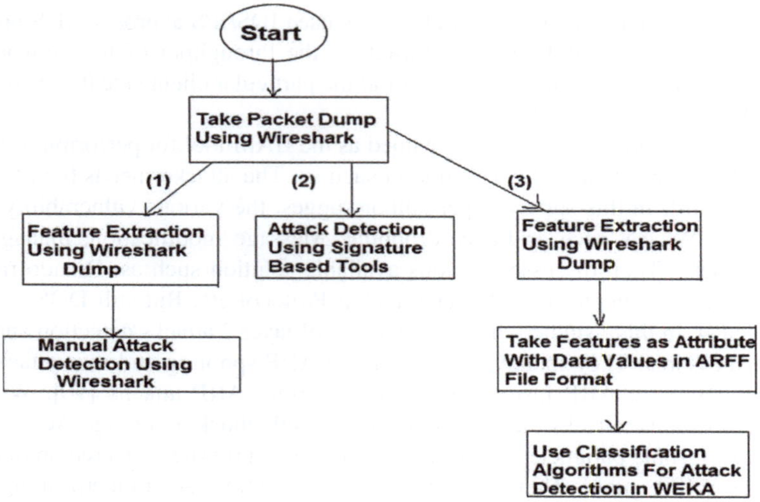

Fig. 1 Proposed approach

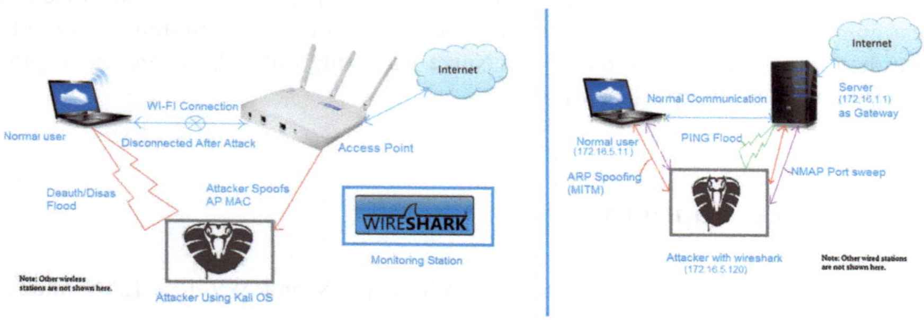

Fig. 2 Wireless LAN and wired LAN setup

4 Setup Environment of Lab

Two setups one for Wired LAN and another for Wireless LAN are mainly used for performing the attacks and analysis of packets. In each setup I capture Wireshark packet dump of 30,000 packets. These packets are firstly manually observed, secondly give in signature based tools and finally apply cross validation approach with various classification algorithms in WEKA tool. The setup is shown Fig. 2.

In the wireless LAN setup there are various clients and APs (access points) communicate to each other through WI-FI. Then attacker comes which spoofs the mac of an particular ap and tried to attempt Deauth or Disassociation flood attack to the particular client or all the clients using Kali OS and I analyze the network traffic using Wireshark in monitor mode, take the packet capture dump for attack detection with 3 proposed approaches. As per the similar way, in Wired LAN setup various clients and server are communicating normally. Then attacker comes, performs the ping flood attack to the server, NMAP (port sweep attack for TCP port 80) and ARP spoofing attack as man in the middle attack. After that monitor all the traffic using Wireshark and take the packet capture dump for attack detection using 3 proposed approaches.

5 Results and Observations

After the following attacks are taking place, detection of these attacks are to be done with the following three ways such as:

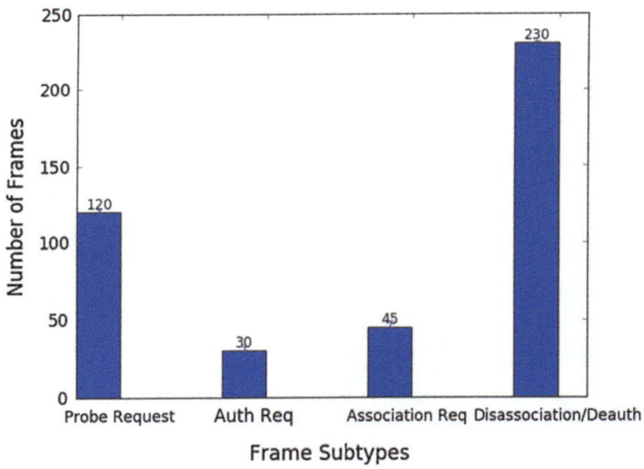

Fig. 3 Packet ratios in the network during wireless LAN attack

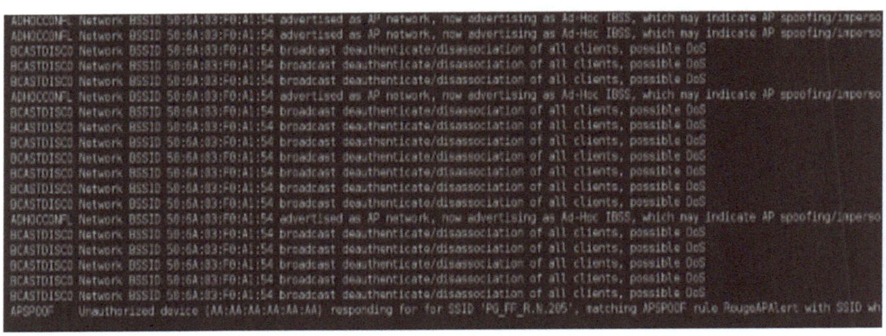

Fig. 4 Wireless LAN attack detection using Kismet

5.1 Manually Through Wireshark

When Deauth, Disas attack are taking place along with AP spoofing. Then I observe the packet capture dump using Wireshark. There is various management frames are seen as probe request, authentication, association frame and Deauthentication or disassociation frames. But the Deauth or Disas frames are exponentially increased irrespective of other frames within 2 min in the network. It is due to the attack as shown in Fig. 3.

For the ping flood attack I observe the Echo (ping) request, Echo (ping) reply information with the ICMP packet header identification number. For NMAP port sweep myself see the particular source address with a port number sending various syn packet to range of an IP address for checking port number 80 and get a response for the open ports. For ARP spoofing (mitm) attack I observe the ICMP redirect error, the mac address of the attacker and mac address of the gateway. Both the MAC address of the attacker and gateway is same after ARP spoofing attack. As the result all the information sending to the gateway by the client also captured by the attacker.

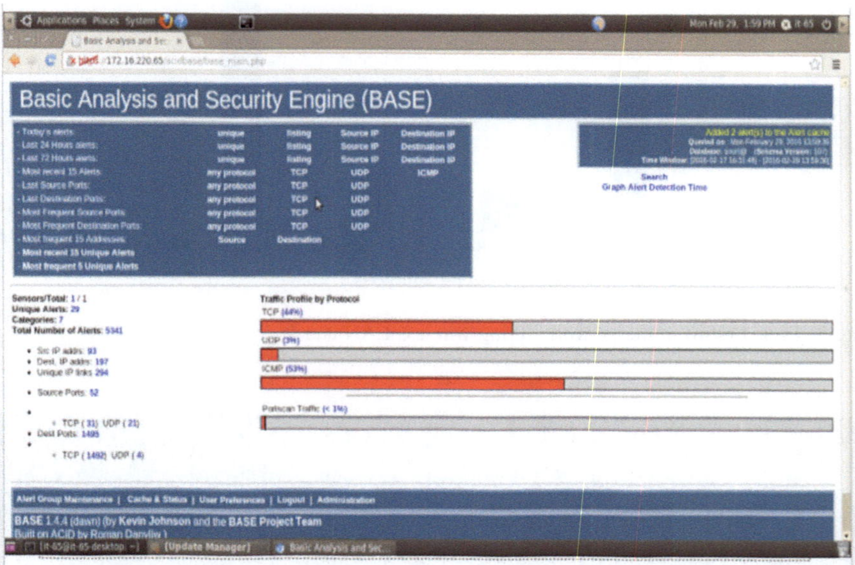

Fig. 5 Wired LAN attack detection using snort

5.2 *Using Signature Based Tool*

For Wireless LAN attack I used the tool named as Kismet. Kismet has various alert rules for detecting the DOS and AP spoofing attack. This tool detects these types of attacks based on the various signatures of these attacks and the threshold value of time such as 5 packets/min for Disassociation attack. The results are shown in Fig. 4.

For Wired LAN attack I used the tool named as Snort implemented as ACID-BASE. For PING scan this tool gives us alert as ICMP PING *NIX, ICMP PING, ICMP Echo Reply. For NMAP scan this gives alert as PORT SWEEP and finally for ARP spoofing, have the ICMP redirect host alert. In the Fig. 5 we represent the ICMP, UDP, TCP, PortScan alert.

5.3 *Using Machine Learning Tool*

In machine learning tool WEKA, my mainly focus on the two algorithms of classification as Naive Bayes and J48 tree for these types of attack detection. The parameters based on which myself defined attack is occurred or not is already given in their respective tables. These parameters are taken collectively for detecting the attacks. The Wireshark packet dump is given as a csv file in the input of the WEKA tool.

In the below WEKA table I compare the location of rouge AP and true AP, also see the secure attribute option which is false it means rouge AP does not use WPA2 protocol for packets, channel at which the original AP is worked is different from rouge AP channel and also see the various parameters differ from the original one. So, WEKA predict these packets as a yes, no for attack detection (Table 1).

Table 1 Wireless LAN attack detection using tool WEKA

Parameters considered	Predicted attack as yes	Predicted attack as no
Source MAC address	Netgear_61:ad:da	Netgear_61:ad:da
Destination MAC address	Xiaomi_7f:04:4f	Xiaomi_7f:04:4f
Range of sequence no.	0–0	712–718
Latitude of AP	28.663615	28.663405
Longitude of AP	77.233468	77.233700
Frame length	34	56
Secure	False	True
Signal strength	90	80
Channel no.	11	6
Reason code	Unspecified reason	MICFLeaving
Frame control flags	0x00	0x08
Other information	DeauthDiasas	DeauthDiasas

Table 2 PING scan attack detection using tool WEKA

Parameters considered	Predicted attack as yes	Predicted attack as no
Source IP address	172.16.5.120	172.16.1.1
Destination IP address	172.16.1.1	172.16.5.11
Frame length	74	74
Identification number for ICMP	0x0100	0x0100
Time to live	64 (for Linux)	128 (for Windows)
Packet count	10	2
Other information	PING reqreply	PING reqreply

Table 3 NMAP scan (portsweep) attack detection using tool WEKA

Parameters considered	Predicted attack as yes	Predicted attack as no
Source IP address	172.16.5.120	172.16.1.1
Destination IP address	172.16.1.1-15	172.16.5.11-25
Frame length	58	66
Source port	3128	33422
Destination port	80	80
Window size	1024	8192
Other information	SYN	SYN

In the PING scan or PING flood and NMAP portsweep attack, if administrator (172.16.1.1) scans the network for testing the IP addresses or testing the specific services is working or not, then its not an attack otherwise for rest of the users which want to perform such an action these parameters detected them as an attack. Admin (Linux sever) always send 2 PING request but attacker send any no. of ping request for performing ping flood. My server is on Linux. So, I mainly look TTL for Linux to more security purpose. For Portsweep admin always use fixed size of window and particular source port no. to scanning the services. As shown in Tables 2 and 3.

In the Table 4 I show the ARP spoofing as source address is attacker address and destination address is gateway, both having the same mac address and ICMP redirect error. So, the parameters predicted these as an attack. On the other hand both source and destination having different mac address, protocol uses as TCP rather than ICMP or UDP and working as normal user, gateway. So, its not an attack.

As everyone see that from the below graph accuracy I have met is quite sufficient with taking the parameters of mine in the machine learning tool WEKA using any of the classification algorithms Naive Bayes or J48 tree (Fig. 6).

Table 4 ARP spoofing attack detection using tool WEKA

Parameters considered	Predicted attack as yes	PREDICTED ATTACK AS NO
Source IP address	172.16.5.120	172.16.5.120
Destination IP address	172.16.1.1	172.16.1.1
MAC address of source	00:26:b9:22:4b:8a	00:26:b9:22:4b:8a
MAC address of destination	00:26:b9:22:4b:8a	a0:48:1c:a5:b1:9e
Protocol	ICMP, UDP	TCP
Other information	Redirect (ICMP error)	Normal data

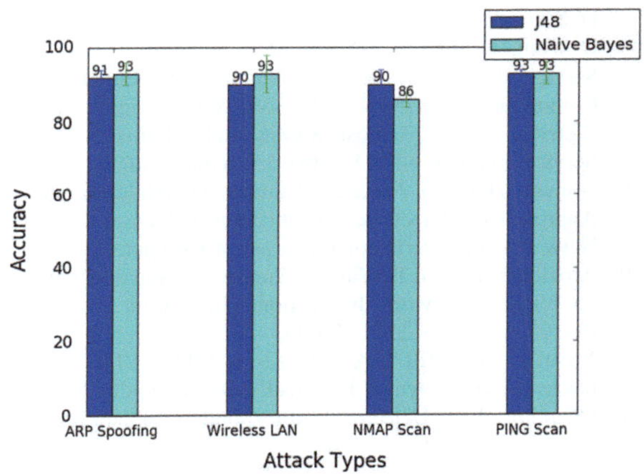

Fig. 6 Accuracy comparison of attacks using Naive Bayes and J48 tree

6 Conclusion and Future Work

As I see that there are various wireless LAN attack and wired LAN attack which are very dangerous for any network environment. So, myself need to detect these types of attacks. My mainly concern on the 3 ways for detecting these kinds of attacks. First one is the manually inspection of packets for these kinds of attacks, then use the signature based and machine learning tool for detection of these. As I see the signature based tools detect these kinds of attacks with less accuracy. So, I need some more parameters as mention in machine learning tool for high accuracy rate. As you see machine learning tool WEKA give me very high satisfactory results using my parameters. For the future work, I combine the signature based parameters and machine learning parameters for very high accuracy results along with more parameters. My mainly next step is concentrate on the prevention strategy of these kinds of attacks.

References

1. Mitchell, Changhua He John C.: Security Analysis and Improvements for IEEE 802.11 i, In: 12th annual network and distributed system security symposium, NDSS05 (2005).
2. Farooq, Taimur, David Llewellyn-Jones, and Madjid M.: MAC Layer DoS Attacks in IEEE 802.11 Networks, In: The 11th Annual Conference on the Convergence of Telecommunications, Networking and Broadcasting, PGNet, Liverpool, UK, (2010).
3. Ratnayake, Deepthi N., et al.: An intelligent approach to detect probe request attacks in IEEE 802.11 networks, In: Engineering Applications of Neural Networks, Springer Berlin Heidelberg, pp. 372–381, (2011).
4. Bellardo, John, and Stefan S.: 802.11 Denial-of-Service Attacks: Real Vulnerabilities and Practical Solutions, USENIX security, (2003).
5. Bernaschi, Massimo, Francesco Ferreri, and Leonardo V.: Access points vulnerabilities to DoS attacks in 802.11 networks, Wireless Networks14.2, pp. 159–169, (2008).
6. B. Vani, L.: Framework to Detect and Prevent Medium Access Control Layer Denial of Service Attacks in WLAN, International Journal of Computer Networks and Wireless Communications, ISSN: 2250-3501 Vol .3, No 2, April (2013).
7. Agarwal, Mohini, Santosh Biswas, and Sukumar N.: Detection of Deauthentication Denial of Service attack in 802.11 networks, India Conference, INDICON, IEEE, (2013).
8. Noman, Haitham Ameen, Shahidan M. Abdullah, and Haydar Imad M.: An Automated Approach to Detect Deauthentication and Disassociation Dos Attacks on Wireless 802.11 Networks, In: International Journal of Computer Science Issues, IJCSI 12.4 pp. 107 (2015).
9. Arockiam, L., and B. Vani: A Survey of Denial of Service Attacks and its Counter measures on Wireless Network, International Journal on Computer Science and Engineering Vol. 02, No. 05, pp. 1563–1571 (2011).
10. Yusuf B.: LAYER 2 ATTACKS & MITIGATION TECHNIQUES. http://www.sanog.org/resources/sanog7/yusuf-L2-attack-mitigation.pdf (2005).
11. OConnor, T. J.: Detecting and responding to data link layer attacks, SANS Institute InfoSec Reading Room, Oct 13 (2010).
12. Tao, Kai, Jing Li, and Srinivas S.: Wise guard-MAC address spoofing detection system for wireless LANs, Second International Conference on Security and Cryptography, Barcelona, Spain, pp. 140–147 (2007).
13. Korck, Michal, Jaroslav Lmer, and Frantisek J.: Intrusion Prevention/Intrusion Detection System (IPS/IDS) For Wifi Networks, International Journal of Computer Networks and Communications 6.4, pp. 77, (2014).
14. Nevlud, Pavel, et al.: Anomaly-based Network Intrusion Detection Methods, Advances in Electrical and Electronic Engineering 11.6, pp. 468, (2013).

Enhancing WSN Lifetime Using TLH: A Routing Scheme

Gaurav Bathla and Rajneesh Randhawa

Abstract A Wireless Sensor Network (WSN) consists of sensor nodes which are distributed over the network, where each sensor node composed of sensing unit along with limited computational power, low storage capacity and limited non-rechargeable battery source, resulting WSN to be energy constrained. In the proposed work a heterogeneous network is considered, in which uniformly placed nodes are divided into two levels of heterogeneity on the basis of energy. For the sake of long network lifetime, transmission to BS is done via Cluster Heads (CH's) resulting in faster drainage of CH node. So to manage the load on CH's, a new static clustering technique gained into its inception which includes CH's of higher energy, that in addition to increase network lifetime, also will reduce hop count to BS. The simulation result shows a significant improvement in stability as well as network lifetime of the proposed technique.

Keywords Routing · Heterogeneity · Clustering · Wireless sensor networks · WSN

1 Introduction

WSN contain large number (from hundreds to thousands) of small nodes which are capable of sensing data from environment and can send the data with required amplification to the next node after some computation. Also these nodes are equipped with sensing, computing, aggregating and wireless communications capability. Classical approaches either do not ensure equal distribution of load

G. Bathla (✉)
Department of Computer Engineering, Punjabi University, Patiala, India
e-mail: gauravbathla86@gmail.com

G. Bathla
Chandigarh University, Mohali, India

R. Randhawa
Department of Computer Science, Punjabi University, Patiala, India

G.M. Perez et al. (eds.), *Networking Communication and Data Knowledge Engineering*, Lecture Notes on Data Engineering and Communications Technologies 3, https://doi.org/10.1007/978-981-10-4585-1_3

among sensor network or reflect expenditure of high transmission power, resulting faster drainage of energy of highly used nodes; thus making them non-operational in the network. So, an ardent care needs to be taken in order to provide energy efficiency while developing any WSN protocol.

One major improvement in routing was done by aggregating the data at CH level (where cluster is a part of sensing area with some sensor nodes along with one aggregating point called CH), resulting less energy consumption of nodes [1–3]. By means of clustering, communication bandwidth is conserved because it limits the inter-cluster interactions between CH's, thus avoiding redundant exchange of messages [4]. Scalability of nodes in the network also affects the result of the employed clustering techniques. Use of heterogeneous nodes for real time scenario (i.e. sensor nodes of different capacity in terms of storage, power or processing capacity) in WSN is an effective way to increase network lifetime and reliability [5, 6] along with improved throughput [7].

In this paper, a routing strategy is proposed as a solution to tackle lifetime versus reporting time by using heterogeneity (in terms of battery power) among sensor nodes; which are deployed uniformly in the sensor field. Only 4% of the total nodes are deputed as CH's which leads to the origin of the concept **T**wo **L**evel **H**eterogeneity (TLH), where nodes are categorized in two categories; one are those nodes having higher energy (CH) and rest are normal (Non-CH) nodes. Each normal node will keep on transmitting the sensed data to its CH after certain time intervals. CH's further transmits the aggregated data to BS for further processing.

2 Literature Survey

In [4] Heinzelman et al. proposed energy efficient communication protocol named Low Energy Adaptive Clustering Hierarchy (LEACH) which distributes load evenly to all nodes of the network. In addition to CH rotation, received information at every CH was processed locally, aggregated and in turn transmitted BS for further processing. This protocol achieves 8x improvement over earlier techniques.

In [8] Manjeshwar et al. presented TEEN protocol, which was based on energy variation defined by threshold values according to the type of the application user needs. These were Hard Threshold (HT) and Soft Threshold (ST) energy levels. The nodes kept on sensing the environment continuously and transmitting when conditions meets the HT criteria, after that node transmits based on the ST, i.e. any change in the environment w.r.t. ST would be intimated.

In [9] Lindsey et al. proposed PEGASIS, a flat based routing protocol, focused to improve lifetime of sensor network by forming a chain among sensor nodes so as to transmit data to the node next in the path towards BS. CH's were chosen without any approach (like minimum distance or highest energy) resulting in failure of CH from random locations over the field.

In [10] Tan et al. proposed PEDAP, a minimum spanning tree-based protocol along with its power-aware version (PEDAP-PA). Both these methods were centralized algorithms governed by the BS which were used for gathering of network data and was also responsible for computing routing information. Since healthy amount of energy is consumed by intermediate nodes of the tree with respect to overall network consumption, so few nodes relapses their energy soon. As a result after some certain number of rounds, BS re-computes routing information to exclude dead nodes.

Xin Liu et al. [11] proposed tiering structure based on distance of nodes from BS. Nodes with higher tier IDs used multi hop transmission to transmit data to BS via lower tier IDs. Since transmission of data by nodes is directly proportional to square of distance between nodes, so the transmitter node's energy drainage was faster than receiving one's.

In [12] Gaurav Bathla et al. were trying to minimize energy usage in each round by creating Minimum spanning Multi-Tier protocol (MSMTP), in which concept of Minimum Spanning Tree with tiering was used along with defining a threshold energy level. Advantage of the technique was even load distribution on the nodes of the network and improvement in stable lifetime of the network.

In [13] Navid Hassanza deh et al. concentrated on the sensor network with the mobility of the nodes for data sources as well as sinks in a direction. They extended the existing work with few added mechanisms to make data collection more efficient for mobility purpose only. Also their simulation doubles the packet delivery ratio along with consuming lesser energy.

In [14] Jau-Yang Chang et al. proposed an algorithm for energy efficient routing based on uniform clustering algorithms where they proposed for a centralized cluster tree routing structure that was formed to reduce the transmission distances. Energy of each sensor node was accounted for selecting their CH and thus reducing the transmission distance by employing multi-hop approach.

In [15] Gaurav Bathla proposed a routing structure based on minimum spanning tree connection to transmit packet to a relay node which transmits aggregated data to BS. This flat routing scheme along with a heterogeneous node reduces energy requirement by almost 50% corresponding to its homogeneous counterpart but resulting a long time to reach data to BS.

In [16] R. Sudarmani et al. proposed a minimum spanning tree scheme for clustered Heterogeneous Sensor Networks with Mobile Sink, in which High Energy (HE) nodes were used as CH and Low Energy nodes as Non-CH. Data is gathered with the help of a mobile sink circulated around the network. This scheme transmits data in 5 hops to BS.

3 Problem Statement

WSNs are based on batteries which are neither chargeable nor replaceable. Sensor nodes sense/collect and pass the data further into the network for further use. Rest of the time they are in standby mode to save energy so passing and receiving of data consumes most of the energy of the nodes.

For effectively utilizing power of SN's, aside from putting them on standby mode (during their Non-operational period), proper energy consumption of sensor nodes should also be taken care for designing their working model. Malfunctioning of nodes also affects the network data and might also cause in loss of the data. So it is necessary to design a routing strategy where nodes with set threshold will participate as the part of working system.

Many routing techniques are proposed till now out of them few are categorized into direct transmission technique in which data to BS was transmitted in 1–2 hops by sensor nodes. This category is fast operational but possesses short network lifetime. While other categorization is multi-hop technique, in which data is transmitted in form of chain resulting slow data approaching BS. This technique is slower than direct transmission but possesses more lifetime.

This paper proposes a new method for clustering in the sensor network which is divided into two phases; one as set up phase and other is transmission phase. In set up phase all nodes will be associated with a CH and remain dedicated (for transmitting the sensed data). In transmission phase, sensed data will be transmitted by sensor nodes to respective CH's at certain regular intervals, after which further transmission of aggregated data to BS takes place until the network remains operational.

4 Proposed Work

As per the problem stated in the last section, to balance nodes energy with respect to network lifetime, details of the proposed routing scheme are given below

4.1 Two Level Heterogeneity (TLH) Scheme

In TLH all nodes are distributed uniformly across the network area which is divided into 4 equal square shaped clusters. Only one node (central node of cluster) per cluster is selected as a head, which will send aggregated data to BS (as shown in Fig. 1).

For efficient utilization of limited energy of SN's, two levels in energy are used. The higher energy nodes are CH's and rest are cluster members. CH's are of higher battery power due to their redundant use and additional tasks (like data aggregation and transmission to BS). By using such energy configured network the stable lifetime of the network is optimized. Figure 1 shows the proposed model, where darker nodes are CH's and rest are cluster members. By following this proposal, results are much better among existing techniques (like LEACH, PEGASIS, PEDAP and MSMTP).

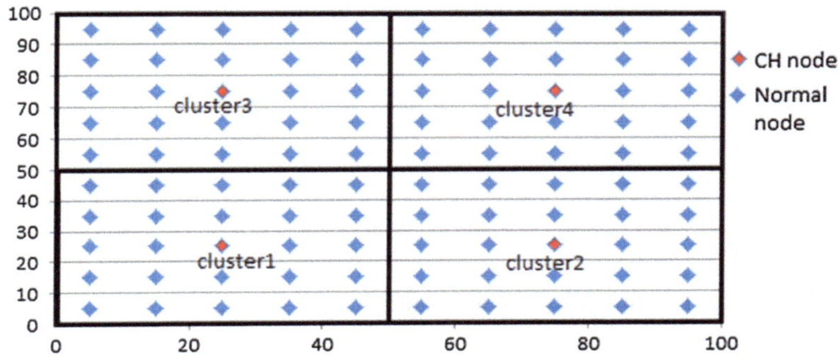

Fig. 1 Proposed scheme: uniform placement + CH at centralized position

4.2 Sensor Radio Model

Energy of sensor nodes is key factors in WSN by pursue of which their circuitry will work. This one time energy of nodes is required to be utilized effectively. Major energy consuming factors are

- **Transmission of data** [14]

$$\mathrm{E_{Tx}(k,d) = E_{elec} * k + E_{amp} * k},$$

  where E_{amp} is the amplification energy so total transmission cost is elaborated as:

$$E_{Tx}(k,d) = \begin{cases} E_{elec} * k + \varepsilon_{fs} * k * d^2 & \mathrm{d \le d0 \text{ within the network}} \\ E_{elec} * k + \varepsilon_{mp} * k * d^4 & \mathrm{d > d0 \text{ to the BS}} \end{cases}$$

  where $\varepsilon_{fs}, \varepsilon_{mp}$ are free space and multi path space coefficients

- **Receiving the data**

$$E_{Rx}(k,d) = E_{elec} * k$$

- **Aggregating the data**

$$E_{Da}(k) = 5 * 10^{-9} * k$$

Here k is the message size in bits between the nodes separated by distance d.

Let $\mathbf{E_{round,\ i}}$ is the **Energy consumed by a cluster$_\mathbf{i}$** for a round, then it is calculated by:

$$E_{round,i} = E_{CH} + \sum\nolimits_{j=1}^{J} E_{non-CH,J}$$

E_{CH} is energy consumed by a CH and consists of receiving (E_{Rx}) cost + aggregating (E_{Da}) + transmitting (E_{Tx}) data to BS, J depicts non-CH nodes in a cluster.

Network Energy consumption in a round

$$E_{round} = E_{CH} + \sum\nolimits_{m=1}^{4} E_{round,i}$$

It is well known that most of the energy is consumed by transmission of data. Larger the size of cluster more will be the distance for transmission to CH and hence energy required by SN's. So it's a better to reduce cluster size to minimize the transmission energy.

4.3 Workflow of Simulation

The Simulation is executed after initializing the starting values (like energy, position of SNs) then is repeated until any node is capable of transmitting data to their reporting head. Steps are:

i. Distribute SN's uniformly over the field of interest.
ii. Divide network into four equal rectangular partitions (clusters) and select the center most node in each cluster as CH.
iii. CH is the node having higher energy than remaining nodes and is responsible for aggregating the received data from SN's and then in turn transmit to BS.
iv. In every round, all SN's sense and transmit the data to their CH where it gets aggregated and transmitted to BS.
v. For every round energy deduction is done for:
   a. Non-CH nodes, for transmission cost to BS.
   b. CH nodes, for receiving cost from Non CH nodes + aggregation cost + transmission cost to BS.
vi. The round count is implemented for the network until nodes possess energy more than dead energy along with keeping track of round number when a node depletes its energy below dead energy.

5 Simulation Study

This section elaborates the simulation environment of the proposed technique and output in form of graph and tables. Graphs depict the comparison of proposed scheme with existing one's illustrating network stable lifetime. Simulation is done in C++.

5.1 Network Set Up

Few considerations while implementing the proposed scheme are:

i. All nodes are considered to be fixed and uniformly distributed over the network.
ii. Two types of nodes are present in the network, one are CH's having much more energy and other are non-CH nodes of less energy, second type of nodes that just sense and transmit data to CH.
iii. Sensor nodes sense and transmit data to their CH's on regular intervals.
iv. Sensor nodes switch their transceivers off after transmitting to CH, as per their TDMA schedule.

5.2 Simulation Environment

Simulation parameters for the network set up are depicted in Table 1 shown below:

Table 1 Simulation environment

Parameter	Value
E_{elec}	50nj/bit
$E_{Data\ Aggregation}$	5nj/bit/signal
$E_{free\ space}$	10pj/bit/m^2
$E_{multi\ path}$	0.0013pj/bit/m^4
Packet size	2000 bits
Sensing area	100 * 100
BS(X, Y)	50,175

6 Result and Discussion

Based on the values of Table 1, simulation is executed and results obtained are depicted in Fig. 2, showing the round number corresponding to which a node is died.

It is easily observed from Fig. 2 that proposed technique maintains good stable life time. Proposed routing technique is designed so as to stabilize the network as much possible. There is no extra load on any node except CH, which is gathering data from its cluster members. Remaining nodes are homogeneous and are just sensing and transmitting data to their respective CH's.

Stability of a network is of prime concern and is defined by the time until all the deployed nodes are operational. Up to stable time, all the deployed nodes are doing their task, so data is reliable and proper decision can be taken on this behalf. Proposed scheme improves network stable lifetime by factor of more than 12. Result comparisons of proposed scheme (TLH) with respect to previous protocols are shown in Figs. 3 and 4, based on the simulation parameters already defined in Table 2.

Figure 3 shows the comparison between in **F**irst **N**ode **D**ied versus Round number in which node is died in the proposed technique with respect to previous techniques. Results of previous techniques are referred from [14]. Graph is self-explanatory illustrating the improvement in stability in the proposed technique.

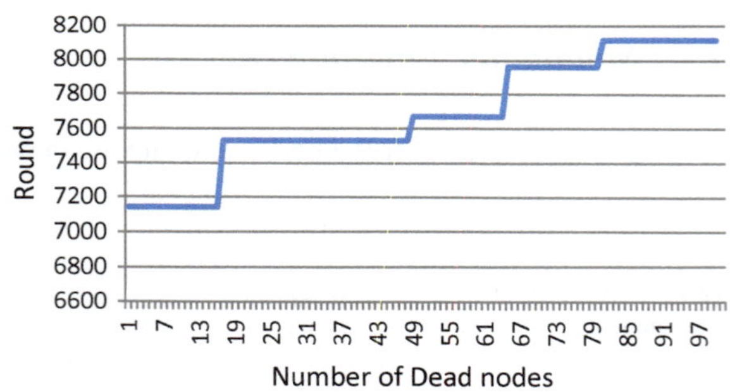

Fig. 2 Sensor node dead versus round number

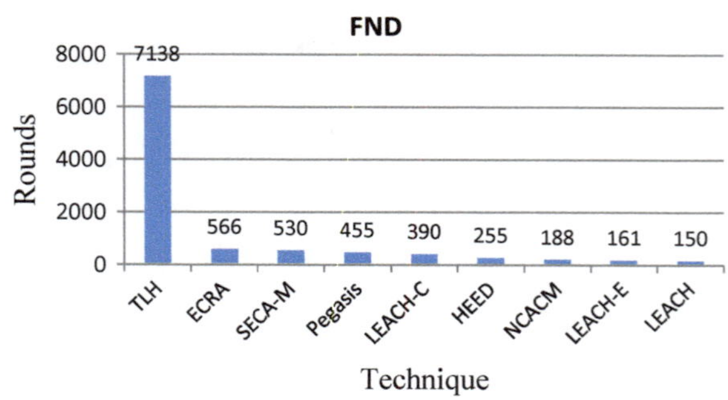

Fig. 3 First sensor node dead

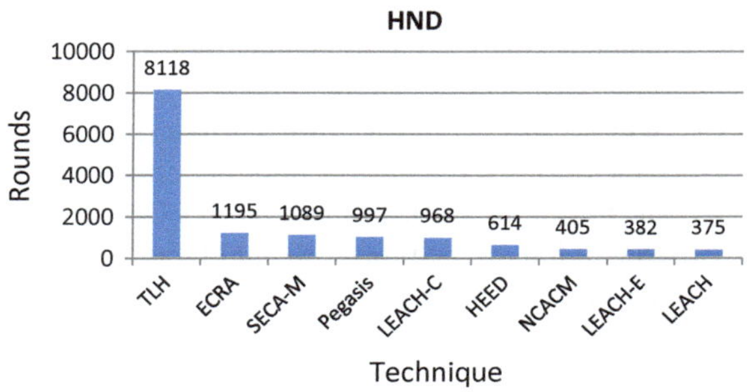

Fig. 4 Half of the sensor nodes dead

Table 2 Parameters for FND and LND comparison

Parameter	Value
$E_{\text{initial Energy}}$	Non-CH nodes 0.83 J/node
	CH nodes Node no. (22, 27) = 35.43 J Node no. (72, 77) = 26.52 J
No. of nodes (N)	100
Total network energy	Just under 200 J

Similarly Fig. 4 explains the network stage when half of the nodes are dead. This shows some intermediate stage of results in terms of stability, because only 50% of the nodes are operational at this stage.

Results of existing schemes are observed to be much lesser than proposed scheme in terms of number of rounds the network data is transmitted to BS. This is because of low transmission cost to CH every time. All the nodes know where to transmit the data and at last consolidated packet is transmitted to BS by CH's. So the sensed information is reached at final destination only within 2 hop counts without using much energy.

6.1 Advantages of Proposed Scheme with Flat Routing Heterogeneous Model

The proposed scheme is compared with a heterogeneous model proposed in [15] in which all the nodes are connected with Minimum Spanning Tree with nearby same energy level of the network, proposed scheme is obtaining nearby results (in terms of lifetime of the network) as were proposed in [15] but is having several advantages, which are

i. Scheme proposed in [15] is flat based scheme while proposed scheme is as fast as direct transmission. Also in [16] authors are able to send data to sink in 5 hops, along with tracking overhead of mobile sink.

ii. [15] contains multilevel in heterogeneity (all nodes with different energy level) which is very hard to implement in real time environment, while TLH requires only two level hierarchy.
iii. Scheme of [15] is very difficult to execute in real time scenario, as it will be typical to place nodes of different energy at their specific position, while no such issue in TLH, because except CH all nodes are homogeneous.
iv. Due to longer path traveling and more traffic, chances of data loss are more in flat based routing as in [15] while this is not such a problem in cluster based schemes as proposed.

7 Conclusion

In this paper a routing protocol based on two level heterogeneity of energy for sensor nodes is proposed. Through simulation the results are proven to be better than existing homogeneous as well as heterogeneous counterparts who claim themselves near to optimal routing protocol. In each round of communication, low load on all cluster member nodes and more load on high energy CH's leads towards increasing the stable lifetime of the network. Higher energy backup of CH's has a great impact on the system lifetime.

Further, the time to reach data to BS is much faster as compared to the chain based protocols; because in chain based protocols, data packet traverses through the nodes in a linear fashion and being aggregated at many points; thus increasing the transmission delay. In the proposed technique the data is transmitted to BS in just two hops via CH (i.e. node → CH → BS). So in only two hops, data of network is reached to BS for further processing. This is just twice time of direct transmission. So advantage of proposed protocol is fast operating as well as energy efficient.

8 Future Scope

In terms of future work Reliability can be added to proposed work by taking back up of data received at CH by any other node, so that in case of any type of failure of CH (physical hazard or any other reason), data of whole cluster will be preserved at back up node. This back up can be discarded when it reaches at BS.

One can improve this work by adding the network aggregator node which collects data from the CH's and transmit the data to BS.

References

1. Tillapart, Piyakul, Sanguan Thammarojsakul, Thanachai Thumthawatworn, and Pratit Santiprabhob. "An approach to hybrid clustering and routing in wireless sensor networks." DOI:10.1109/AERO.2005.1559523, In Aerospace Conference, IEEE, pp. 1–8, IEEE, 2005.
2. Ryu, J-H., S. Song, and D-H. Cho. "Energy-conserving clustering scheme for multicasting in two-tier mobile ad-hoc networks." Electronics Letters, vol. 37, issue 20, pp. 1253–1255, 2001.
3. Hou, T-C., and T-J. Tsai. "Distributed clustering for multimedia support in mobile multihop ad hoc networks." IEICE transactions on communications, vol. 84, issue 4, pp. 760–770, 2001.
4. Heinzelman, Wendi Rabiner, Anantha Chandrakasan, and Hari Balakrishnan. "Energy-efficient communication protocol for wireless microsensor networks." In System sciences, Proceedings of the 33rd annual Hawaii international conference, pp. 1–10, IEEE, 2000.
5. Younis, Mohamed, Moustafa Youssef, and Khaled Arisha. "Energy-aware management for cluster-based sensor networks." Computer networks, doi:10.1016/S1389-1286(03)00305-0, vol. 43, issue 5, pp. 649–668, 2003.
6. Pignaton de Freitas, E., Tales Heimfarth, Carlos Eduardo Pereira, Armando Morado Ferreira, Flávio Rech Wagner, and Tony Larsson. "Evaluation of coordination strategies for heterogeneous sensor networks aiming at surveillance applications." In Sensors, doi:10. 1109/icsens.2009.5398314, pp. 591–596. IEEE, 2009.
7. Corchado, Juan M., Javier Bajo, Dante I. Tapia, and Ajith Abraham. "Using heterogeneous wireless sensor networks in a telemonitoring system for healthcare." Information Technology in Biomedicine, IEEE Transactions, vol. 14, issue 2, pp. 234–240, 2010.
8. Manjeshwar, Arati, and Dharma P. Agrawal. "TEEN: a routing protocol for enhanced efficiency in wireless sensor networks." In Parallel and Distributed Processing Symposium, International, vol. 3, pp. 30189a–30195a, IEEE Computer Society, 2001.
9. Lindsey, Stephanie, and Cauligi S. Raghavendra. "PEGASIS: Power-efficient gathering in sensor information systems." In Aerospace conference proceedings, vol. 3, pp. 1125–1130, IEEE, 2002.
10. Tan, Hüseyin Özgür, and Ibrahim Körpeoğlu. "Power efficient data gathering and aggregation in wireless sensor networks." ACM Sigmod Record, vol. 32, issue 4, pp. 66–71, 2003.
11. Liu, Xin, Quanyu Wang, and Xuliang Jin. "An energy-efficient routing protocol for wireless sensor networks." In 7th World Congress Intelligent Control and Automation (WCICA), pp. 1728–1733. IEEE, 2008.
12. Bathla Gaurav, Khan Gulista "Minimum Spanning Tree based Routing Strategy for Homogeneous WSN", International Journal on Cloud Computing Services and Architecture (IJCCSA), vol. 1, issue 2, pp. 22–29, August 2011,
13. Hassanzadeh, Navid, Olaf Landsiedel, Frederik Hermans, Olof Rensfelt, and Thiemo Voigt. "Efficient mobile data collection with mobile collect." In 8th International Conference on Distributed Computing in Sensor Systems (DCOSS), pp. 25–32. IEEE, 2012.
14. Chang, Jau-Yang, and Pei-Hao Ju. "An energy-saving routing architecture with a uniform clustering algorithm for wireless body sensor networks." Future Generation Computer Systems, vol. 35, pp. 128–140, 2014.
15. Bathla, Gaurav. "Minimum Spanning Tree based Protocol for Heterogeneous Wireless Sensor Networks." i-Manager's Journal on Wireless Communication Networks vol. 1, issue 4, pp. 12–22, 2013.
16. Sudarmani, R., and R. Vanithamani. "Minimum spanning tree for clustered heterogeneous sensor networks with mobile sink." In 2015 IEEE International Conference on Computational Intelligence and Computing Research (ICCIC), pp. 1–6. IEEE, 2015.

IP Traceback Schemes for DDoS Attack

Vipul Mandhar and Virender Ranga

Abstract Nowadays the Internet is exposed to a span of web threats. In the modernized era, multifarious types of attacks are discovered on the Internet, along with the utmost disastrous attack, Distributed Denial of Service (DDoS) attacks. In such course of attacks, an immense number of settle arrangement tie in with one another to make the services baseless for honest users. These composed systems frequently mask their existence by counterfeit technique. IP traceback is a way used to catch the real path of web packets in such scenario. This paper provides a schematized investigation of various IP traceback approaches with their fruitful domain and doorway for forthcoming research in this thrust expanse of IP traceback.

Keywords DDoS attack · IP traceback · Spoofed IP address
Packet marking · Packet logging · Ant colony system · Star colouring algorithm

1 Introduction

In the public eye, the Internet is growing swiftly now-a-days because of its exceptional features, availability and low cost. It has given a boost to good fortune in every track we can think of, but like every coin has two sides, Internet has its own comfort and harm. Many illicit activities are committed on the internet. DDOS attack is one of such activities. This attack involves flooding of the bandwidth of true users by the attacker. DDoS web attacks are imaginable due to susceptible architectonics of TCP/IP protocol batch where a packet is sent from source to destination without confirming its source address. To unsettle the avail of a server, the schooled attackers may launch a DDoS attack. The originator IP address attached to the packet can be used by the attacker in order to mask him-self from

---

V. Mandhar (✉) · V. Ranga
Department of Computer Engineering, National Institute of Technology,
Kurukshetra, Haryana, India
e-mail: vipulmandhar130793@gmail.com

G.M. Perez et al. (eds.), *Networking Communication and Data Knowledge Engineering*, Lecture Notes on Data Engineering and Communications Technologies 3, https://doi.org/10.1007/978-981-10-4585-1_4

discerning. IP spoofing is used to form hosts difficult to resist against an amplification attack. For such reasons, developing an architecture to strike the actual source of attacks is the main challenge in DDoS. IP traceback is an impressive approach to finding the actual path of network packets from where it is actually provoked. In this paper, we accept the said challenge and provide a systematized investigation of different IP traceback approaches to tackle DDoS attack.

The paper is further organized as follows: Sect. 2 describe the background study on DDoS attack and present scenario of DDoS attack in India and abroad. Section 3, discusses the different basic IP traceback approaches with their pros and cons. Further in this section, we discuss the new messaging IP traceback approaches which have overcome the weakness of the previous messaging schemes i.e. DPM (Deterministic Packet Marking) and DFM (Deterministic Flow Marking). Hybrid approaches are discussed with open research issues related to IP traceback problems. Section 4 describes the comparison table for all discussed techniques. Finally, Sect. 5 concludes the paper and highlights the scope for forthcoming research work.

2 Background Details & Related Work

During an amplification attack, the actual paths through which the attack comes toward the legitimate user get poorly affected resulting in a massive amount of work ignominy to the overall network. This makes the detection and filtration of attack traffic at the earliest a very important task for the legitimate users as well as internet service providers.

Present statistics of DDoS attack [1]: India is 4th largest (7.43%) victim of DDoS attack. In the first quarter of 2015, India accounts for 6.93% of all DDoS attacks in the world, which is increased to 7.43% in the second quarter of 2015. With comparison with Q2 of 2014, DDoS attacks all over the world have increased by 132.43%. China (37.01%) is the top victim of DDoS attack as shown in (see Fig. 1).

3 IP Traceback Schemes

As outlined earlier finding the attacker's origin is most crucial since the attackers may spoof someone else's IP address for doing malicious attacks to mask their identity. After the malicious DDoS attack, our on hand aim is to identify the root of the packets from where the attackers generate them. The process of seeking the attackers is not that much easy because an attacker may use some others IP address or there are so many paths through which attacker is able to do this attack. The continuous flood of spurious packets may be sent from the different-different zone. ISPs are bound for providing the Internet to the users. Two types of internet domains are used for communication between the routers under same ISP or different ISP. The intra-domain link is liable for the communication process between the

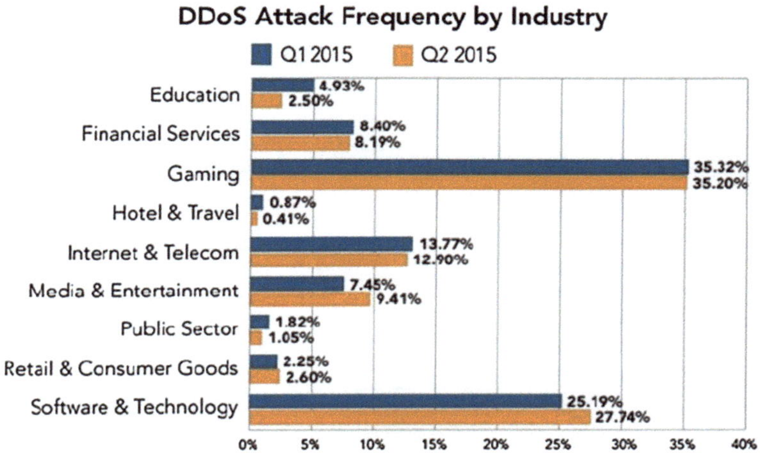

Fig. 1 DDoS attack frequency by industry [1]

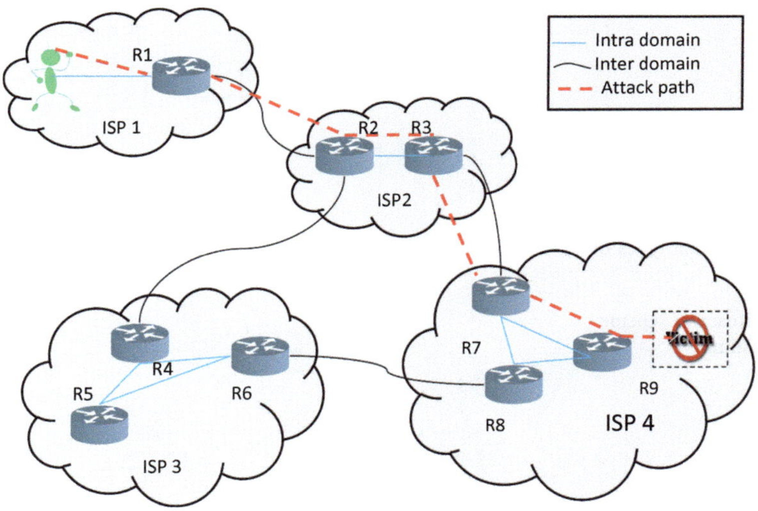

Fig. 2 Attack paths with inter and intra domain links

routers under the same ISP, whereas Inter-domain link is used for settling connection between two routers of distinct ISP. The path followed by the packet from its root to the destination is referred as attack path. Typically, amplification attack may compromise count of attack path to reach the terminal over the Internet. The Intra and Inter-domain links constitute attack path. This can be illustrated in (see Fig. 2) where the path traversed by the attack packets from origin to the destination is shown.

IP traceback approaches can be categorized according to the taxonomy as shown in Fig. 3. These can be used for the compilation of information of trace. These

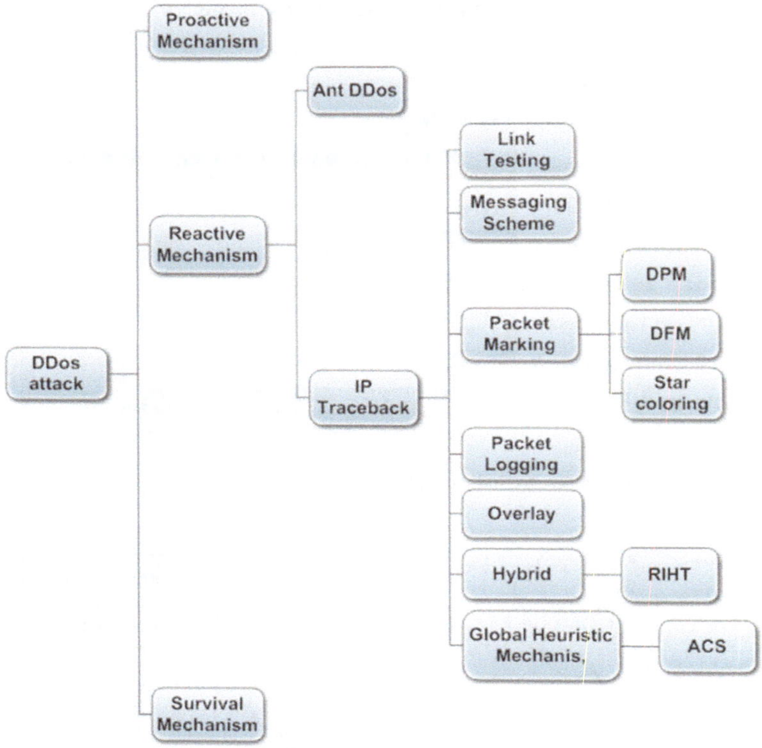

Fig. 3 Different types of trace-back schemes

Fig. 4 Reactive scheme classification

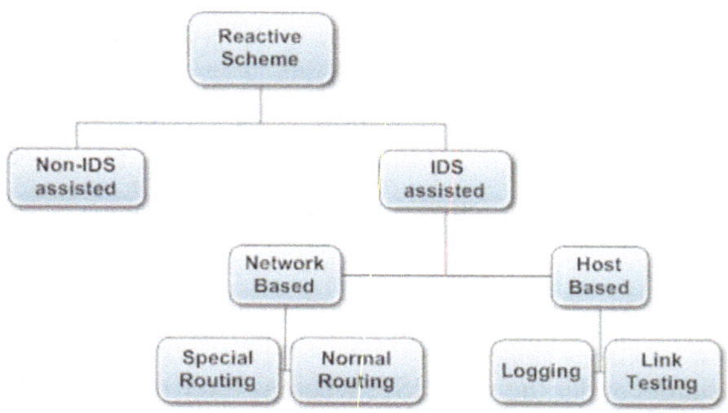

approaches are distinct as per their deployment strategies, storage requirements information collecting, algorithm, etc.

Further, there are countless schemes poses for traceback which can be classified basically along two dimensions as reactive and pro-active schemes [2]. Figures 4 and 5 provide an elaborate classification of various existing schemes according to their functionality.

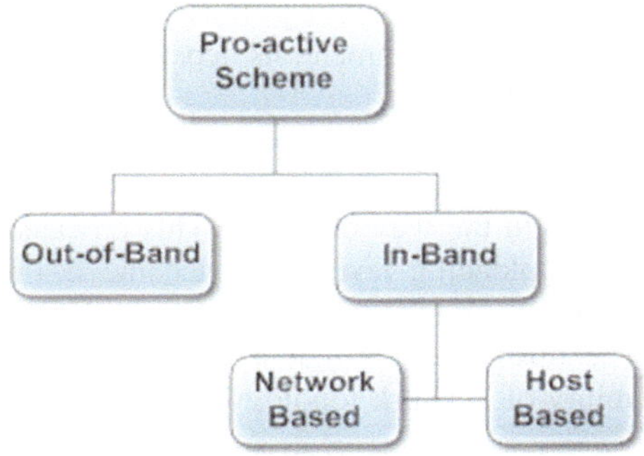

Fig. 5 Pro-Active scheme classification

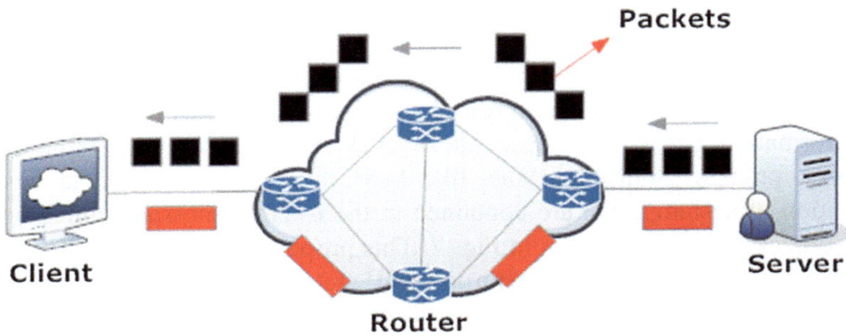

Fig. 6 Link testing

3.1 Reactive Mechanism Based Traceback Schemes: This Section Shows the Different Reactive Mechanism Based Traceback Schemes

3.1.1 Link Testing

In this proposed approach developed by Singh et al. [3], the traceback process begins from the router that is nearest to the legitimate users. All the upstream links are analyzed in a periodic manner till the source of the packet or the attacker is found this approach will give a path to the router that is nearest to the attacker as shown in (see Fig. 6). The two existing link testing scheme are: Input debugging and another one has controlled flooding.

Input debugging. For the specific packet, every router has some potential to settle its ingress links. Upstream routers are used to aggregate the information of the victim under an amplification attack and can put up an attack packet signature and send it to uploading routers. Then router can consequently identify the attacker by periodically investigating its upstream links against the received attack packet signature. This scheme has the pros of being consistent with extant protocols and infrastructure, along with granting good guide for incremental formatting and the little bandwidth hanging on network traffic. The limitations include incompatibility for DDoS environment, its dependency on assistance among ISPs and it works only during the attack.

Controlled flooding. An existing ISP map is used in controlled flooding. Victim monotony flood packets to its uploading routers and synchronously determines switch in the way of attack. This repeated process can record the attack source at each uploading level. This scheme has advantages of being logical with actual protocols and infrastructure and it supports easy and cumulative implementation. The obstruction is that it works only during ongoing attack and required prior ability of network topology.

3.1.2 Messaging

In proposed approach [3], the messaging basically uses ICMP. This packet is probabilistically generated by each router known as–Trace message or trace packet. The information carried by these packages is then used as an input for the trace-back process. Some constants like MAC address, previous and next hop information, timestamp, etc. are contained in the i–Trace message which helps in the trace-back process as shown in Fig. 7. This proposed scheme has advantages to support incremental distribution with low ISP (Internet service providers) cooperation, Consistent with current protocols and infrastructure, and allows the post-attack inquiry. The limitation is that it is easily mistreated by attackers if proposed scheme lacks in certification support.

3.1.3 Packet Marking

A thought-provoking solution to traceback problem is packet marking. The complete encrypted route information is stored in the packet. The victim can spot all the incoming packets from the attacker and can use this information i.e. sent by the attacker to trace the actual origin of the packet. This proposed approach needs a

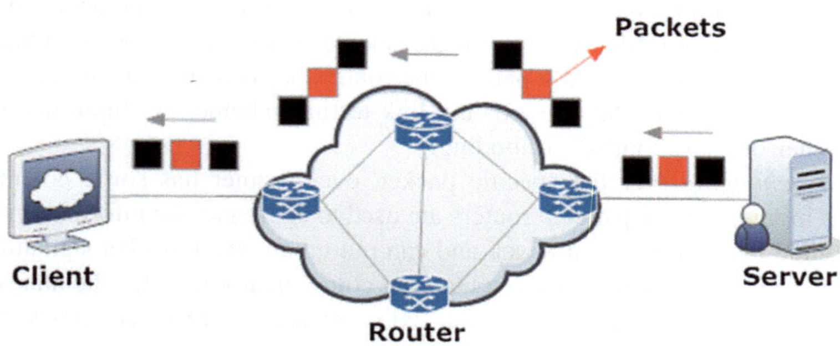

Fig. 7 Messaging scheme

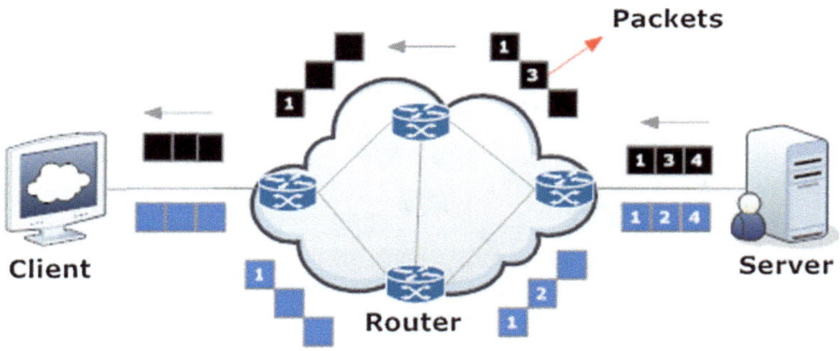

Fig. 8 Packet marking scheme

mark encoding algorithm to mark the packets properly as shown in Fig. 8. The proposed scheme has advantages of being in sync with current network structure. Mostly current trace-back methods belong to the network group [4, 5]. The motive of these techniques is to identify the attack path entirely or partially. Smooth and formable discharge in comparison to other approaches and Gutter (Least) ISP support required. The limitation is that packet discontinuity issues occurs due to an exhaust of description domain in the traceback approach and sometimes may result in high false positive. Moreover, the total count of the marked packet received by the victim node is used for traceback process. Some tactics for marking based scheme as follows:

- Node appending.
- Node sampling.
- Edge sampling.

Node appending: The complete path is traced by using these tactics with the help of a single attack packet. The traceback data is attached to the authentic IP packet header.

Node sampling: IP address of the router is responsible for the specifically marked information at each node. Each router has some specific parameters such as marking function, color or identity number assigned as per traceback schemes.

Edge sampling: It involves encrypting edge information, like start, end, etc. The distance field is responsible for constructing the attack path and it is independent of Internet topology.

Two promising schemes for tracing cyber-attacks are shown in Fig. 9.

Deterministic Packet Marking (DPM): DPM trace the path up to the end router closer to the attacker. In (DPM) the main key feature is that the end router closest to the attacker is marked [6]. So this approach will help in traceback the origin of the packet in a proper way because every packet is marked on the same basis.

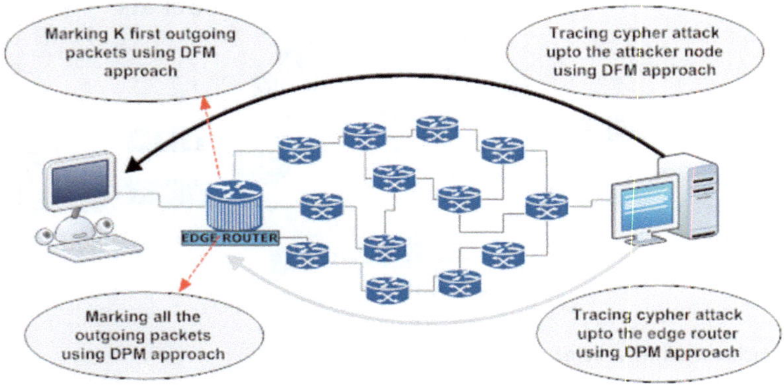

Fig. 9 Two promising schemes for tracing

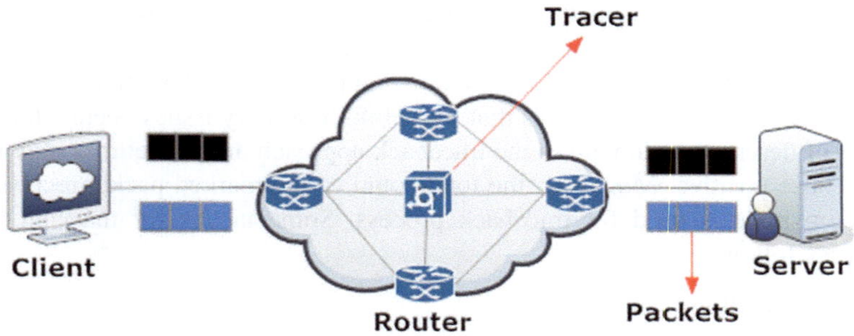

Fig. 10 Packet logging

Deterministic Flow Marking (DFM): DFM allow the legitimate user to trace the source from where the actual attacker initiated, even if the attack is performed behind the NAT (Network address translation). The main advantage of DFM is that it will not mark the every single packet i.e. coming from the attacker side it will mark K first packets of each flow. This will help it to contain the feature of both DPM and PPM (probabilistic packet marking). The aim of DFM is to trace the source node (s) of the attack which may be located on the LAN existing at the back of edge routers. For detail, see [7].

3.1.4 Packet Logging

In this proposed approach every single packet that traverses in the network plays an important role, so information of every packet (a hashed value of IP header fields) is stored in the router through which it passes. The stored information in every router is used to trace the packets origin. This approach is powerful since a single packet is able to trace the attacker's position as shown in Fig. 10.

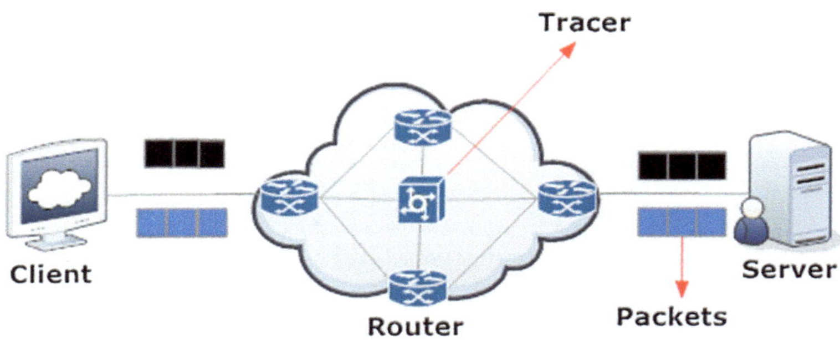

Fig. 11 Overlay

3.1.5 Overlay

In the approach [3], tracking router is a special type of router is used in overlay trace-back scheme this tracking router is responsible for monitoring the traffic in the network. A command is issued whenever an attack is detected. This command will direct the traffic to pass through the tracking router. Then router will examine the traffic and collect the important information useful for traceback scheme as illustrates in Fig. 11. This scheme has many advantages that it provides perfect trace-back results, cogently handle DDoS attacks. It has the limitation that it shows high implementation cost and an attacker may also target the tracing routers while DDos attack.

3.2 Hybrid Traceback Schemes

Whenever any two different traceback schemes merged to work together for traceback mechanism, it becomes a hybrid traceback scheme. This approach could yield more efficient and good result to find the root of the attacker because this will use two different approaches together for the single type of work. In hybrid traceback scheme, Yang et al. [8], mark router's interface number and in order to deal with this logging and marking issues packet logging is integrated with a hash table for IP traceback. In this scheme, only the interface used to transfer the packet is marked by the router and this will help to traceback. Marking field is logged into the hash table due of the lack of field space in the packet. Finally, the index table is only stored in the packet. This logging and marking process we rerun until the required destination receives the packet. To traceback the path of the origin we reverse such processes. In RIHT, router connected to other router and the local network will send the packet to the core router. The host sends the packet through the closest router to it and then this router will send the packet to other router and so on till it reached the destination as shown in Fig. 12. RIHT has scheme has following properties:

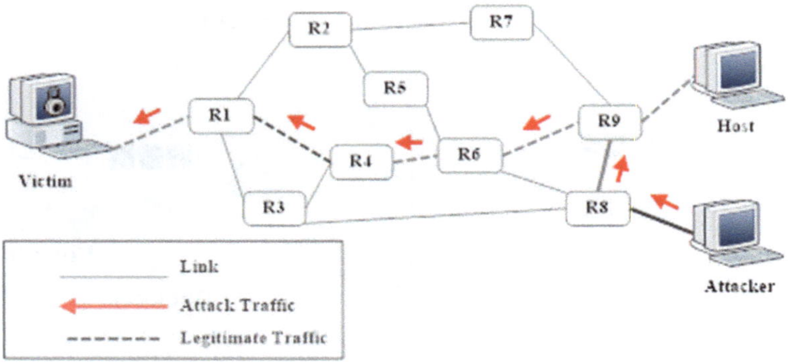

Fig. 12 RIHT

- Arbitrary router's storage requirement is edged above the number of paths leading to router, and the routers do not require refreshing there logged tracking information.
- This strategy helps to acquire the zero false negative rates in reconstruction of the path from victim to the attacker.
- This path reconstruction scheme has high efficiency.
- It can censor the attack traffic.

3.3 Special IP Traceback Schemes

It offers to refer colors to the routers, these preferred color are used as a mark and attached to the en-router IP packet. Finally, the attack path is constructed using these mark. Any graph G of star colouring is a vertex colouring of G so no path of length 3 is bi-colored. Most of the existing packet marking execution requires a large count of packets to pool on the attack path but in this style, it reduces the aerial (overhead). Star chromatic no. Of the graph G, denoted as Xs(G) which is minimum no of colors required for star color G. Colors are pre-mentioned to each router and (worn) used as marks for attack path construction [9]. To trim the IP-header the colors are reused but at the same time, we can distinguish between routers by using some action of its unique Ip address to traceback the actual source of the attack. The actual attack path built up by the legitimate user is thus a flow of colors. Given a flow of colors, Star coloring satisfy that from any distinct node the attack path can be variously traced.

Attack Path Construction: In this approach, each packet on their path to the destination is marked by the routers with probability. All the information collected in the packets from the traversed node of the attack path is stored and set by the victim. If

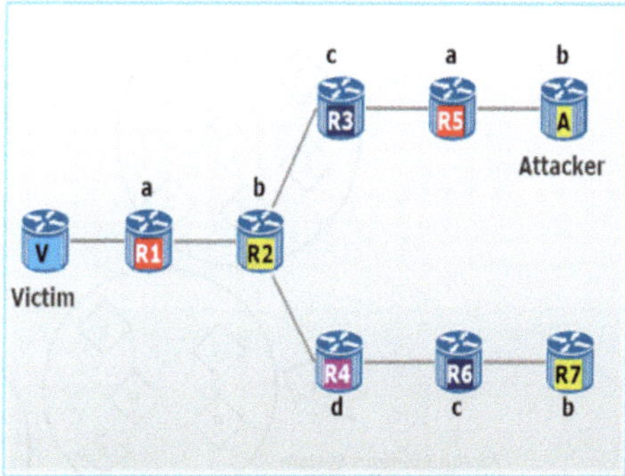

Fig. 13 Star coloring

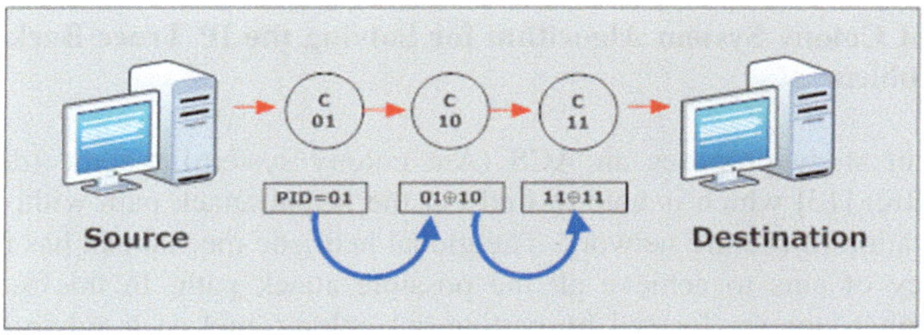

Fig. 14 Construction of path identifier

any router is d hop distance away from the victim then the point of victim swing the router will be inclined with the probability (1-)d-1, where d is maximum node degree which is less than n, (d << n).The range for the total number of colors is (Ω (d);O (d2) [10]). In star coloring algorithm, the single color can be repeated in an attack path and hence truthful separate the packets placed on the basis of their color and their count. If we see the actual attack path a-b-c-a-b (as shown in Fig. 13) the victim will accept the packets from router R1 and router R5 and from router R2 and router A pose same color data. To build up the attack path there must be the least number of packets received from the router i.e. closest to the attacker. The actual attack path will be constructed by distinct marks sorting in decreasing order from the victim's distance as shown in Fig. 14. The TTL field is used during reformation of the attack path [11]. If two nodes lie on two different paths then their TTL [12] value must be same, but if they are on same track then their value will be different. The main problem with existing packet marking technique is successfully addressed by imprinting the fingerprint of the path in each packet. In this process, a path identifier referred as (path ID) is embedded in the packet by the en-routers along with their color. To keep the path ID field size compatible with color field we can either XOR or can use one's complement to compute the sum at each en-router. Victim when finally receives the packet, it receives XOR summation as a mark with that packet.

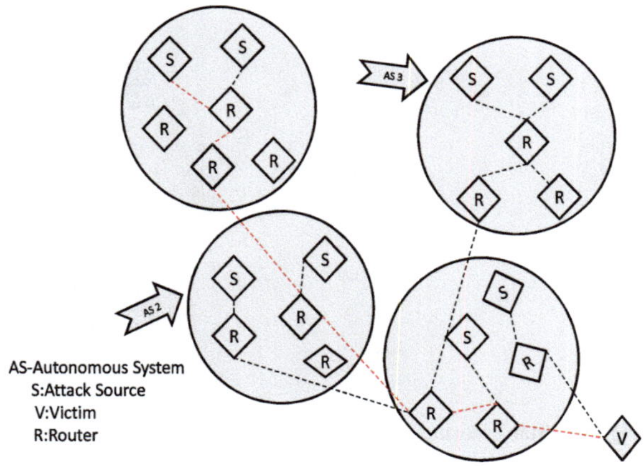

Fig. 15 ACS algorithm

3.3.1 Ant Colony System Algorithm for Solving the IP Trace-Back Problem

The present study proposes an ACS (Ant colony system) scheme (denoted as ACS-IPTBK) [13] which is able to find out the actual attack path without having the information of routing network. The global heuristic mechanism has reinforced the potency of ants to achieve all the possible attack path. In this mechanism, multiple subgroups are formed by parting ant colony, and each subgroup has its own essence updating rule. The performance of the ACS–IP Traceback algorithm in regenerating the actual attack path that is constructed using the algorithm can be inspected using NS2 simulations as shown in Fig. 15.

This scheme has multiple advantages that the positive feedback accounts for the swift discovery of solutions, distributed computation avoids imperfect convergence. The limitation is that it has slow convergence and also performs poorly for TSP problems larger than 75 cities and there is no centralized processor to lead the AS toward the solution [14].

4 Comparison of Different Proposed IP Traceback Schemes

It has been observed that the link testing scheme is set side by side with other traceback schemes like marking, overlay and hybrid has large overhead of the transmitted packets because in link testing scheme the authors mainly consider the connection between the legitimate user and the attacker but in other schemes they use both the link and the router information for traceback process. As shown in Table 1 logging, overlay, hybrid and ACS need less time for traceback as compared to link testing, marking and messaging schemes due to less number of packets. In

Table 1 Comparison between different categories of IP traceback scheme

		Comparison of IP traceback schemes					
Approach evaluation metrics	Link testing	Messaging scheme	Marking scheme	Logging scheme	Overlay	Hybrid schemes	Ant colony system
Packets	Large	Medium	Medium	Very less	Very less	Very less	Very less
Time for traceback	Long	Short	Short	Medium	Short	Very short	Very short
Handling packet transformation	Good	Yes	Fair	Yes	Yes	Yes	Yes
Security	Possible	Possible	Possible	Difficult	Possible	Possible	Possible
DDoS handling capability	No	Good	Good	Good	Fair	Fair	Fair
False positive	High	Low	Low	Very low	Low	Low	Low
ISP privacy	No	Yes	Rarely	Yes	Yes	Yes	Yes

link testing scheme there are no DDoS handling capabilities compared with other schemes because there is less use of routers in trackback process. Link testing scheme has a large false positive rate as compared to all other proposed schemes because there is no fixed path or routers used for traceback process. Link testing scheme has no ISP privacy but other proposed schemes have ISP privacy because they allow both pre and post-analyze traceback.

5 Conclusions and Future Direction

In this paper, we discussed the DDoS attacks and how to traceback the attacker's origin after the attack. Different IP Traceback approaches like link testing, messaging, packet marking, packet logging are used to traceback the attack, we have also explored some advanced traceback schemes like DFM and DPM which reduces the computational overhead, bandwidth overhead, memory overhead, traceback rate, false positive rate. DFM phase out the spoofed marking, planted by the compromised router in the attack trail. For future work turn up with the added version of RIHT that use16-bit marking field to bypass the problems induced by packet wound. We have also explored some methods which solve the IP Traceback problems like Star Coloring and ACS (Ant Colony System). In the future work, the offline form of star coloring algorithm will be refined. In ACS schema, the ability of the ants to find all the likely attack paths with in the unfolding space was boost by dividing the portion of the ant colony into the subgroups, so that each subgroup handles a different track revise rule. For the future purpose in ACS method, the ants will pretend to revolution and share their judgment. However, this may not be achievable in the physical world because of IP attack.

References

1. Akamai, December 2015, Akamai State of the Internet- Security Report, http://trak.in/tags/business/2015/08/21/india-ddos-attack-growth-web-app-gaming-target/.Last accessed (6-Oct-2-16).
2. Santhanam, Lakshmi, Anup Kumar, and Dharma P. Agrawal. "Taxonomy of IP traceback." *Journal of Information Assurance and security* 1, no. 1 (2006): 79–94.
3. Singh, Karanpreet, Paramvir Singh, and Krishan Kumar. "A systematic review of IP traceback schemes for denial of service attacks." *Computers & Security* 56 (2016): 111–139.
4. Savage, Stefan, David Wetherall, Anna Karlin, and Tom Anderson. "Network support for IP traceback." *IEEE/ACM Transactions on Networking (TON)* 9, no. 3 (2001): 226–237.
5. Song, Dawn Xiaodong, and Adrian Perrig. "Advanced and authenticated marking schemes for IP traceback." In *INFOCOM 2001. Twentieth Annual Joint Conference of the IEEE Computer and Communications Societies. Proceedings. IEEE*, vol. 2, pp. 878–886. IEEE, 2001.
6. Belenky, Andrey, and Nirwan Ansari. "IP traceback with deterministic packet marking." *IEEE communications letters* 7, no. 4 (2003): 162–164.
7. Foroushani, Vahid Aghaei, and A. Nur Zincir-Heywood. "TDFA: traceback-based defense against DDoS flooding attacks." In 2014 IEEE 28th International Conference on Advanced Information Networking and Applications, pp. 597–604. IEEE, 2014.
8. Yang, Ming-Hour, and Ming-Chien Yang. "RIHT: a novel hybrid IP traceback scheme." IEEE Transactions on Information Forensics and Security 7, no. 2 (2012): 789–797.
9. Roy, Sangita, Anurag Singh, and Ashok Singh Sairam. "IP traceback in star colored networks." In *2013 Fifth International Conference on Communication Systems and Networks (COMSNETS)*, pp. 1–9. IEEE, 2013.
10. Muthuprasanna, Muthusrinivasan, G. Manimaran, Mansoor Alicherry, and Vijay Kumar. "Coloring the Internet: IP traceback." In *12th International Conference on Parallel and Distributed Systems-(ICPADS'06)*, vol. 1, pp. 8-pp. IEEE, 2006.
11. Wang, Ping, Hui-Tang Lin, and Tzy-Shiah Wang. "An improved ant colony system algorithm for solving the IP traceback problem." Information Sciences 326 (2016): 172–187.
12. Yaar, Abraham, Adrian Perrig, and Dawn Song. "FIT: fast Internet traceback." In *Proceedings IEEE 24th Annual Joint Conference of the IEEE Computer and Communications Societies.*, vol. 2, pp. 1395–1406. IEEE, 2005.
13. Lai, Gu Hsin, Chia-Mei Chen, Bing-Chiang Jeng, and Willams Chao. "Ant-based IP traceback." *Expert Systems with Applications* 34, no. 4 (2008): 3071–3080.
14. Gu Hsin Lai, Chia-Mei Chen, Bing-Chiang Jeng, Willams Chao, "Ant-based IP traceback", Expert Systems with Applications 34 (2008) pp. 3071–3080.

Adaptability of Various Mobility Models for Flying AdHoc Networks—A Review

Kuldeep Singh and Anil Kumar Verma

Abstract Researchers worldwide have immensely contributed towards the field of MANETs. In recent years, Unmanned Aerial Vehicles (UAVs) are being extensively used to create ad hoc networks due to their ability to be used in tactical as well as civilian areas. Interestingly, the research focus is being extended from one large UAV based network to multi-UAV ad hoc network consisting mini or micro UAVs, commonly known as Flying Ad hoc Network (FANET). There are various mobility models that are being used in MANETs and it is the right time to explore the capabilities of those mobility models for their use in FANETs. The underlying mobility model plays a vital role in simulating the performance of a routing protocol. In this paper, an analysis of various mobility models have been carried on parameters such as dependencies, connectivity metrics and real time applications with an eye on the adaptability of these models in FANETs.

Keywords FANETs · Adhoc network · Mobility models · UAV
Connectivity matrices

1 Introduction

Although mobile ad hoc networks have a large number of applications, but there are certain situations such as drowning or military combat field where traditional MANETs is not very useful. So there is a need of advance technology which can be used to establish a communication network in such scenario. Researchers came up with an idea called FANET. A Flying ad hoc network (FANET) belongs to a particularly new sub-class of Mobile ad hoc networks (MANET). FANET consist a group of homogeneous or heterogeneous flying agents called unmanned aerial vehicles (UAVs) which are capable to communicate with each in a group, and interact with their neighbors to acquire valuable information [1]. FANET allowed sending

K. Singh (✉) · A.K. Verma
Department of Computer Science Engineering, Mobile Computing and Communication (MC2) Research Group, Thapar University, Patiala, India
e-mail: kuldeepisp@gmail.com

G.M. Perez et al. (eds.), *Networking Communication and Data Knowledge Engineering*, Lecture Notes on Data Engineering and Communications Technologies 3, https://doi.org/10.1007/978-981-10-4585-1_5

information speedily and accurately in circumstances, where common ad hoc networks are not able to perform. Popularity of FANETs is increasing rapidly because of their huge applications and characteristics such as flexibility, easy to deploy and relatively less expenses [1]. The usage of unmanned aerial vehicle in FANETs promises new advantages for its applications such as disaster monitoring, border surveillance, traffic monitoring, military battlefield, relay for ad hoc networks, flooding, wildfire and earthquakes etc. Single UAVs has been used for many applications such as surveillance and monitoring purpose from last many years, but instead of using a single UAV based system, multi-UAV systems have many advantages. However, FANETs have many other challenges such as high mobility, frequent topology change, routing and security. As high mobility in FANET nodes is a big challenge and create several other challenges for FANETs to perform efficiently and accurately [2]. So In this paper, we have done an analysis of different mobility models for FANET. The simulation uses the synthetic environment for experiments and provides a very easy way to monitor the network and also uses reproducible scenarios in files that can be easily modified. Therefore, simulation is used by the majority of researchers because it offers a doable option. Simulation environment has several features, mobility models are one of them. Mobility models define the path through which a network node will move. So they play a crucial role in network topology change and subsequently in evaluation of ad hoc network performance [3, 4].

2 Mobility Models

Nowadays the main focus of ad hoc network researchers is in the development and evaluation of FANET routing protocols. For the assessment of FANET routing protocols, researchers are using new test beds and simulators such as UAVRF, RAVEN, GRASP, Simbeeotic, UAVSim etc. and synthetic models for mobility of nodes [1]. The movement of nodes in the ad hoc networks is based on the mobility model used in the simulation. Mobility model defines the moving pattern, velocity and acceleration of a node with the time [5]. Mobility models emulate moving pattern of network nodes in real life application in a logical way [6]. We have described and analyze different mobility models for flying ad hoc network to the facet of real life applications [7]. Characterization of mobility models is described in Fig. 1. Based on the different characteristics such as dependencies, randomness and restriction, mobility models can be classified into different classes [8].

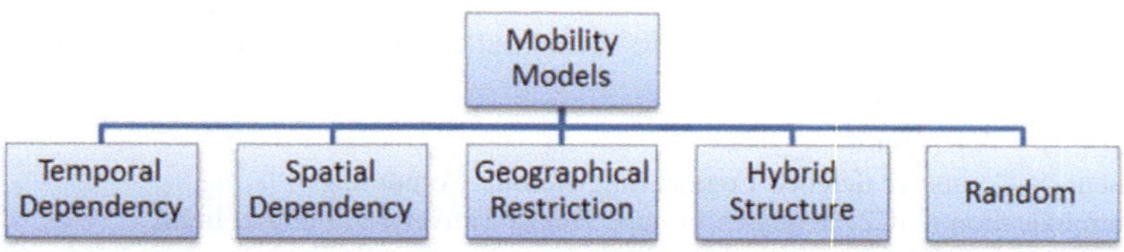

Fig. 1 Classification of mobility models

2.1 Temporal Dependency

In this class of mobility models, the current movement of a node is influenced by the past movements, i.e. the current velocity of any mobile node is dependent on the prior velocity of that node [8]. Degree of temporal dependence is defined as the amount of relationship between the velocities of two mobile nodes at two time slots which are not too distant from each other.

$$Temporal\ Degree(x, t, t') = RD(\bar{v}_x(t), \bar{v}_x(t')) * SR(\bar{v}_x(t), \bar{v}_x(t')) \ldots \quad (1)$$

In Eq. 1, $\mathrm{Temporal\,Degree}(x, t, t')$ describe the temporal degree of node x at time instance t and t'. $\mathrm{RD}(\bar{v}_x(t), \bar{v}_x(t'))$ refers to the relative direction of node x at time instances t and t'. $\mathrm{SR}(\bar{v}_x(t), \bar{v}_x(t'))$ refers to the speed ratio of node x at time instances t and t'. Degree of temporal is high when both nodes move in the same direction with almost similar speed in a certain time interval.

2.1.1 Gauss-Markov

In this model, initially a position, speed and direction are assigned to each node [9]. After a fixed interval of time n, speed and direction of each node is updated. Speed and direction of a node at nth instance is calculated with the help equations given below [10]:

$$S_n = xS_{n-1} + (1-x)\bar{S} + \sqrt{(1-x^2)S_{x_{n-1}}} \ldots\ldots\ldots\ldots \quad (2)$$

$$D_n = xD_{n-1} + (1-x)\overline{D} + \sqrt{(1-x^2)D_{x_{n-1}}} \ldots\ldots\ldots \quad (3)$$

In Eqs. 2 and 3, S_n and D_n are the fresh speed and direction of the mobile node at time interval n. ‘x’, $0 \leq x \leq 1$, is a tuning parameter for randomness. $\bar{S}$ and $\overline{D}$ representing mean of speed and direction respectively. $S_{x_{n-1}}$ and $D_{x_{n-1}}$ Gaussian distribution random variable [11]. At each time interval new position of the node is calculated with the help of current location, speed and direction.

$$x_n = x_{n-1} + s_{n-1} \cos d_{n-1} \ldots\ldots\ldots \quad (4)$$

$$y_n = y_{n-1} + s_{n-1} \sin d_{n-1} \ldots\ldots\ldots \quad (5)$$

In Eqs. 4 and 5, (x_n, y_n) and (x_{n-1}, y_{n-1}) are the x and y coordinates of the mobile node at nth and $n-1$th time instance. And s_{n-1} and d_{n-1} are the speed and direction of the mobile nodes at $n-1$th time instance.

2.1.2 Smooth Random

Due to memory less nature of the Random Waypoint model, it sometimes shows unrealistic movements. Bettstetter [12] proposed mobility model that changes speed and direction of node smoothly. This model considers mobile nodes have a preferred set of velocities instead of uniformly distributed velocities.

$$P_v(v) = \begin{cases} P(v) = 0\delta(v)v = 0 & \\ P(v = 0.5V_{max})\delta(v - 0.5V_{max})v = 0.5V_{max} & \\ P(v = V_{max})\delta(v - V_{max})v = V_{max} & \\ \dfrac{1 - P(v=0) - P(v - 0.5V_{max}) - P(v - V_{max})}{V_{max}} & 0 < V_{max} < 1 \\ 0 & otherwise \end{cases} \quad (6)$$

In Smooth Model [13], the frequency of change in speed is assumed to be a Poisson process and frequency of direction changes is assumed to be exponentially distributed [14]. Whenever there is any directional change, net moment direction is calculated using following probability distribution function: $\mathrm{P}_{\varnothing}(\varnothing) = \frac{1}{2\pi}$, for $0 \le \varnothing < 2\pi$. Where $\Delta\varnothing(\mathrm{t})$ is defined as:

$$\Delta\varnothing(\mathrm{t}) = \begin{cases} \varnothing(\mathrm{t}) - \varnothing(\mathrm{t}') + 2\pi & \text{for } -2\pi < \varnothing(\mathrm{t}) - \varnothing(\mathrm{t}') \le -\pi \\ \varnothing(\mathrm{t}) - \varnothing(\mathrm{t}') & \text{for } -\pi < \varnothing(\mathrm{t}) - \varnothing(\mathrm{t}') \le \pi \\ \varnothing(\mathrm{t}) - \varnothing(\mathrm{t}) - 2\pi & \text{for } \pi < \varnothing(\mathrm{t}) - \varnothing(\mathrm{t}') \le 2\pi \end{cases} \quad \ldots\ldots\ldots \quad (7)$$

2.2 *Spatial Dependency*

Spatial dependency of mobile nodes occurs when nodes move together in a group. Movement of a mobile node possibly influences the mobility of another node. The reference group is one of the approaches to realizing spatial dependency. Degree of spatial dependence is defined as the amount of relationship between the velocities of two mobile nodes which are close to each other [15].

$$Spatial\,Degree(x, y, t) = RD\left(\bar{v}_x(t), \bar{v}_y(t)\right) * SR\left(\bar{v}_x(t), \bar{v}_y(t)\right) \ldots \quad (8)$$

Where, Spatial Degree(x, y, t) describe the spatial degree of node x and y at time instance t. $\mathrm{RD}(\bar{\mathrm{v}}_{\mathrm{x}}(\mathrm{t}), \bar{\mathrm{v}}_{\mathrm{y}}(\mathrm{t}))$ Refer to relative directions between node x and y at time instance t. $\mathrm{SR}(\bar{\mathrm{v}}_{\mathrm{x}}(\mathrm{t}), \bar{\mathrm{v}}_{\mathrm{y}}(\mathrm{t}))$ refer to the speed ratio between node x and y at time instance t. The degree of spatial is high when both nodes move in the same direction with almost similar speed. However, spatial degree gradually decreases if the relative direction of nodes or speed ratio reduces.

2.2.1 Reference Point Group Mobility Model (RPGM)

In RPGM model, nodes move in a group and each node can move randomly within the group. The mobility pattern of group nodes is dependent on the center of the group called reference point [15]. It could be a logical center of the group or a leader node. Movement of center of group specified movement of all other nodes in the group and also determines their speed and direction.

2.2.2 Community Based Mobility Model

As in primeval time, nomadic societies moved from one place to other place. This model follows the same behavior and sometimes known as nomadic mobility model. It represents groups of mobile nodes that jointly move from one location to another [16]. In this model every mobile node is free to roam around a reference point by using entity mobility model such as random waypoint, random walk. As reference point changes, all group nodes travel towards the new area specified.

2.2.3 Pursue Mobility Model

In pursue mobility model, each node in the group chase a node called master node. Pursue model can be seen as a group of policemen following the thief. In pursue mobility model position of each node can be calculated using Eq. 9:

$$N_p = \left[O_p + a(T - O_p) + Random\,Vector\right) \ldots\ldots\ldots \tag{9}$$

N_p and O_p are the new and old position of a mobile node. 'a' is the acceleration and T is position of nodes [3]. Random vector is obtained by mobility model.

2.2.4 Social Network Founded Model

The social network founded mobility model is based upon the centrality of a node, node with higher centrality have a larger probability of strong social relationship. It is also known as interaction indicator. Centrality of a node depends upon the closeness of that node to other node. Initially, all the nodes in the network are grouped into clouds based on their interaction indicator. All the nodes in a cloud moves according to random waypoint model [12].

2.3 Geographical Restrictions

Mobile nodes in a mobility model have spatial and temporal dependency but, there are many situations where it is impractical to assume that mobile node will cover the entire simulation area [8].

2.3.1 Area-Graph Based Model

In this model geographical area is divided into small non overlapping sub-area called cluster [17]. The cluster is made with higher density mobile nodes and has a path in between lower density nodes. The clusters are treated as vertices and paths are treated as the edges of the area graph. Random waypoint model is used to manage the movements in the cluster of area graph model [18]. The Area graph model is very suitable in the scenario like tactical, rescue network requirement.

2.3.2 Map Based Model

In map based mobility model, a restriction is imposed on the movement of node means it have restriction on the movement area geographically on the network node [13]. The model follows the modeling movement in the street of urban area which composed many horizontal and vertical streets. Nodes move along the vertical and horizontal street, according to the map with possibility to turn at the meeting point of the street. This model imposed huge restriction on node movement.

2.3.3 Manhattan Grid Model

Manhattan model is mainly proposed for the movement of traffic in the urban areas, streets on map are organized as horizontal and vertical. Mobile nodes move in the horizontal and vertical street of the geographical area [9]. This model uses a probabilistic approach to decide the node movement at the intersection points of the streets. Probability of left and right turn at the intersection is 0.25 and probability of going straight is 0.50. This model is very suitable for vehicular traffic in the urban area.

2.3.4 Obstacle Model

The obstacle mobility model was designed to manage the moving pattern of nodes in the scenario which look like real world topographies. Buildings, hills and other structures could be a barrier to the wireless communication between mobile nodes as well as to the movement of nodes [19]. This model uses Voronoi Diagram to avoid the obstacle in the way of mobile nodes.

2.4 Hybrid Structure

Hybrid class is attained by integration of all mobility metrics to find optimum performance mobility models. Hybrid models have a combination of structure having relative speed, spatial, temporal dependence and clustering [8].

2.4.1 Free Way Mobility

This model emulates the movement of mobile nodes where spatial, temporal exists between nodes and geographical restrictions also imposed on mobile nodes [10]. In this model velocity of nodes is temporal dependent on its earlier velocity as well as influenced by the other nodes within the neighborhood. So, this mobility model represents movement behavior of nodes which have environmental constraint and temporal and spatial dependency.

2.4.2 Catastrophe-Scenario Model

Similar to real catastrophic situations, this model works well on fast and slow movement nodes. This model could be considered for obstacles, tactical areas, and group movements. Some other scenarios like catastrophe, hostage release as well as the squad scenario are realized to support the requirements of this model [20].

2.4.3 User Oriented Meta Model

To realize complex mobility scenarios, IllyaStepanov et al. proposed user oriented meta model [21, 22]. This model consists of three apparatus; first modeling that considers the environmental conditions, restrictions of mobile nodes and attraction points for simulation area. Next step contains the sequencing of all the movements of the users i.e. attraction points. Third, prepared model should be able to reflect all the temporal and spatial dependencies of mobile user.

2.4.4 Disaster Area Model

This model provide realistic scenario to represent movements in disaster areas. Disaster situations demonstration structured movement based on civil protection maneuvers. This model supports group mobility, diverse area-based movement and movements on the optimal paths avoiding obstacles with node joining/leaving scenario [23]. To simulate this model, simulation is divided into planned areas with the concept of civil protection using both stationary and transport nodes [24].

2.5 *Random*

Random class consists of those models which do not have any dependencies as well as any other restriction. These models are the often used in past years.

2.5.1 Random-Waypoint

The Random Waypoint Model was proposed by Broch et al. [24]. This model is used to build several other mobility models. It involves hiatus before changes in path direction and velocity of a mobile node [9]. Before starting traveling towards to a new target at a selected mobile node stays for a short phase of time i.e. breaks.

2.5.2 Clustered Mobility

In Cluster mobility model, network nodes are divided into clusters and each cluster have member node and head node [25]. Member nodes of a cluster have one hop distance from its cluster head. Cluster head maintains IP addresses of its member nodes and all member nodes accounts their cluster head IP address.

2.5.3 Random Walk

The Random walk model is one of the basic mobility models in which node starts moving from its present location to a new destination by choosing a random direction and a random speed. But the direction and speed are selected from pre-defined ranges $[0, 2\pi]$ and [minimum-speed, maximum-speed] respectively [9]. Random walk mobility model node changes its movement after one of the two parameters satisfies either a fixed distance 'd' or a fixed amount of time 't' is completed.

3 Connectivity Graph Metrics

Since the connectivity graph between mobile nodes have an effect on the performance of routing protocol in the ad hoc networks, so it is very important to analyze the connectivity graph between nodes. These connectivity metrics will help in linking mobility model with routing protocol performance [15].

3.1 Number of Link Changes

The Number of link changes between two nodes x and y depends upon how many times the link between these two nodes change from “down” to “up” and vice versa and can be calculated as:

$$\text{Link Change}(x, y) = \sum_{t=1}^{T} X(x, y, t) \tag{11}$$

where, T is the simulation time and $X(x, y, t)$ is a random variable such that $X(x, y, t) = 1$ iff $\{A(x, y, t-1) = 0 \text{ and } A(x, y, t) = 1\}$ or $\{A(x, y, t-1) = 1 \text{ and } A(x, y, t) = 0\}$

3.2 Link Duration

It defines the average number of link durations between any two nodes x and y. It can also be said that it is the durability of links between two mobile nodes [26]. $A(x, y, t)$ is the variable set as 1 if there is a link among two mobile nodes.

$$\text{Link Duration}(x, y) = \begin{cases} \dfrac{\sum_{t-1}^{T} A(x,y,t)}{\text{Link Change}(x,y)} & \text{if Link Change} \neq 0 \\ \sum\limits_{t-1}^{T} A(x, y, t) & \text{Otherwise} \end{cases} \quad \ldots \tag{12}$$

3.3 Link Availability

It is the time instance t when a path is available between two mobile nodes x, y. $A(x, y, t)$ is variable indicator set as 1 if there is a path between node x and y [10].

$$\text{Path Available}(x, y) = \begin{cases} \dfrac{\sum_{t=\text{start}(x,y)}^{T} A(x,y,t)}{T - \text{start}(x,y)} & \text{if } T - \text{start}(x, y) > 0 \\ 0 & \text{Otherwise} \end{cases} \quad .. \tag{13}$$

Comparisons of Various Mobility Models

Table 1 Comparison of various mobility models

Mobility model	Connectivity metrics			Dependencies			Requirement for real life scenario					
	Average number of links	Average link duration	Average path availability	Spatial	Temporal	Geographical	Obstacles	Tactical areas	Optical paths	Group movements	Units leaves scenario	Heterogeneous velocity
Temporal dependency based												
Gauss-markov	–	L	–	–	M	–	–	E	–	–	–	E
Smooth	A	A	–	–	M	–	–	E	M	–	–	M
Spatial dependency based												
Reference point group	H	H	–	M	E	E	E	E	E	M	E	E
Community based	–	A	A	M	–	–	–	–	M	M	–	E
Pursue	A	A	–	M	–	–	–	–	–	–	–	–
Social network	H	–	–	M	–	–	–	E	–	M	–	E
Geographical dependency based												
Area based	–	H	–	–	–	M	E	M	E	–	–	E
Map based	L	–	L	–	–	M	–	–	–	–	–	E
Manhattan-grid	–	L	–	–	–	M	–	–	–	–	–	E
Obstacle	L	L	–	–	–	M	M	–	–	–	–	E
Hybrid												
Free way	A	H	–	M	M	M	–	–	–	–	–	E
Disaster area	A	–	A	–	M	M	M	M	M	M	M	M
Catastrophe scenario	–	A	–	M	–	M	M	M	–	–	–	M
User oriented meta model	–	H	–	M	M	M	M	–	M	M	–	M

(continued)

Table 1 (continued)

Mobility model	Connectivity metrics			Dependencies			Requirement for real life scenario					
	Average number of links	Average link duration	Average path availability	Spatial	Temporal	Geographical	Obstacles	Tactical areas	Optical paths	Group movements	Units leaves scenario	Heterogeneous velocity
Random based												
Random waypoint	–	H	–	–	–	–	–	E	M	–	–	E
Clustered	H	–	H	E	–	–	–	E	M	–	–	E
Random walk	–	A	L	–	–	–	–	E	–	–	–	E

*L—Low, A—Average, H—High, M—Explicitly modeled, E—Can be Extended

4 Conclusion and Future Scope

In this paper, we have discussed and compared various mobility models on the basis of various parameters viz. connectivity metrics, various dependencies and real-time application parameters. The analysis directs that a mobility model which support good connectivity metrics is likely to be the best candidate for mobility modeling in FANETs and can be easily extended for the real time application. Table 1 indicates that obstacle, map based and disaster area mobility models are the best models among all other models. Therefore, these are the best candidate to adopt for FANET environment. Further, this analysis will support in the design and development of new mobility models to be developed for FANETs.

Acknowledgements This work is supported by the Department of Electronics and Information Technology, Ministry of Communication and IT, Government of INDIA as a part of VISVESVERYA PHD Scheme.

References

1. Bekmezci, I., Sahingoz, O.K. and Temel, Ş., 2013. Flying ad-hoc networks (FANETs): A survey. *Ad Hoc Networks*, *11*(3), pp. 1254–1270.
2. Sahingoz, O.K., 2014. Networking models in flying ad-hoc networks (FANETs): Concepts and challenges. *Journal of Intelligent & Robotic Systems*, *74*(1–2), pp. 513–527.
3. Singh, K. and Verma, A.K., 2014, May. Applying OLSR routing in FANETs. In *Advanced Communication Control and Computing Technologies (ICACCCT), 2014 International Conference on* (pp. 1212–1215). IEEE.
4. Lim, S., Yu, C. and Das, C.R., 2006, November. Clustered mobility model for scale-free wireless networks. In *Proceedings. 2006 31st IEEE Conference on Local Computer Networks* (pp. 231–238). IEEE.
5. Aschenbruck, N., Gerhards-Padilla, E., Gerharz, M., Frank, M. and Martini, P., 2007, October. Modelling mobility in disaster area scenarios. In *Proceedings of the 10th ACM Symposium on Modeling, analysis, and simulation of wireless and mobile systems* (pp. 4–12). ACM.
6. Lakhtaria, K.I. ed., 2012. *Technological Advancements and Applications in Mobile Ad-Hoc Networks: Research Trends: Research Trends*. IGI Global.
7. Pomportes, S., Tomasik, J. and Vèque, V., 2011. A composite mobility model for ad hoc networks in disaster areas. *REV Journal on Electronics and Communications*, *1*(1).
8. Roy, R.R., 2011. Mobility Model Characteristics. In *Handbook of Mobile Ad Hoc Networks for Mobility Models* (pp. 23–32). Springer US.
9. Sindhuja, M., Selvamani, K., Kannan, A. and Kanimozhi, S., 2015. Mobility Assisted Uncertainty Reduction in MANETS Using Forward Node Optimization. *Procedia Computer Science*, *48*, pp. 408–413.
10. Aschenbruck, N., Gerhards-Padilla, E. and Martini, P., 2008. A survey on mobility models for performance analysis in tactical mobile networks. *Journal of Telecommunications and Information Technology*, pp. 54–61.
11. Kumar, R. and Dave, M., 2014. Mobility models and their affect on data aggregation and dissemination in vehicular networks. *Wireless Personal Communications*, *79*(3), pp. 2237–2269.

12. Bettstetter, C., 2001. Mobility modeling in wireless networks: categorization, smooth movement, and border effects. *ACM SIGMOBILE Mobile Computing and Communications Review*, *5*(3), pp. 55–66.
13. Bouachir, O., Abrassart, A., Garcia, F. and Larrieu, N., 2014, May. A mobility model for UAV ad hoc network. In *Unmanned Aircraft Systems (ICUAS), 2014 International Conference on* (pp. 383–388). IEEE.
14. Musolesi, M., Hailes, S. and Mascolo, C., 2004, October. An ad hoc mobility model founded on social network theory. In *Proceedings of the 7th ACM international symposium on Modeling, analysis and simulation of wireless and mobile systems* (pp. 20–24). ACM.
15. Camp, T., Boleng, J. and Davies, V., 2002. A survey of mobility models for ad hoc network research. *Wireless communications and mobile computing*, *2*(5), pp. 483–502.
16. Bai, F., Sadagopan, N. and Helmy, A., 2003, March. IMPORTANT: A framework to systematically analyze the Impact of Mobility on Performance of RouTing protocols for Adhoc NeTworks. In *INFOCOM 2003. Twenty-Second Annual Joint Conference of the IEEE Computer and Communications. IEEE Societies* (Vol. 2, pp. 825–835). IEEE.
17. Kumar, S., Sharma, S.C. and Suman, B., 2010. Mobility metrics based classification & analysis of mobility model for tactical network. *International Journal of Next-Generation Networks*, *2*(3), pp. 39–51.
18. Stepanov, I., Hahner, J., Becker, C., Tian, J. and Rothermel, K., 2003, October. A meta-model and framework for user mobility in mobile networks. In *Networks, 2003. ICON2003. The 11th IEEE International Conference on* (pp. 231–238). IEEE.
19. Bettstetter, C., 2001, July. Smooth is better than sharp: a random mobility model for simulation of wireless networks. In *Proceedings of the 4th ACM international workshop on Modeling, analysis and simulation of wireless and mobile systems* (pp. 19–27). ACM.
20. Hong, X., Gerla, M., Pei, G. and Chiang, C.C., 1999, August. A group mobility model for ad hoc wireless networks. In *Proceedings of the 2nd ACM international workshop on Modeling, analysis and simulation of wireless and mobile systems* (pp. 53–60). ACM.
21. Musolesi, M. and Mascolo, C., 2006, May. A community based mobility model for ad hoc network research. In *Proceedings of the 2nd international workshop on Multi-hop ad hoc networks: from theory to reality* (pp. 31–38). ACM.
22. Aschenbruck, N., Gerhards-Padilla, E., Gerharz, M., Frank, M. and Martini, P., 2007, October. Modelling mobility in disaster area scenarios. In *Proceedings of the 10th ACM Symposium on Modeling, analysis, and simulation of wireless and mobile systems* (pp. 4–12). ACM.
23. Bittner, S., Raffel, W.U. and Scholz, M., 2005, March. The area graph-based mobility model and its impact on data dissemination. In *Pervasive Computing and Communications Workshops, 2005.PerCom 2005 Workshops. Third IEEE International Conference on* (pp. 268–272). IEEE.
24. Broch, J., Maltz, D.A., Johnson, D.B., Hu, Y.C. and Jetcheva, J., 1998, October. A performance comparison of multi-hop wireless ad hoc network routing protocols. In *Proceedings of the 4th annual ACM/IEEE international conference on Mobile computing and networking* (pp. 85–97). ACM.
25. Tian, J., Hahner, J., Becker, C., Stepanov, I. and Rothermel, K., 2002, April. Graph-based mobility model for mobile ad hoc network simulation. In *Simulation Symposium, 2002. Proceedings. 35th Annual* (pp. 337–344). IEEE.
26. Johansson, P., Larsson, T., Hedman, N., Mielczarek, B. and Degermark, M., 1999, August. Scenario-based performance analysis of routing protocols for mobile ad-hoc networks. In *Proceedings of the 5th annual ACM/IEEE international conference on Mobile computing and networking* (pp. 195–206). ACM.

Reliable Packet Delivery in Vehicular Networks Using WAVE for Communication Among High Speed Vehicles

Arjun Arora, Nitin Rakesh and Krishn K. Mishra

Abstract This paper by way of simulation based on NS3 and SUMO shows multi-channel working of WAVE which in turn helps in defining communication among high speed vehicles and their surrounding environment. NS3 being a popular tool for simulation supports WAVE and its extension modules. The main objective is to lower the delivery time of packet among connected network of vehicles. The roads are divided based on the assumption of traffic density i.e. high and low respectively. The vehicles share their knowledge of position and fastest possible delivery path. Dijkstra based calculations are used for shortest path with weights linked to traffic density. For simplification of calculation two strategies are used to find shortest path, one relays information to the nearest road with high density, second is to simply carry and forward the information to the destination. Prior gained knowledge by a vehicle about its surrounding is used for selection of best path. The paper shows the reliable packet delivery to conventional methods such as Vehicle Assisted Data Delivery (VADD) in city traffic conditions.

Keywords VANET · NS3 · Simulation · WAVE · Vehicular communication

A. Arora (✉)
Department of CSE, UPES, Dehradun, India
e-mail: a.arora@ddn.upes.ac.in; arjunarora06dit@gmail.com

N. Rakesh
Department of CSE, ASET, Amity University Campus, Noida, India
e-mail: nrakesh@amity.edu

K.K. Mishra
Department of CSE, MNNIT, Allahahbad, India
e-mail: kkm@mnnit.ac.in

G.M. Perez et al. (eds.), *Networking Communication and Data Knowledge Engineering*, Lecture Notes on Data Engineering and Communications Technologies 3, https://doi.org/10.1007/978-981-10-4585-1_6

1 Introduction

Vehicular Ad hoc networks (VANETS) are simply a collection of on-board units (OBU) and road side units (RSUs) which communicates using Dedicated Short Range Communications (DSRC) utilizing 75 MHz in 5.9 GHz frequency band for WAVE. The Fig. 1 shows frequency channel layout of WAVE which is divided into 7 channels with one channel for service safety applications and the remaining for service non-safety applications [1]. The nature of VANET environment being dynamic it is most likely that packet forwarding over varied time intervals may be disconnected causing network delays. In such a case the message containing information needs to be with the current vehicle until new node becomes active.

In order to effectively utilize the seven channels IEEE 1609.4 and 802.11p provides co-ordination methods and channel access mechanisms. Extensive research has undergone in the area of VANETs for new utilization of these channels. The usage of the medium for communication may change with time of the day or vary over period of same days. Large number of studies have addressed to the problem of data aggregation in case of vehicular environments [2]. Now if every vehicle includes the complete path of itself and its neighbours it consumes large amounts of bandwidth which makes use of network resources, however this problem can be subdued by data encoding and compression.

But the question that comes is that what part of the path should be shared by the vehicle. Simply stating every vehicle may limit its sharing of path knowledge over a limited area around its location. As mentioned previously the WAVE has seven channels one for safety service and remaining for non-safety service both being controlled by time division [3]. With time being synced at default sync intervals of 100 ms. when the specific channel intervals begin it always starts with a guard interval which is used for accounting radio switching and inaccuracies in timing among different vehicles. The WAVE MAC interval architecture describes transit operations, data queuing, prioritization, channel selection and channel routing as shown in Fig. 2.

For reliable packet delivery conventional method used in case of VANET was Vehicle Assisted Data Delivery (VADD) which segments the roads and fastest route is selected around the current vehicle location as well as destination. The current vehicle considers adjacent vehicles in the same road segment to select the

Channel number	172	174	176	178	180	182	184

Channel type	SCH	SCH	SCH	CCH	SCH	SCH	SCH

Frequency Range (GHz)	5.855~5.865	5.865~ 5.875	5.875~5.885	5.885~5.895	5.895~5.905	5.905~5.915	5.915~5.925

Fig. 1 Channel layout of WAVE system with 1 control channel (CCH) and 6 service channel (SCH)

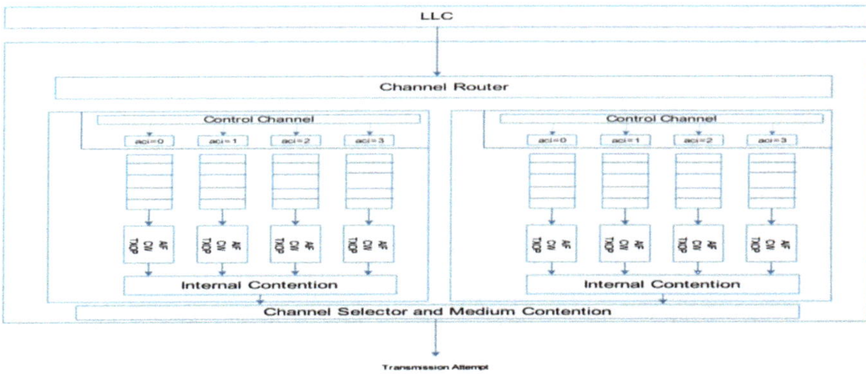

Fig. 2 Internal architecture of WAVE MAC

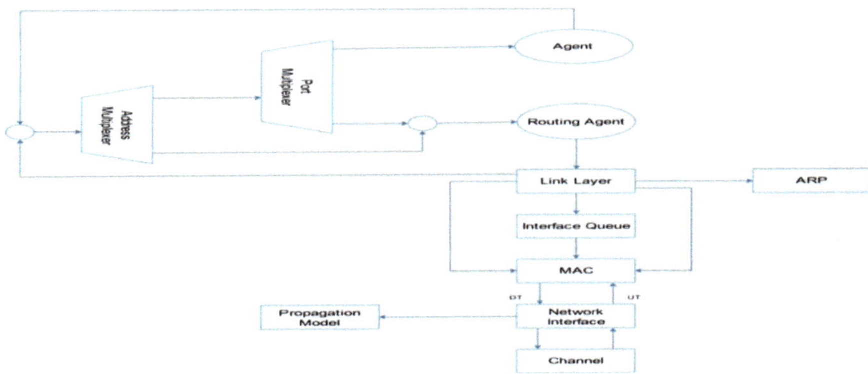

Fig. 3 Structure of mobile node

best path as next carrier [4]. In this paper a new approach is simulated in which time sharing is the main focus for every vehicle in order to carry a given message. Each vehicle is aware of high density segments in four opposite directions through its digital map. The aggregation includes time needed to reach any high density segment of the road. In the aggregation the probable vehicle trajectory is not important a vehicle does not act as a relay [5]. On the contrary in the case of relay after receiving the packet, relay mechanism uses location based broadcasting to relay packet towards destination.

Extensive studies have been conducted on propagation of messages in vehicular networks. Although the conventional methods are mainly focussed on homogenous traffic and high, low density in traffic is not considered. Figure 3 shows the mobile node structure in NS-3 which has been designed to support 802.11 MAC with a single channel [6, 7]. However this mobile node is not appropriate for WAVE multi-channel operation. Therefore we need to have every single MAC with its own

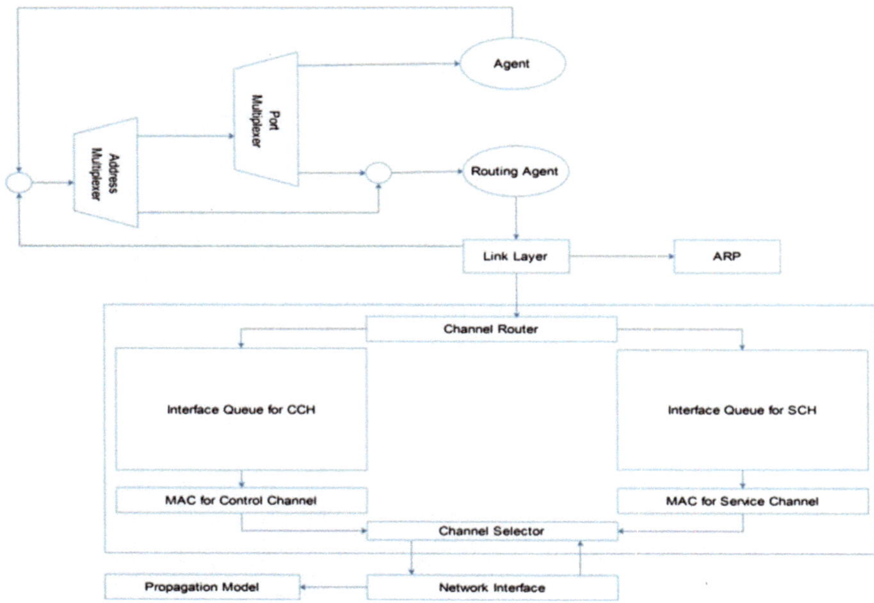

Fig. 4 Structure of modified mobile node

priority based packet buffering. Figure 4 shows modified mobile node structure in order to support the multi-channel co-ordination as required by WAVE.

2 Packet Delivery in Vehicular Ad Hoc Networks

In case of VANETs the current location of the vehicle with corresponding geographical elements need to be considered in order to achieve high rate of reliable packet delivery. In this paper an approach is discussed to combine as well as distribute knowledge gained by a node regarding the best paths and is forwarded to interested nodes. While such information in case of VANETS is time bound i.e. if path information is not delivered within a few minutes it becomes useless and distribution of such information would simply occupy network resources [8]. Therefore the goal of the proposed algorithm is to provide information in a reliable fashion to areas of interest rather than to specific nodes.

An assumption is taken into consideration that most of the commuters follow a repetitive path about 60% of the time. The same assumption can be used to statistically exploit such repetitive patterns between vehicles. This pattern can be used to calculate the degree to which the given information is useful to a vehicle when it receives from its neighbours [9, 10]. This can be used as a measure to whether this information can be used and/or forwarded by using the channel provided by

Table 1 Creation time and sending time of packets

Packet no.	Ch. no.	Packet creation time (s)	Packet sending time (s)
T1	CCH	11.03	11.0302
T2	SCH	11.03	11.0632
T3	CCH	11.06	11.1022
T4	SCH	11.07	11.0702
T5	CCH	11.14	11.1402
T6	SCH	11.13	11.1641

WAVE. In order to verify maximum channel utilization and interval access operation we use the same simulation which is explained in details in further sections of the paper. In the simulation nodes are created using the modified mobile node structure and one set of nodes are configured for sending and the other set for receiving packets. The nodes sending packets generate a control packet (CCH) and a signalling packet (SCH) which is sent to either CCH or SCH [11]. For proper transmission a packet in CCH interval moving towards SCH must wait for SCH interval. Packet creation and sending times are extracted from the trace files generated by NS3 are explained in Table 1. It can be clearly seen that a given packet T1 with CCH interval is formed at 11.03 s and immediately sent at 11.0302 s. Similarly packet T2 with SCH interval formed at 11.03 s was delayed because of the CCH interval and was sent at 11.0632 s with SCH interval.

Each vehicle stores its own path traversed based on which future path is predicted. When two or more vehicles encounter they exchange information regarding previous knowledge of path traversed by each vehicle, the information also contains knowledge about time taken by a packet for delivery reliably. Every delay in measurement is not same as a result of which the variation in delays may be large [12, 13]. Hence information exchange should be considered as an aggregation of information similar to previously captured information. Since storage is limited and recording multiple hops would lead to large amounts of information, the proposed algorithm's main objective is to combine the various encounters and measure delay in message delivery towards popular paths. Advantage of this process is that it results in greater degree of precision.

3 Performance Evaluation of Proposed Algorithm

In this section the proposed algorithm's performance evaluation is explained in detail. However before going into the details certain assumptions that were made need to be discussed, which are that each node has a copy of information that is being distributed which means more than one copy of a message may be moving around in the network. The simulation is setup in such a manner that each segment of the road is part of an intersection. A general assumption is taken that for a given road segment the vehicle in that segment is identifiable as being at one place [14].

On longer road segments there is a possibility that cars may not meet however this proposed algorithm only functions when a given vehicles are within the range of communication. The vehicle is assumed to be moving along a road segment i.e. the vehicle is either covering the whole road segment or partially covering it before moving out of the network.

For the sake of keeping things simple a given road segment is considered to be having exclusive density characteristics with respect to traffic. Thus each vehicle on a given road segment would identify same density of traffic. Although these features vary with time the inter-vehicle distances have an exponential distribution which has a mean distance of $1/\alpha_k$ where α_k is simply the density of vehicles on a given road segment ‘k’. Now as assumed that traffic density sensed by each vehicle in given road segment is uniform, the same assumption is used to define l_k which is simply the delay in packet forwarding for a given road segment and equals to:

$$l_k = \left(1 - e^{-P\alpha_k}\right)\frac{Q_k}{P}b + e^{-P\alpha^k}\frac{Q_k}{S_k} \ldots\ldots\ldots\ldots \quad (1)$$

Density of vehicles α_k in a given road segment ‘k’ can be computed using previously known data or by using the count of neighbours in a given road segment k. the traffic on a given road segment k could be high or low. Traffic in a given road segment is considered to be high if delay of relay in a packet is lesser than that for actually carrying it by a factor of ‘ρ’:

$$\left(1 - e^{-P\alpha_k}\right)\frac{Q_k}{P}b < \rho e^{-P\alpha_k}\frac{Q_k}{S_k} \quad 0 < \rho < 1 \ldots\ldots\ldots\ldots \quad (2)$$

Where P is the range of transmission and Q_k is length of a given road segment ‘k’. ‘b’ is a constant which approximates the delay for packet relay for a given pair of neighbours. S_k is simply the travel speed in a road segment ‘k’. The same categorization of road segments is used later on as well in order to select path with lowest delay in packet relay. Realistic simulations of vehicular traffic makes it clear that uniform distribution of traffic is not present [15]. High traffic road segments show that a single connected mesh is created with time or smaller meshes may be formed of low traffic. Each pair of source and destination can either be surrounded by high traffic road segments or separated by a road segment. In both scenarios nodes relaying information could end up with selecting the best path, including segments of high traffic.

By using knowledge of traffic characteristics roads can be categorised as to having high or low traffic however such methods as also used by VADD for selecting paths with lowest packet delay are not effective when traffic density is very low in which finding appropriate path with lowest packet delay becomes problematic. In case of VADD it selects the next best possible path on the basis of currently available vehicles in range [16, 17]. However lacks knowledge of future candidate vehicles for information relay. In such cases VADD ends up selecting a bad candidate for relay resulting in greater delay in packet transmission and also increases un-reliability. The proposed algorithm however proves to be more effective than VADD as it is able to find better paths with lowest delay as well as it is able to utilize knowledge gained for future path prediction maintaining high levels of privacy [18, 19]. A heuristic approach is used to reduce number of possible path selections in a few decisions. These decisions are taken in such a manner that the path selected would have road segments with shortest path not necessarily directly pointing towards destination but road segments of high traffic density. The advantage of doing so is that high density road segments provide mesh type connectivity among vehicles with lower delay in packet relay.

Now as mentioned earlier 60% of vehicle trips are on previously traversed paths and each vehicle keeps a copy of knowledge it has gained. Therefore the same knowledge of path can be used for future path prediction as well as computing forwarding time which is expected for a given road segment 'k' referred to as M_{f_k}:

$$M_{f_k} = \begin{cases} \frac{Q_k}{S_k}, & for\, k \in high\, density\, traffic \\ l_k, & for\, k \in low\, density\, traffic \end{cases} \quad \ldots\ldots\ldots\ldots \quad (3)$$

The equation mentioned above is based on the fact that vehicles in high density road segments are able to quickly relay packets towards their destinations. As this approach is over and above VADD, which acts as the basis for forwarding time computation for a given expected path and is relative to node density for a given road segment [20, 21]. In order to remove the requirement of global knowledge on travelling time, each vehicle calculates delay in delivery for a message. Dijkstra's shortest path calculation is used to compute delay in delivery for a road segment.

The expected time of travel for a given road segment 'k' is defined by keeping in mind the heavy traffic segments surrounding the segment in question as $G^{u}_{exp}(k)$. This value is initially kept equal to the summation of expected times of travel for different vehicles, starting with the current segment in question till that vehicle is reached which is present on the first heavy traffic road segment [22]. With each iteration for a given road segment the initial value is changed based on the experienced time of travel and is given by $G^{v}_{exp}(k)$ [23, 24]. Each moment of time in which a vehicle is able to locate a new surrounding vehicle it tries to synch it's time of travel that it has experienced. Whenever a new time is sent from a surrounding vehicle given as $G^{n}_{exp}(k)$ an update is made as per the following:

$$G_{exp}(k) = \beta(\alpha G^{n}_{exp}(k) + (1-\alpha)G^{o}_{exp}(k) + (1-\beta)T^{V}_{exp}(k) \ldots\ldots\ldots \quad (4)$$

B being the weighing factor used to adjust vehicle time in order to compute overall travel time:

$$\beta = e^{-\log_2 \frac{\gamma^2.neigbouring\, vehicles\, count}{\gamma^2.average\, of\, vehicle\, times}} \quad \ldots\ldots\ldots \quad (5)$$

Each vehicles broadcasts its new time of travel in the high traffic road segment if there is a vehicle which has not got updates. This process is done periodically and occurs at every Z_{int}. As explained in the previous section the rate at which information is synched is based on the fact of traffic sensed in the background [25]. If network resources are heavily consumed and time taken to sync is too long then the sync broadcasts are not sent at all. Time sync can be computed as E_{sn} where:

$$E_{sn} = e^{-A.C.B} \dots\dots\dots\dots \tag{6}$$

Such that

$$C = \frac{1}{C_{pp}\frac{1}{n}\sum_{i=1}^{n}\frac{1}{\left[\frac{C_{rc}(i)}{C_{su}}\right] + \in}} \dots\dots\dots\dots \tag{7}$$

And

$$B = \frac{\sum_{j=1}^{m}\frac{F_j}{X_j}}{C_{su}} \dots\dots\dots\dots \tag{8}$$

Where 'A' is the adjustment factor, 'C' is the urgency in time required for sending sync broadcast with C_{rc} as remaining time, C_{su} as time interval between updates, C_{pp} is period of prediction, 'B' acting as utilization factor for the network and F, X representing bits per transmission and rate of transmission respectively [26].

4 Simulation Parameters and Results

In order to validate the results of the proposed algorithm it is compared with VADD. In the research community VADD is considered as the basis for implementation in a large amount of VANET based applications. NS-3 is used as a simulator for broadcasting and syncing transmissions among vehicular nodes. The traces generated by SUMO are recorded and are used to simulate in NS-3 which acts as the urban simulator for mobility. The screenshot of SUMO is shown in Fig. 5 which depicts the road map created. The trace files contain large amounts of data so in order to simplify only a randomly selected selection from the trace file is simulated in NS-3 [27, 28]. The performance results of the proposed algorithm are compared with VADD. Important parameters used in the simulation are mentioned in the Table 2.

In the setup 150 vehicles are considered which are flowing in from both direction of the two roundabouts and the cross-section of the high speed highway traffic density is considered to be from low to high with vehicles moving in high traffic road segments as well as low traffic road segments. Comparison is done on the basis of delivery rate, delay and overhead.

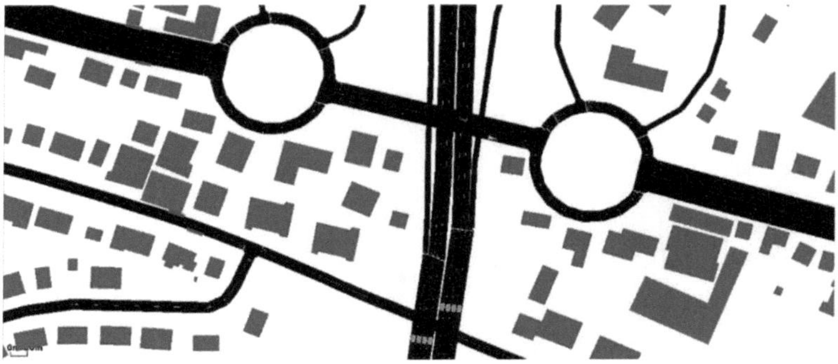

Fig. 5 Roadmap created for the simulation

Table 2 Simulation Parameters Used

Parameter	Value
Network interface	802.11 p
Number of nodes	150
Data rate	1 Mbps
Sync interval	2.5 s
Contact time	20 s
Transmission range	100 m
Packet size	5 kb
Packet rate	0.2/s

Fig. 6 Data delivery rate versus number of iterations

In the Fig. 6 delivery rate is shown as a function of total number of iterations. With each iteration the knowledge gained by the vehicle is adjusted based on previous iterations. The timer stops even if delivery to vehicle continues in case of the desired vehicle receiving its packet. The graph shows that initially the proposed

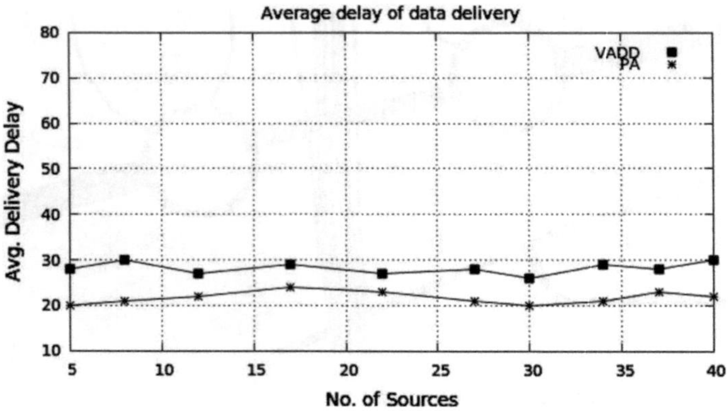

Fig. 7 Average delay of delivery versus number of sources

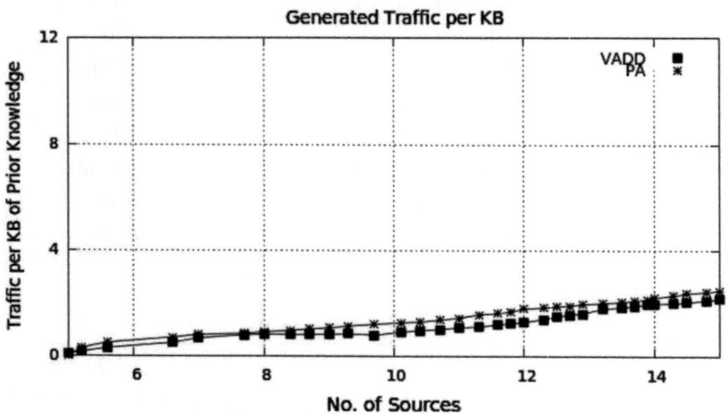

Fig. 8 Generated traffic per KB versus number of sources

algorithm (PA) has a very low delivery rate but with time by making use of previous knowledge and adjustment factor it shows better performance than VADD. The reason for this better performance is due to the usage of prior gained knowledge for a given road segment for each vehicle which is broadcasted and synced with other vehicles in the same road segment.

In the Fig. 7 the average delay in delivery is shown, the proposed algorithm performs better than VADD in case of high traffic and low traffic road segments this is because as mentioned earlier the PA uses relay which reduces packet delay significantly even in case of large number of packets being generated in the network for distribution. In Fig. 8 the effect of overhead distribution of prior gained knowledge is shown. The generated data is compared with control packets at the time of full network utilization. Comparative to VADD the proposed algorithm shows better results this is because the PA keeps a check on the rate of sync between vehicular nodes which does not result in over utilization of network resources which if occurs is bad as VANETs have mainly been used for safety based applications.

5 Conclusion

The proposed algorithm shows functioning of environmental aware distribution method which can be used in VANETs. The algorithm is based on the assumption that most vehicles over a period of time follow a repetitive path which can be used as knowledge gained over a period of time to predict the future path of the vehicle. The main objective is to simply provide a high rate of data delivery among vehicles moving from source to destination with reliability. In order to provide better results low density traffic road segments need to be considered with relaying. The results show promising improvements in delay of packets with contained overhead. For future work more performance based comparisons can be shown with larger network diversity and different traffic conditions. Also improvement is required in data packet queuing when packets are carried by a vehicular node. Certain traffic conditions are not considered such as traffic jams and accidents during the execution of simulations however that can be included in the future.

References

1. Di Felice, M., Ghandour, A. J., Artail, H., and Bononi, L. 2012. "On the Impact of Multi-channel Technology on Safety-Message Delivery in IEEE 802.11 p/1609.4 Vehicular Networks". In Computer Communications and Networks (ICCCN), 2012 21st International Conference on (pp. 1–8). IEEE.
2. A. Bessani, M. Correia, B. Quaresma, F. Andr, and P. Sousa, "Dep-Sky: dependable and secure storage in a cloud-of-clouds," Proc. Sixth conference on computer systems (EuroSys'11), ACM, 2011, pp. 31–46.
3. S. Sehrish; G. Mackey; P. Shang; J. Wang; J. Bent, "Supporting HPC Analytics Applications with Access Patterns Using Data Restructuring and Data-Centric Scheduling Techniques in MapReduce," Parallel and Distributed Systems, IEEE Transactions on, vol. 24, no. 1, pp. 158–169, Jan. 2013.
4. K. George; V. Venugopal, "Design and performance measurement of a high-performance computing cluster," Instrumentation and Measurement Technology Conference (I2MTC), 2012 IEEE International, vol., no., pp. 2531–2536, 13–16 May 2012.
5. N. Khaitiyakun, T. Sanguankotchakorn, and A Tunpan. "Data dissemination on manet using content delivery network (cdn) technique". In *Information Networking (ICOIN), 2014 International Conference on*, pp. 502–506, Feb 2014.
6. Jinwoo Nam, Seong-Mun Kim and Sung-Gi Min, "Extended wireless mesh network for VANET with geographical routing protocol," *11th International Conference on Wireless Communications, Networking and Mobile Computing (WiCOM 2015)*, Shanghai, 2015, pp. 1–6.

7. L. Ramakrishnan, R. S. Canon, K. Muriki, I. Sakrejda, and N. J. Wright. "Evaluating Interconnect and virtualization performance for high performance computing", ACM Performance Evaluation Review, 40(2), 2012.
8. Y. Bai, D. Xie, S. Wang and M. Zhong, "Multi-path transmission protocol in VANET," *2015 International Conference on Connected Vehicles and Expo (ICCVE)*, Shenzhen, 2015, pp. 204–209.
9. I. B. Jemaa, O. Shagdar, F. J. Martinez, P. Garrido and F. Nashashibi, "Extended mobility management and routing protocols for internet-to-VANET multicasting," *2015 12th Annual IEEE Consumer Communications and Networking Conference (CCNC)*, Las Vegas, NV, 2015, pp. 904–909.
10. K. C. Lee, U. Lee, and M. Gerla, Oct. 2009. "Survey of routing protocols in vehicular ad hoc networks". *Advances in Vehicular Ad-Hoc Networks: Developments and Challenges, IGI Global.*
11. I. O. Chelha and S. Rakrak, "Best nodes approach for alert message dissemination in VANET (BNAMDV)," *RFID And Adaptive Wireless Sensor Networks (RAWSN), 2015 Third International Workshop on*, Agadir, 2015, pp. 82–85.
12. D. Jiang, & L. Delgrossi, (2008). "IEEE 802.11p: towards an international standard for wireless access in vehicular environments". In Proceedings of 67th IEEE vehicular technology conference on vehicular technology (pp. 2036–2040), May 2008.
13. F. Granelli, G. Boato, and D. Kliazovich, 2006. "MORA:A movement-based routing algorithm for vehicle adhoc networks". *1st IEEE Workshop on Automotive Networking and Applications (AutoNet 2006)* 2006. San Francisco,USA.
14. J. Zhao, and G. Cao, 2008. "VADD: Vehicle-Assisted Data Delivery in Vehicular". *Vehicular Technology, IEEE Transactions on*.57.3.1910-1922.
15. G. Yan, D.B. Rawat and B.B. Bista, "Towards Secure Vehicular Clouds," Proc. Complex, Intelligent and Software Intensive Systems (CISIS), 2012 Sixth International Conference on, 2012, pp. 370–375.
16. Sandeep Pratap Singh, Shiv Shankar P Shukla, Nitin Rakesh, Vipin Tyagi, "Problem reduction in online payment system using hybrid model", International Journal of Managing Information Technology (IJMIT) Vol.3, No.3, pp. 62–71, August 2011.
17. Praveen K Gupta, Nitin Rakesh, "Different job scheduling methodologies for web application and web server in a cloud computing environment", 2010 3rd International Conference on Emerging Trends in Engineering and Technology (ICETET), pp. 569–572, 2010, India.
18. S. Pearson and A. Benameur, Privacy, "Security and Trust Issues Arising from Cloud Computing" Proc. Cloud Computing Technology and Science (CloudCom), 2012 IEEE Second International Conference on, 2012, pp. 693–702.
19. T. Atéchian, L. Brunie, J. Roth, and J. Gutiérrez. "DGcastoR: direction-based geo-cast routing protocol for query dissemination in VANET". *IADIS International Telecommunications, Networks and Systems.* 2008.
20. A. Festag, (2009). "Global standardization of network and transport protocols for ITS with 5 GHz radio technologies". In Proceedings of the ETSI TC ITS workshop, Sophia Antipolis, France, February 2009.
21. R.K. Shrestha, S. Moh, I. Chung, and D. Choi 2010. "Vertex-based multi-hop vehicle-to infrastructure routing for vehicular ad hoc networks". In *IEEE proceedings of 43rdHawaii International Conference on System Sciences (HICSS).*
22. L. Kristiana, C. Schmitt and B. Stiller, "Survey of angle-based forwarding methods in VANET communications," *2016 Wireless Days (WD)*, Toulouse, 2016, pp. 1–3.
23. Nitin Rakesh and Vipin Tyagi, "Linear Network Coding on Multi-Mesh of Trees using All to All Broadcast," International Journal of Computer Science Issues, vol 8, no 3, pp. 462–471, 2011.
24. Nitin Rakesh and Vipin Tyagi, "Parallel architecture coding: link failure–recovery mechanism (PAC: LF—RM)," Springer International Journal of System Assurance Engineering and Management, vol 4, no 4, pp. 386–396, 2013.

25. S. Grafling, P. Mahonen, & J. Riihijarvi. "Performance evaluation of IEEE 1609 WAVE and IEEE 802.11 p for vehicular communications". In Ubiquitous and Future Networks (ICUFN), 2010 Second International Conference on (pp. 344–348). IEEE.
26. Praveen K. Gupta and Nitin Rakesh, "Different Job Scheduling Methodologies for Web Application and Web Server in a Cloud Computing Environment", 2010 3rd International Conference on Emerging Trends in Engineering and Technology, IEEE, pp. 569–572, Goa, India, Nov. 19–21.
27. Nitin Rakesh and Vipin Tyagi, "Linear-code multicast on parallel architectures," Elsevier Advances in Engineering Software, vol. 42, pp. 1074–1088, 2011.
28. Nitin Rakesh and Nitin, "Analysis of Multi-Sort Algorithm on Multi-Mesh of Trees (MMT) Architecture", Springer Journal of Supercomputing, vol 57, no 3, pp. 276–313, 2011.

Integration of IEC 61850 MMS and IEEE 802.22 for Smart Grid Communication

Vasudev Dehalwar, Akhtar Kalam, Mohan Lal Kolhe, Aladin Zayegh and Anil Kumar Dubey

Abstract The reliability of Smart Grid depends on two-way communication between substation and utility. IEC 61850 is an international standard defined to ensure interoperability between Substation Automation System (SAS). IEC 61850 services are mapped on the Manufacturing Message Specification (MMS) especially for meter data management system. MMS is the OSI protocol that run over TCP/IP or OSI networks to support IEC 61850 services. Presently, the MMS uses Ethernet as the layer 2 protocol. However, for long distance communication in remote areas wireless communication is the prefered mode. The Cognitive Radio based IEEE 802.22 is next generation standard for Wireless Regional Area Network (WRAN) that can support long distance wireless communication for low-latency, high-volume, reliable and secure communication. This paper shows the potential to integrate IEC 61850 MMS with IEEE 802.22 for long distance Smart Grid communication.

Keywords IEEE 802.22 · IEC 61850 · Cognitive Radio · Smart Grid

---

V. Dehalwar · A. Kalam · A. Zayegh
College of Engineering and Science, Victoria University, Melbourne, Australia
e-mail: vasudev.dehalwar@live.vu.edu.au

A. Kalam
e-mail: akhtar.Kalam@vu.edu.au

A. Zayegh
e-mail: aladin.Zayegh@vu.edu.au

M.L. Kolhe
Faculty of Engineering and Science, University of Agder, 4604 Kristiansand, Norway
e-mail: mohan.l.kolhe@uia.no

A.K. Dubey (✉)
Government Engineering College Ajmer, Ajmer, Rajasthan, India
e-mail: anildudenish@gmail.com

G.M. Perez et al. (eds.), *Networking Communication and Data Knowledge Engineering*, Lecture Notes on Data Engineering and Communications Technologies 3, https://doi.org/10.1007/978-981-10-4585-1_7

1 Introduction

The communication technology performs an important task of collecting information from various stages of energy generation, transmission, and distribution [1]. The high-level architecture of today's electric grid allows monitor and control of transmission and distribution substation remotely, thereby enhancing operational efficiency of the grid. The recommendation of National Institute of Standards and Technology (NIST) breaks the Smart Grid into seven domains for data interchange and smart judgments as described in Fig. 1 [2, 3]. The domain comprises of customers, markets, service providers, operations, bulk generation, transmission, and distribution. The distribution side of the model comprises of Home Area Networks (HANs), Building Area Networks (BANs), and Industrial Area Networks (IANs) [4, 5]. In the utility network, WAN is referred as core network that connects utility's infrastructure called Field Area Networks (FANs) with the control center [1]. FAN is a part of distribution grid which acts as an interface, connecting end nodes IEDs (Intelligent Electronic Device) of a microgrid with Control Center. The applications supported by the Smart Grid are Advanced Metering Infrastructure, Demand Response, Dynamic Pricing, Distributed Energy Resources (including PHEV charging), etc. Building Automation and Control Networks (BACnet), Home Energy Management Systems (HEMS), etc. also oversees demand in the networks. The utility can prefer either wired or wireless network on the basis of requirements in the utility.

The fast and timely response is the essence of Smart Grid. The distribution and transmission management functions that are critical in grid management are Supervisory Control and Data Acquisition (SCADA), Flexible AC Transmission

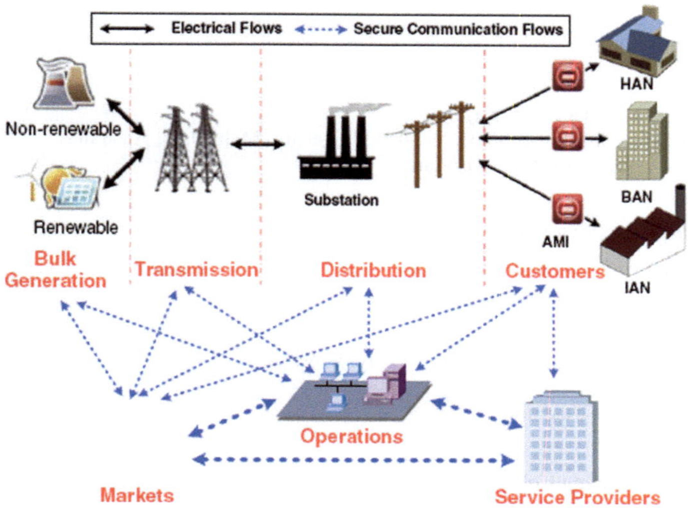

Fig. 1 Conceptual architecture of NSIT [2, 6]

System (FACTS), Energy management system (EMS), Digital Fault Recorder (DFR), Wide-area situational awareness and control (WASA&C), Dynamic Line Rating (DLR), etc. Some advance Smart Grid functions are Fault Detection, Isolation, and Restoration; Conservation Voltage Regulation; Volt/Var control; Power Quality controls, etc. all of them requires fast and timely response [7].

The main contributions of this study are investigating the suitability of IEEE 802.22 for synchrophasors transfer and teleprotection in Smart Grid and determining the compatibility of IEC 61850 with IEEE 802.22 for integration and data exchange. Section 2 presents the communication architecture of Smart Grid. Section 3 discusses features and standards of IEC 61850. Section 4 describes IEEE 802.22 standards relevant to this study. Section 5, discuss the integration of IEC 61850 with IEEE 802.22 with an objective to provide long distance and reliable communication. Section 6, present the final communication model. Section 7, conclude the study.

2 Smart Grid Communication

Distributed microgrids are interconnect by cluster routers/gateways to Core networks of large distribution and transmission grid. The edge networks are the distributed substations in the Smart Grid. The merging of the data traffic at the substation is shown in Fig. 2 [1]. The cluster routers also called WAN router (WR) are the interface between core network of Wide Area Network (WAN) and distributed substation. Substation automation was first introduced in the 1970s through the first generation of SCADA systems [7]. The control panel of SCADA comprises of control, automation and protection devices which are located at one place in the building. With the advancement in technology, SCADA substation automation transformed its communication entity into Ethernet-based since 2004. The Ethernet-based cable communication symbolizes a revolution in the data exchange. However, the SCADA developers started introducing proprietary protocols for data exchange between the control centers and substations.

The data and control center (DCCs) host the operations of utility on their servers. Data acquisition, control, and other operations from edge network are collected on servers through the cluster router. The core-edge network illustrated in Fig. 2 shows the aggregation of traffic at DCCs for energy management, meter data management and host of other applications. There is a considerable increase in the volume of data exchange between core and edge network. The large volumes of data are produced by modern substations at a regular interval. The data produced at regular interval are (i) change in voltages and currents status sampled at a very high frequency (approx. 50/60 times in a second) called phasor measurement and (ii) change in the state of any switch [8]. Furthermore, the data communicated to DCCs had to be processed for smart decision making at enterprise level. The entire processes generate large volumes of data classified as Big Data which requires

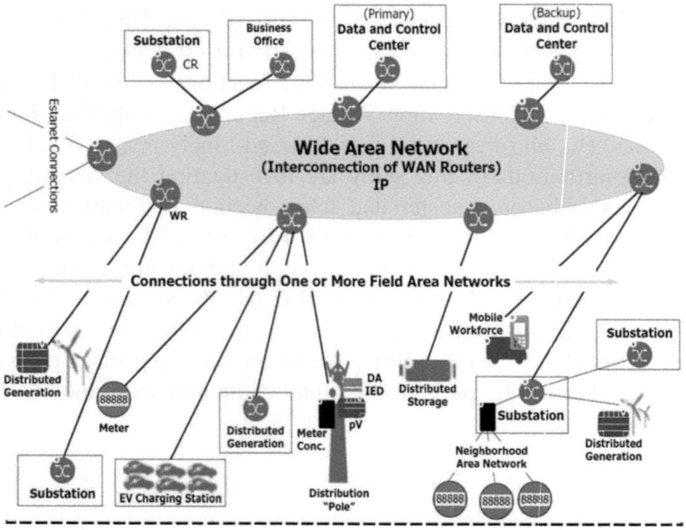

Fig. 2 Core-edge network topology in smart grid [1]

better communication facilities. The Cognitive Radio based IEEE 802.22 Wireless Regional Area Network (WRAN) have the potential to transfer data wirelessly to long distance in core-edge network of the Smart Grid.

3 IEC 61850

IEC 61850 protocol suite was designed to address the issue of interoperability between Intelligent Electronic Device (IEDs) of different vendors within substation automation systems. IEDs are microprocessor based controller that receives inputs from CT, VT, switchgear, etc. to perform protection, control and associated functions. The basic approach in designing IEC 61850 happens to partition a physical device into logical devices, that is further portioned into Data Objects, Logical Nodes, and Data Attributes as shown in Fig. 3 [9, 10].

The different types of performance classes in IEC 61850 are mapped to corresponding communication protocols with an idea to assist the particular requirement of different types of messages. The different class of mappings are GOOSE (Generic Object Oriented Substation Event), Time Sync and GSSE (Generic substation status event), SV (Sampled Values), MMS (Manufacturing Message Specification) shown in Fig. 4 [9, 10]. The message types having similar requirements are clubbed together and mapped onto the same protocol. For example, the time critical messages like type 1 and 1A are mapped to GOOSE and directly embedded into Ethernet to cut processing delay.

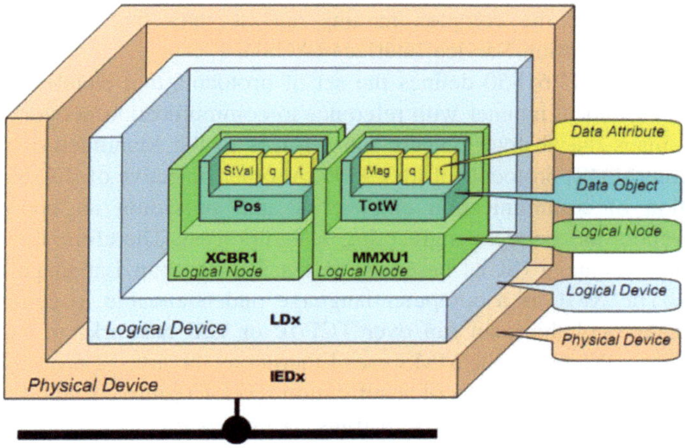

Fig. 3 IEC 61850 data modeling [9, 10]

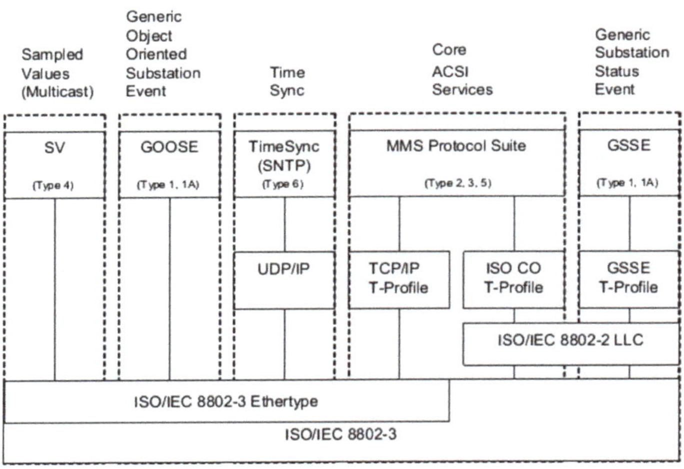

Fig. 4 Mapping of various protocols [10, 11]

The SV and GOOSE are mapped directly onto Ethernet to reduce processing time, but direct mapping of these protocols on Ethernet has a disadvantage. Since they are not encapsulated with higher layer protocol such as TCP/UDP they cannot be routed over Wide Area Network (WAN) which is necessary for teleprotection and transmitting synchrophasors. Instead, other application like smart meter reading through Meter data management system (MDMS) are not time critical which can be mapped to MMS. The MMS is a protocol that supports transfer of supervisory control information and process data between networked devices and control center.

MMS also supports the complex naming and services that are mapped onto Abstract Communication Service Interface (ACSI).

The ACSI of IEC 61850 defines the set of protocols that enables all IEDs to behave in an identical manner with reference to communication network. The IEC 61850-8-1 maps the abstract objects and services to the Manufacturing Message Specification (MMS) protocols of ISO 9506. Another objective of IEC 61850 is the simplification of communication engineering and providing an opportunity to support a vendor independent engineering design process. Therefore, the substation configuration language (SCL) based on Extensible Markup Language (XML) is developed. The XML is a computer language understandable to computers [7]. MMS is a protocol than can run over TCP/IP or OSI networks to support IEC 61850 services. Presently, the MMS uses Ethernet as the layer 2 protocol which is favorable communication protocol at substation level. Cognitive Radio based MAC layer protocol can replace Ethernet for transferring supervisory control information and real-time process data between networked devices and control center.

4 IEEE 802.22 WARN

WRAN can provide low-latency, high-volume, reliable and secure communication over long wireless network. The IEEE 802.22 standard are developed for WRAN that uses television white space (TVWS) [9, 12, 13]. The secondary user is allowed to use unused spectrum opportunistic if it is not being utilized by the primary user. The next generation ingenious wireless communication system called Cognitive Radio (CR) is cognizant of its surrounding, reads the situation and adapts to statistical changes in the incoming RF stimuli [14]. The important features of Cognitive Radio are learning, intelligence, awareness, adaptability, efficiency and reliability. The major tasks associated with Cognitive Radio are depicted in Fig. 5. The tasks include (a) Radio-scene analysis, (b) Channel identification and predictive modeling of radio scene and (c) Spectrum management and transmitted power control [14].

IEEE 802.22 can be used to exchange and transfer of utility data to Data and Control Center through coordinated efforts. The end nodes in edge network are Customer Premise Equipment (CPE) that functions as Network Control and Management Systems (NCMS) at both Cognitive Base station (CBS) and Customer Premise Equipment (CPE) [9, 12]. The BS and CPEs shall store and collect the managed objects in the format defined by WRAN Management Information Base (MIB). The Network Control System contains the service flow and the associated QoS information that is pre-populated in service classes at the BS and instantiated when a CPE requests services. The services supported by NCMS also include: Authentication, Authorization, and Accounting (AAA) services, Radio Resources Management (RRM) services, Location-Based services (LBS) management, Network Management services, Service Flow Management services, and Security services.

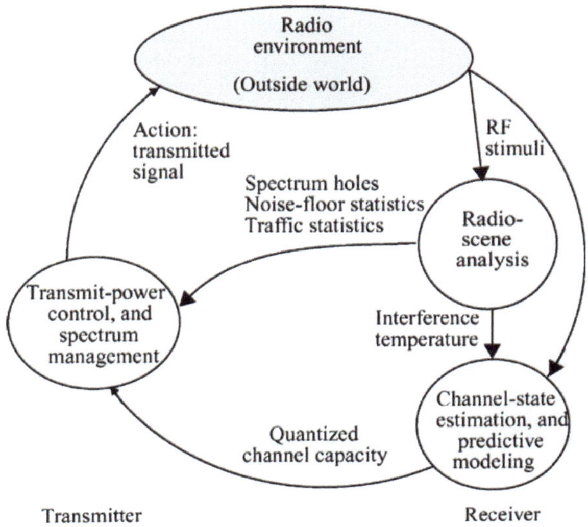

Fig. 5 Basic cognitive cycle [14]

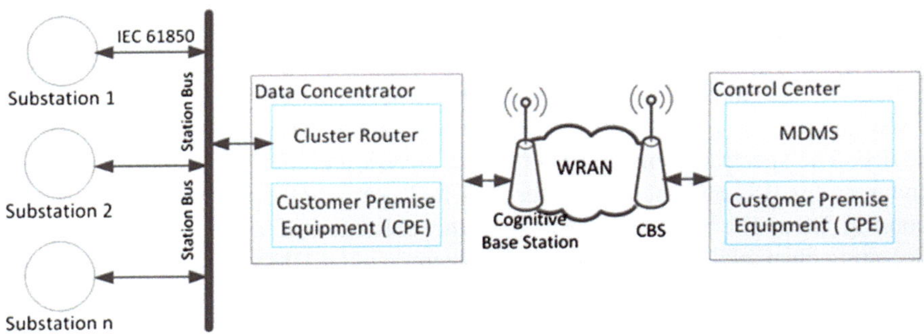

Fig. 6 Communication model of IEC 61850 and IEEE 802.22

5 Integration of IEC 61850 MMS with IEEE 802.22

The design and configuration of the Cognitive Radio network for the data transfer with the external agents/control centers is depicted in Fig. 6. The IEC 61850 support client–server and publisher–subscriber communication principles [7].

The MMS traffic model is designed and implemented based on the technical specifications of the IEC 61850. The MMS is an application layer protocol that defines how the messages have to be transported by underlying protocols. Figure 7 depicts layered architecture of MMS stack and MAC layer of IEEE 802.22.

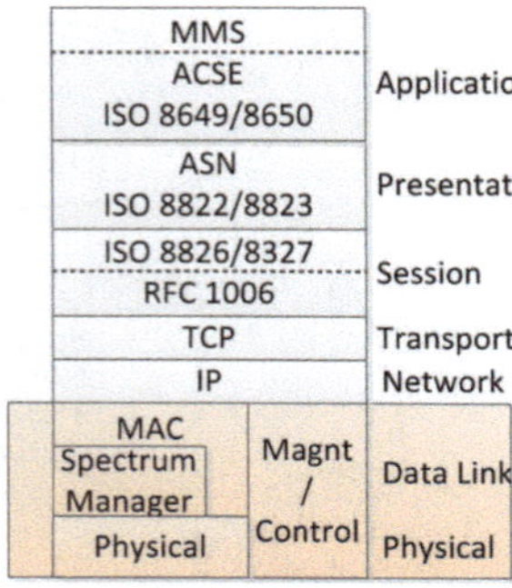

Fig. 7 Layered architecture of protocol integration

6 Communication Model

The state transition diagram of Fig. 5 explicitly mentioned the phases of model implementation. When the CPE is powered up, it needs to be admitted into the network by the BS depending on the capabilities of CPE negotiated by the BS. This may involve many tasks (e.g., geolocation and sensing channels) and handshaking between the CPE and the BS, and the procedure is called initialization and network association. The incumbent safe initialization and network association process is critical for Cognitive Radio systems. The process of spectrum sensing requires observing the RF spectrum for its occupancy by either incumbents or other WRANs device. The spectrum sensing is performed by BS and CPE by scanning all set of channels for BSs and incumbent services and synchronizes the network to neighbouring BSs. The CPE will recognize the existence of a BS transmission and operating channels. The BS shall send concentrated OFDM symbols composed of a superframe preamble, a frame preamble and an SCH once every superframe in its operating channel. The channels N and N $\pm$ 1 are scan by CPE to pass the sensing and timing requirements. If these channels pass the selection criteria, BS and CPE perform initial ranging, authentication and key exchange functions. The database of used and unused channels is maintained by BS for a particular service area. The Spectrum Manager (SM) at the BS receives an initial list of available channels from the database service. If there is no database service, the SM initially considers all channels available. Spectrum Sensing Automaton (SSA) and Spectrum Sensing Function (SSF) are an important entity of all devices (BS and CPEs). Spectrum sensing function is processed by SM through the SSA interfaced with SSF. The sensing behaviour of SSA is normally controlled by BS. The SSA keeps track of channels that are engaged by incumbents and the channels that are vacant which can be used for WRAN transmission. In order to avoid interference to incumbents the SSA allows dynamic frequency selection on a real-time basis. The status of network is updated frequently by the BS through the superframe control header (SCH) for in-band sensing during the quiet periods defined by the SM.

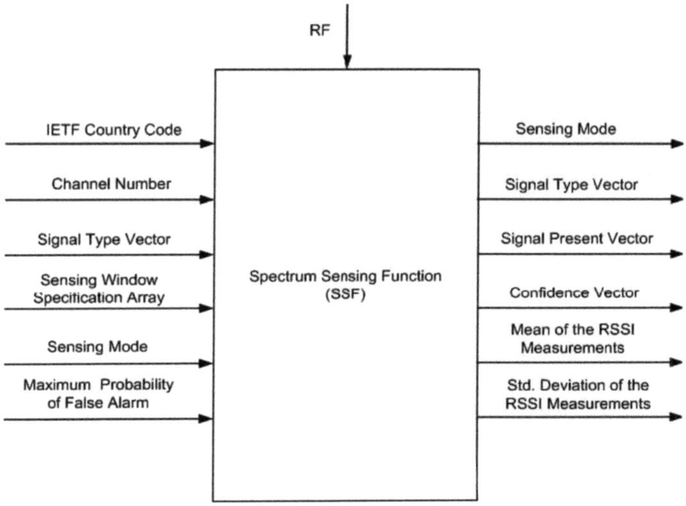

Fig. 8 Spectrum sensing function [12]

For each channel, an RSSI measurement shall be performed on the WRAN signal path and attempt shall be made to detect a WRAN superframe header or a CBP burst. If an SCH is captured and the level of the RF signal on the WRAN signal path is sufficiently high, attempt shall be made to acquire the frame header, the broadcast PDUs sent by the BS to advertise the WRAN service for BS and CPE initialization. The list of channels prohibited from incumbent operation obtained from the base station by the IPC-UPD management MAC message. If an SCH can be acquired but the signal level is insufficient or a CBP burst can be acquired but cannot be decoded, the presence of a WRAN signal shall be recorded along with the channel number and the measured RSSI. If an SCH or a CBP burst cannot detect the presence of broadcast incumbents then their signal type can be determined through RF signal sensing and signal classification. The result of the measurement and the signal classification shall be provided to the SM at the BS or stored locally at the CPE. The channel shall be incremented and the above initial sensing shall be repeated. The RF spectrum is continuously observed by SSF both at the base station and the CPEs and updated regularly. The SSF is part of MAC management and the inputs and outputs functions of the SSF are illustrated in Fig. 8.

7 Conclusions

The complexities of Smart Grid make it difficult for real-time communication of data. The prevailing wireless technologies are not capable to transmit Big Data of Smart Grid to control center due to bandwidth restrictions. The new IEEE 802.22

WRAN uses Cognitive Radio for communication, which has the potential to support high bandwidth, fast and reliable communication. The performance classes of IEC 61850 can use Manufacturing Message Specification (MMS) for transmitting synchrophasors and teleprotection over WAN. The superframe structure of IEEE 802.22 is capable to send data on upstream burst of uplink frequency. The integration of IEC 61850 with IEEE 802.22 can provide better communication facility in Smart Grid.

References

1. J. G. D. Kenneth C. Budka, Marina Thottan, Communication Networks for Smart Grids: Springer-Verlag London, 2014.
2. G. Locke and P. D. Gallagher, "NIST framework and roadmap for smart grid interoperability standards, release 1.0," *National Institute of Standards and Technology,* p. 33, 2010.
3. Y. Ye, Q. Yi, H. Sharif, and D. Tipper, "A Survey on Smart Grid Communication Infrastructures: Motivations, Requirements and Challenges," *Communications Surveys & Tutorials, IEEE,* vol. 15, pp. 5–20, 2013.
4. W. Meng, R. Ma, and H. H. Chen, "Smart grid neighborhood area networks: a survey," *Network, IEEE,* vol. 28, pp. 24–32, 2014.
5. Y. Rong, Z. Yan, S. Gjessing, Y. Chau, X. Shengli, and M. Guizani, "Cognitive radio based hierarchical communications infrastructure for smart grid," *Network, IEEE,* vol. 25, pp. 6–14, 2011.
6. J. Gao, Y. Xiao, J. Liu, W. Liang, and C. L. P. Chen, "A survey of communication/networking in Smart Grids," *Future Generation Computer Systems,* vol. 28, pp. 391–404, 2012.
7. B. M. Buchholz and Z. A. Styczynski, *Smart grids: fundamentals and technologies in electricity networks*: Heidelberg: New York: Springer, [2014], 2014.
8. A. Bose, "Smart Transmission Grid Applications and Their Supporting Infrastructure," *Smart Grid, IEEE Transactions on,* vol. 1, pp. 11–19, 2010.
9. V. Dehalwar, A. Kalam, M. L. Kolhe, and A. Zayegh, "Review of IEEE 802.22 and IEC 61850 for real-time communication in Smart Grid," in *2015 International Conference on Computing and Network Communications (CoCoNet)*, 2015, pp. 571–575.
10. I.-T. Ed.2, "Communication networks and systems for power utility automation – Part 1: Introduction and Overview," 2012.
11. R. E. Mackiewicz, "Overview of IEC 61850 and Benefits," in *Power Systems Conference and Exposition, 2006. PSCE '06. 2006 IEEE PES*, 2006, pp. 623–630.
12. IEEE, "ISO/IEC/IEEE International Standard – Information technology – Telecommunications and information exchange between systems – Local and metropolitan area networks–Specific requirements – Part 22: Cognitive Wireless RAN Medium Access Control (MAC) and Physical Layer (PHY) specifications: Policies and procedures for operation in the TV Band," *ISO/IEC/IEEE 8802-22:2015,* pp. 1–678, 2015.
13. P. Chang-Woo, Z. Xin, S. Chunyi, Z. Ming-Tuo, and H. Harada, "A new standard activity in IEEE 802.22 wireless regional area networks: Enhancement for broadband services and monitoring applications in TV whitespace," in *Wireless Personal Multimedia Communications (WPMC), 2012 15th International Symposium on*, 2012, pp. 108–112.
14. S. Haykin, "Cognitive radio: brain-empowered wireless communications," *Selected Areas in Communications, IEEE Journal on,* vol. 23, pp. 201–220, 2005.

KRUSH-D Approach for the Solution to Node Mobility Issue in Underwater Sensor Network (UWSN)

Nishit Walter, Nitin Rakesh, Rakesh Matam and Shailesh Tiwari

Abstract The most frequently experienced challenge in Underwater Wireless Sensor Network (UWSN) is the mobility issue i.e. the nodes present underwater changes their position from one place to another, therefore affecting the entire communication in the network. In our previous work, we proposed a 2D-based solution namely "Arc moment" for the node mobility. This approach used the Euclidean 2D distance formula and various other assumptions to efficiently maintain the communication. In this paper, we have proposed a 3D based approach named KRUSH-D, which is a further enhancement to our previous 2D approach. The proposed approach that has now been put forward rectifies the mobility issue, and also maintains the communication in the network at the same time. It brings in use the Euclidean 3D distance calculation and the famous KRUSKAL algorithm for path selection. The main objective here is to provide a solution for maintaining reliable communication in UWSN. The proposed work is examined using example (s) in order to provide readers clarity.

Keywords UWSN · Mobility issue · Communication network · Distance

N. Walter (✉) · N. Rakesh · R. Matam · S. Tiwari
Department of Computer Science and Engineering, Amity University, Noida, Uttar Pradesh, India
e-mail: nishitwalter1@hotmail.com

N. Rakesh
e-mail: nitin.rakesh@gmail.com

R. Matam
e-mail: matamrakesh@gmail.com

S. Tiwari
e-mail: shail.tiwari@yahoo.com

R. Matam
Indian Institute of Information Technology, Guwahati, India

S. Tiwari
A.B.E.S Engineering College, Ghaziabad, India

G.M. Perez et al. (eds.), *Networking Communication and Data Knowledge Engineering*, Lecture Notes on Data Engineering and Communications Technologies 3, https://doi.org/10.1007/978-981-10-4585-1_8

1 Introduction

Wireless Sensor Network (WSN) is a distributed network that comprises of autonomous sensors. It is used for the purpose of monitoring physical and environmental conditions, parameters such as temperature, sound etc. WSNare motivated by defense applications such as analyzing battlefield etc. Nowadays WSN plays a very important role in various areas [1]. A WSN comprises of nodes and sensors that are connected to each other, where the counting of nodes can vary depending upon the size of the area that is to be covered and each node is connected with one or more than one sensors. Underwater wireless communication network is a technique that is used for the purpose of sending and receiving information underwater with the help of some sensory devices and underwater autonomous vehicles. For the application of underwater networking, acoustic networking is best suited. It consists of multiple sensory devices that are deployed underwater for monitoring and accumulating data [2]. There is a very large variety of areas where WSN plays a very crucial role such as health care, disaster management and prevention, landslide detection, water quality monitoring etc. [3].

In this paper, we propose a 3D based solution named KRUSH-D which provides an effective and reliable approach for maintaining communication despite some of the nodes getting drifted away [4] from their location leading to node mobility problem [5] and thus establishing a network [6] with a path i.e. fast in formation for transferring data/information [7] from one place to another and thus preventing the network from breaking off [8–10]. The paper is divided in five sections. First section is about introduction of wireless sensor network. The second section comprises of the proposed approach for solving the node mobility issue. Section third is about the dry run of the proposed approach. The result has been given in the fourth section. Fifth section presents the conclusion and scope of future work.

2 Proposed Approach

Considering the 3 Dimensional nature of underwater environment [11], we are considering three axis x, y and z to propose the solution so that each and every aspect of mobility in the underwater environment is addressed. The proposed approach consists of two parts, i.e. first is calculating distance between nodes using 3D distance formula (after calculating the distance, arranging the distance in ascending order) and second is applying KRUSKAL algorithm afterwards, in order to obtain the path for passing the information from source to the destination establishing a communication network (Use the path from source to destination created by application of algorithm; If path is not formed due to mobility of more nodes, then again repeat the steps to obtain path). Thus, preserving the communication network and preventing it from failure even after experiencing node mobility problem.

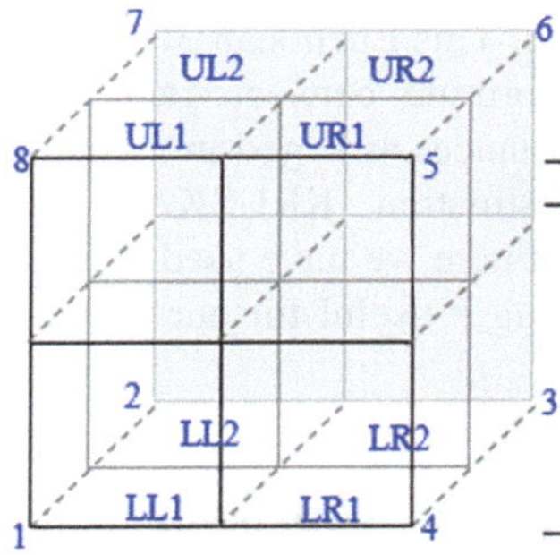

Symbols	Description
LL1	Lower Left 1
LL2	Lower Left 2
LR1	Lower Right 1
LR2	Lower Right 2
UL1	Upper Left 1
UL2	Upper Left 2
UR1	Upper Right 1
UR2	Upper Right 2

Fig. 1 8 quadrants in underwater 3D environment with their names

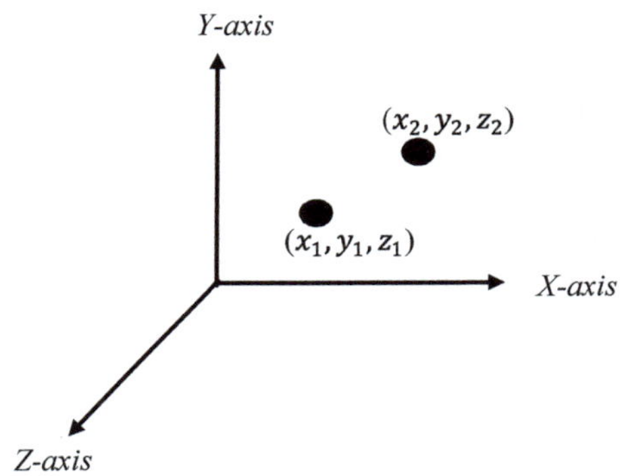

Fig. 2 Two nodes in x, y and z axis

After the execution of the above mentioned method, a path can be formed which in turn results in maintaining the communication network even after suffering node mobility issue. If the termination condition is not satisfied, the approach repeatedly executes till the termination condition is reached, resulting in the formation of a network. As we are dealing with 3D environment of underwater, for better understanding and reestablishment of communication network, we have divided the area into 8 quadrants, so that it can be clearly distinguished whether the nodes are still in their area of work i.e. sensing zone; or have drifted into another quadrant i.e. the domain of quadrant of some other zone. This way, making use of quadrants for maintaining communication in the network can be done efficiently (some details regarding the quadrants is discussed under Sect. 4). Figure 1 shows the arrangement of 8 quadrants. Here the Euclidean 3D distance formula between two points is used to calculate the distance between nodes in 3D environment of underwater. The 3D distance formula is as in Fig. 2.

$$d = \sqrt{(x_2 - x_1)^2 + (y_2 - y_1)^2 + (z_2 - z_1)^2},$$

Here d denotes the distance between nodes. Here (x_1, y_1, z_1) are the coordinates of first node and (x_2, y_2, z_2) are the coordinates of second node among whom the distance is to be calculated. This Euclidean 3D distance formula is the extension of Euclidean 2D distance formula between two points and the KRUSKAL [12] algorithm make use of distances arranged in ascending order and forming the path from source to target/destination. KRUSKAL algorithm is primarily used for making spanning trees but here we have used it functionality in forming path from source to target and making it useful for our proposed paper.

3 Dry Run of Proposed Approach

A communication is set among nodes and sensors underwater. In Fig. 4 Let the line between nodes shows the nodes involved in the flow of information in a communication from X to Y. Due to some disturbance, the node D gets drifted away due to mobility issue (far away then the present nodes). Now, 3D Euclidean distance formula is used to find out the distance among nodes and latter finding the path using KRUSKAL algorithm for establishing fast and efficient path for communication (Fig. 3).

Fig. 3 Communication network

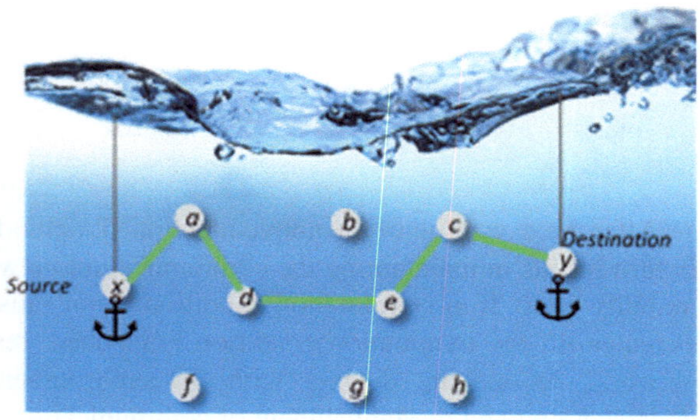

Fig. 4 Node connectivity using KRUSKAL

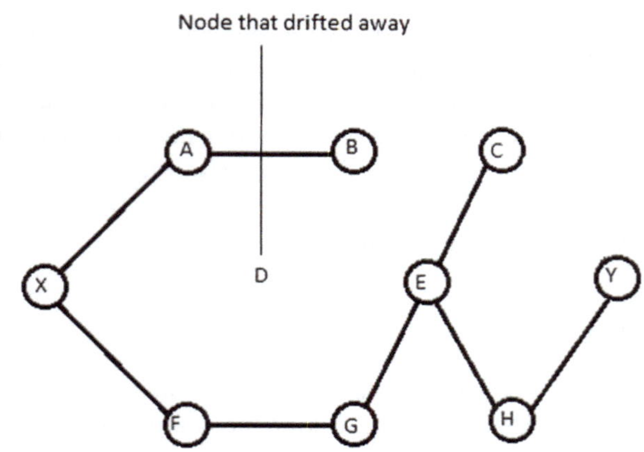

Table 1 Distance calculation

Distance between	Coordinates of first node	Coordinates of second node	Euclidean 3D distance
A to B	1, 3, 2	3, 3, 2	$\sqrt{(3-1)^2+(3-3)^2+(2-2)^2}$ = **2**
B to E	3, 3, 2	5, 2, 1	$\sqrt{(5-3)^2+(2-3)^2+(1-2)^2}$ = **2.4**
E to G	5, 2, 1	3, 1, 1	$\sqrt{(3-5)^2+(1-2)^2+(1-1)^2}$ = **2.2**
G to F	3, 1, 1	1, 1, 1	$\sqrt{(1-3)^2+(1-1)^2+(1-1)^2}$ = **2**
C to Y	6, 3, 2	9, 2, 1	$\sqrt{(9-3)^2+(2-3)^2+(1-2)^2}$ = **3.3**
X to F	0, 2, 1	1, 1, 1	$\sqrt{(1-0)^2+(1-2)^2+(1-1)^2}$ = **1.4**
X to A	0, 2, 1	1, 3, 2	$\sqrt{(1-0)^2+(3-2)^2+(2-1)^2}$ = **1.7**
E to C	5, 2, 1	6, 3, 2	$\sqrt{(6-5)^2+(3-2)^2+(2-1)^2}$ = **1.7**
G to H	3, 1, 1	6, 1, 1	$\sqrt{(6-3)^2+(1-1)^2+(1-1)^2}$ = **3**
H to Y	6, 1, 1	9, 2, 1	$\sqrt{(9-6)^2+(2-1)^2+(1-1)^2}$ = **3.1**
E to H	5, 2, 1	6, 1, 1	$\sqrt{(6-5)^2+(1-2)^2+(1-1)^2}$ = **1.4**

Now calculating distance for neighboring nodes. We are calculating a few distances, with the help of which the following example can be shown; otherwise all the distance needs to be calculated. Here we have used Euclidean distance formula, which is well known. Euclidean distance formulas can be applied to 2D space as well as to 3D space. Here we are using the 3D distance formula as we are dealing with underwater 3D environment, so we have x, y and z coordinates for the representation of node. After applying Euclidean 3D distance formula we get distance between nodes. After getting distance we can apply KRUSKAL algorithm that makes use of these distances that we had calculated in formation of path (Table 1).

We have taken above 11 distances into consideration to show our example. Arranging the distances in ascending order: XF = 1.4, EH = 1.4, XA = 1.7, EC = 1.7, AB = 2, GF = 2, EG = 2.2, BE = 2.4, GH = 3, HY = 3.1, CY = 3.3. Now applying KRUSKAL algorithm on the basis of above 11 distances evaluated, for establishing a path. Figure 4 shows the connectivity of nodes (not considering D) after application of KRUSKAL algorithm and Fig. 5 shows the path established from source to destination for the flow of communication.

Figure 5, shows the establishment of communication, after D being got drifted away. Lines show the flow of communication. In case any other node (or many nodes) gets drifted away, then in that case the distance between all the nodes is

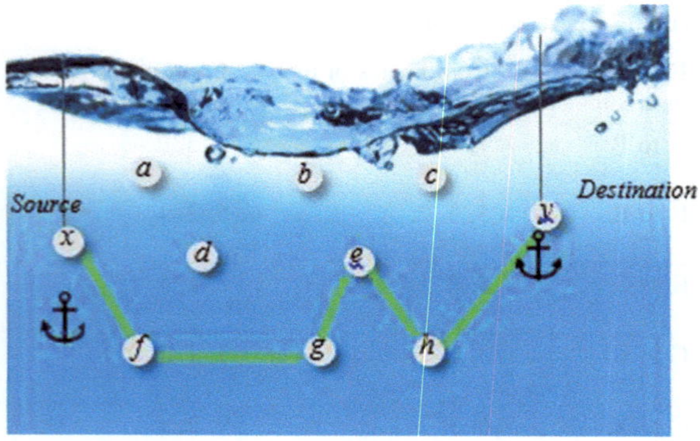

Fig. 5 New path after D drifted away

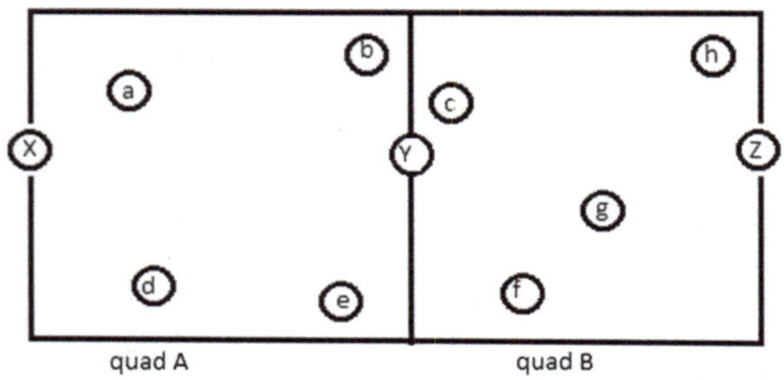

Fig. 6 Two out of eight quadrants of 3D perspective for reference

needed to be calculated and the same procedure needs to be followed for maintaining network. Due to node mobility, there are times when nodes from other quadrant comes closer, in that case we have to consider those as well because there is a possibility that the approaching node/nodes might be much closer to the source or destination. So, they will also be evaluated for Euclidean 3D distance in order to form a path much efficiently from source to destination and thus preserving communication in underwater wireless sensor network. There could be a possibility for nodes of different quadrants approaching near, as shown in the Fig. 6.

In Fig. 6, we consider X, Y and Z as targets and a, b, c, d, e, f, g and h as nodes. In quadrant A, if 'd' or 'e' gets drifted away due to node mobility, the flow of information through the network will take place via nodes a, b and c, thus forming a path x-a-b-c-y, as 'c' is near both 'b' and 'y', hence it will also be taken into consideration while forming the path for the flow of communication in the network. Above shown figure is just an example of two consecutive quadrants, while there exists eight such quadrants in real case scenario. Also nodes from all the quadrants that approach near the network will be considered if they come close enough near the already existing nodes in a network.

4 Results

Table 2, shows the difference that how our proposed work is different from previous work. Dimension used in our work is 3D used for underwater 3D environment, where area is divided into 8 quadrants. Used Euclidean 3D distance formula (d = $\sqrt{(x_2-x_1)^2+(y_2-y_1)^2+(z_2-z_1)^2}$ followed by KRUSKAL algorithm that helps in formulating the path for communication network from source to destination effectively. On the other hand previous work has considered 2D dimension for representation of underwater that is divided into four quadrants. Used Euclidean 2D distance formula ($d=\sqrt{(x_2-x_1)^2+(y_2-y_1)^2}$).

In the above Figs. 7a and 8b, we have shown the difference between the two graphs. One is 2D graph that was used in previous work for representation of nodes in 2D plane that used two values for representation of coordinates in the manner (x_1, y_1) and one is 3D that is used in this paper for the representation of nodes in 3D environment underwater that uses three values for representation of coordinates in the manner (x_1, y_1, z_1).

In the above Fig. 8a and b, we have shown the difference between the two representations of quadrants. In previous work there are four quadrants that were used for the division of underwater plane in four quadrants namely upper left, upper right, lower left and lower right and in this paper we have used 8 quadrants for division of underwater plane namely upper left 1, upper left 2, upper right 1, upper right 2, lower left 1, lower left 2, lower right 1 and lower right 2.

In Fig. 9a–c we have shown the difference between the node representation in previous work and in our work. In previous work nodes were represented in four quadrants namely UL, UR, LL and LR, while in our work nodes were represented in quadrants, for that purpose we have used two Fig. 9b and c for representation. Figure 9a is used for representation in above four quadrants namely UL1, UL2,

Table 2 Difference between proposed work and previous work

Proposed work	Previous works
Dimension taken into account: 3D	Dimension taken into account: 2D
Area: underwater 3D environment	Area: underwater but 2D representation
Quadrants used: 8, because of 3D nature of underwater	Quadrants used: 4
Used Euclidean 3D distance formula	Used Euclidean 2D distance formula
Used KRUSKAL algorithm for selection of path for communication	No such algorithm used
The approach shows a working example	No
Better approach as it considers real problem of 3D underwater environment The approach gives a realistic solution and a better understanding of the node mobility issue Provides a full practical solution to our problem	Doesn't consider real time issue The 2D approach is only good enough to understand the problem on paper Provides only a partial solution to our problem

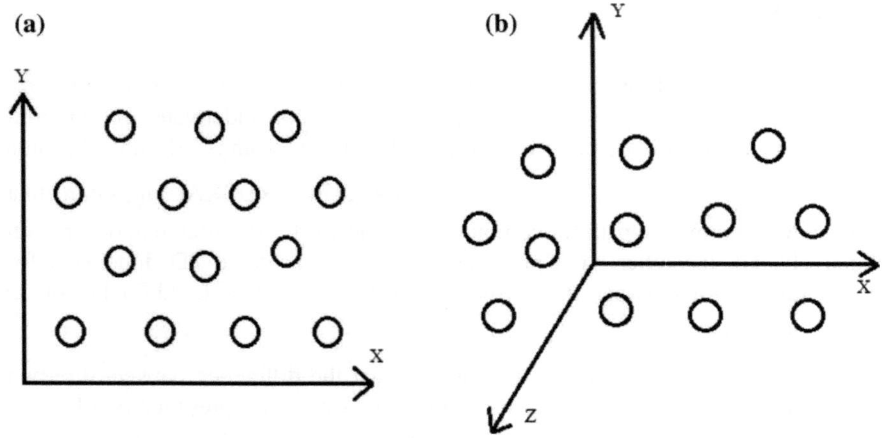

Fig. 7 **a** 2D Graph and **b** 3D Graph

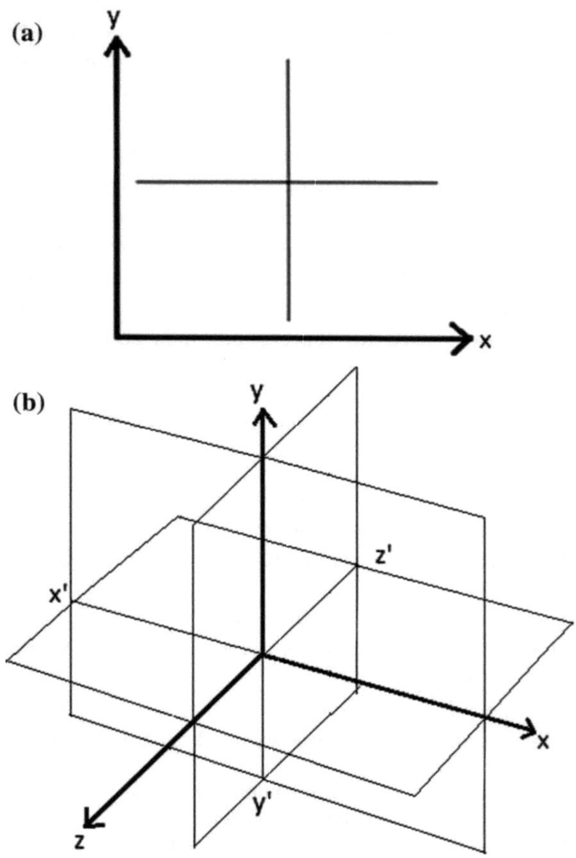

Fig. 8 **a** 4 Quadrants of 2D representation and **b** 8 Quadrants of 3D representation

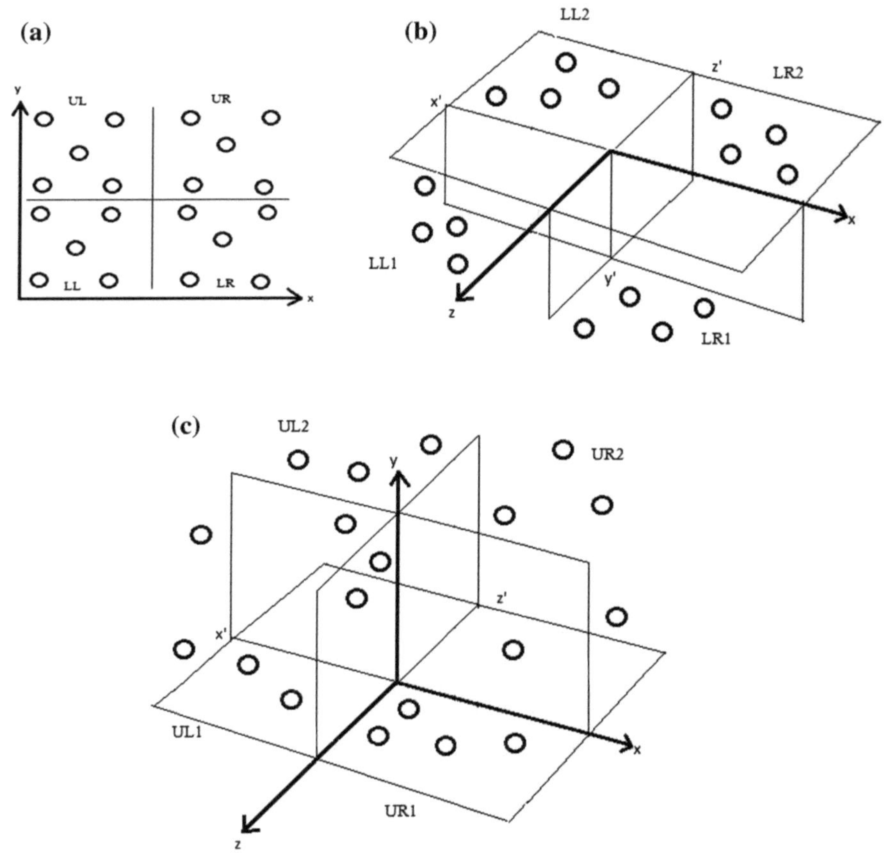

Fig. 9 **a** Nodes in 4 quadrants of 2D perspective and **b** Nodes in *lower* 4 out of 8 quadrants of 3D perspective, **c** Nodes in *upper* 4 out of 8 quadrants of 3D perspective

UR1 and UR2. Figure 9c is used for representation in lower four quadrants namely LL1, LL2, LR1 and LR2. In previous work formula used was Euclidean 2D distance formula [13] and in our work we have used Euclidean 3D distance formula.

Pros of our proposed approach are as follows: Application of our presented approach is explained with an example that makes it understandable with easy application and the approach is easily understandable as there is no complex methodology. Cons of our proposed approach are as follows: Complexity increases as the number of nodes increases and approach is simple in terms of process and application, but time consuming when it comes to calculating the distances among all the nodes again and again after suffering mobility issue problem.

5 Conclusion and Future Work

The 2D approach which was previously proposed is having limitation of real time problem of node mobility. In this paper we have proposed 3D perception (using a total of 8 quadrants) which enables robust approach for real time environment underwater using Euclidean 3D distance along with our proposed KRUSKAL algorithm. The proposed approach provides effective communication which is maintained under dynamic underwater environment without affecting/losing the network communication. Furthermore, the proposed approach also resolves the node mobility issue as this approach exploits the disadvantage of node mobility to its advantage. In our future work, we will bring into consideration the real world scenario and implement our approach in real time scenario to further evaluate and compare the effectiveness of our work.

References

1. Chinna Durai, M; Mohan Kumar, S; Sharmila, "Underwater wireless sensor networks", Compusoft journal, volume- IV, ISSUE- VII, 1899–1902, 2015.
2. Jan Erik Faugstadmo; Subsea, Kongsberg Maritime, Horten, Norway; MagnePettersen; Jens M. Hovem; Arne Lie, "Underwater wireless sensor network", IEE, 422–427, 2010.
3. J. Heidemann; Inst. of Inf. Sci., Southern California Univ., CA; Wei Ye; J. Wills; A. Syed, " Research challenges and applications for underwater sensor networking, IEEE, 228–235, 2006.
4. Ian F. Akylidiz, Dario Pompili, Tommaso Melodia, "Challenges For Efficient Communication in Underwater Acoustic Sensor Network", ACM, 2004.
5. U. Devee Prasan and Dr. S. Murugappan, "Underwater Sensor Networks: Architecture, Research Challenges and Potential Applications" International Journal of Engineering Research and Applications, Vol. 2, Issue 2, pp. 251–256, Mar–Apr 2012.
6. Jun-Hong Cui; Connecticut Univ., Storrs, CT, USA; Jiejun Kong; M. Gerla; Shengli Zhou, " The challenges of building mobile underwater wireless networks for aquatic applications", IEEE, vol 20, issue 3, 12–18, 2006.
7. Jim Partan, Jim Kurose, Brian Neil Levine, "A survey of practical issues in underwater networks", ACM, VOL-11, ISSUE-4, 23–33, 2007.
8. Djauhari, M., & Gan, S. Optimality problem of network topology in stocks market analysis. Physica A: Statistical Mechanics and Its Applications, 419, 108–114, 2015.
9. Dario Pompili, Rutgers and Ian F. Akyildiz, "Overview of Networking Protocols for Underwater Wireless Communications", IEEE Communications Magazine, January, pp 97–102, 2009.
10. Mohamed K. Watfa, Tala Nsouli, Maya Al-Ayache and Omar Ayyash "Reactive localization in underwater wireless sensor Networks", University of Wollongong in Dubai – Papers, 2010.
11. S. Naik and Manisha J. Nene, "Realization of 3d underwater wireless sensor networks and influence of ocean parameters on node location estimation", International Journal of Wireless & Mobile Networks (IJWMN) Vol. 4, No. 2, April 2012.
12. Udit Agarwal, "Algorithms Design and Analysis", Dhanpat Rai & Co., 6th Edition, 2008.
13. Kanika Agarwal, Nitin Rakesh "Node Mobility Issues in Underwater Wireless Sensor Network", International Conference on Computer, Communication, and Computational Sciences (IC4S-2016), Springer 2016, *accepted, in press*.

NS2 Based Structured Network Attack Scrutiny in MANET

Kamal Nayan Chaturvedi, Ankur Kohli, Akshat Kamboj, Shobhit Mendiratta and Nitin Rakesh

Abstract Wireless networks are widely used in almost areas and the urging of users has inspired the development of Mobile Ad Hoc Network (MANET). MANET is a dynamic Network which does not require any backbone or infrastructure network as it deploys peer to peer connection between nodes. There are numerous implementations of MANET ranging from defense to multiuser gaming which necessitates the need to secure our network from various trespassers and attacks. Various attackers use numerous approaches to degrade the network performance. The paper proposes classification to network attacks on MANET by distinguishing them in a structured way with NS2 simulated results. The objective of this paper is to know various attacks and their respective measures to be taken to detect and prevent these attacks.

Keywords MANET · Ad Hoc Network · Routing · Attacks · Mesh

K.N. Chaturvedi (✉) · A. Kohli · A. Kamboj · S. Mendiratta · N. Rakesh
Amity University, Noida, Uttar Pradesh, India
e-mail: kamalnc12@gmail.com

A. Kohli
e-mail: mail2ankurkohli@gmail.com

A. Kamboj
e-mail: mehrokakshat@gmail.com

S. Mendiratta
e-mail: shobhit.mendiratta@gmail.com

N. Rakesh
e-mail: nitin.rakesh@gmail.com

G.M. Perez et al. (eds.), *Networking Communication and Data Knowledge Engineering*, Lecture Notes on Data Engineering and Communications Technologies 3, https://doi.org/10.1007/978-981-10-4585-1_9

1 Introduction

Today the world is growing at a very fast pace; this radical change is due to the influence of the developing communication and network technologies. Wireless network gives flexibility to users to operate at random locations. Ad Hoc Network is the new flourishing technology which supports communication without any kind of a pre-installed or pre-determined infrastructure. Due to its distributed nature, Routing in MANET is trouble free, easy and robust. There are various challenges faced in MANETs because of its dynamic topology, wireless interface and lack of any kind of standard defense mechanism. Because of lack of any kind protection in MANETs it becomes an easy target for the attacker to do any kind of malicious activity and disrupt the communication in the network. The Attacks on MANETs can be combated on the basis of encryption and Authentication and the integrity of the data can be securely maintained.

Existing MANET routing protocols are subject to a variety of attacks that can accede attackers to impact a node's selection of routes or enable flooding attacks. MANET can be secured by use of Security Protocols and solutions as discussed in this paper. As the main objective is "Security for acquiring flawless performance like using least resource forward utmost packet in lesser span of time". In this paper we have proposed all possible viable Attacks on Ad Hoc Network, with their respective operational and well-grounded Solutions. Various Security measures have been listed down to be considered while framing Ad Hoc Network. By outlining these Attacks and their attributes we can design a new Defense measure to protect the Network.

This paper is classified as followed; second section describes about the routing mechanism and about the various routing protocols used along with their respective applications. Section third gives an overview of the various possible attacks on MANETs and the generic solutions that are available for each attack. Section forth is the comparison between the various routing protocols under different types of attacks, section fifth is study of NS2 simulated results and section sixth is conclusions.

2 Routing and Protocols

Wireless network is categorized as Ad Hoc network and Infrastructure network on the basis of resources they require. Both have innumerable applications as Infrastructure network is used for long distance and permanent communication while Ad Hoc Network is more conventional in point to point communication and in various disastrous prone areas for operational activities. Topology of network infrastructure has a central server and fulcrum arrangement of Access Points (AP). These APs are mutually interconnected by some network. It is more exemplary for imperishable network. Infrastructure Network is used for long distance and massive

Transmission comprising of very large number of substratum and framing of Network nodes and terminals. Infrastructure network is in for the long haul and frequently used to aggrandize wired LAN to indulge wireless devices through base stations.

Ad Hoc Network is a disintegrated network in which any device can work as a network node as well as terminal and has peer to peer connection between devices. When Ad hoc network is applied in mobile communication it is called as MANET (Mobile Ad Hoc Network). This network is gaining importance as it gives window of opportunity for unmediated connection between devices. Ad Hoc network do not require any schematic infrastructure and it is task specified and for less span of time. This hierarchy of network has great exceptional implementation in disasters and in deserted locations. Ad Hoc network do not have medial access point. There are uncountable applications of Ad hoc network [1], few of them are following as Military Communications and Operations, Robotic Battlefield, Emergency situations, Search and Operation Services, Dynamic Database Access, Multi User Gaming, On construction Sites and so on.

2.1 Routing

Routing is approach of selecting the shortest and the most secure path in the whole mesh for reaching destination from source. We use various protocols for routing, which lineup the best available path in the network. Routing uses mechanism of selecting next hop IP for transmitting routing table or RREQ packets to its respective destination node. A table is forwarded, carries topological data which leads to discovery of the path with least metric. Routing is much different than network forwarding, Routing perceives the best path and forwarding is act of transmitting the packet to the next hop towards the receiver of that packet. Each router has a forwarding table Generated by routing protocol and Routing pursue routing info to forwarding table. Routing is further classified as shown in Fig. 1 as Proactive routing, Reactive routing and Hybrid routing;

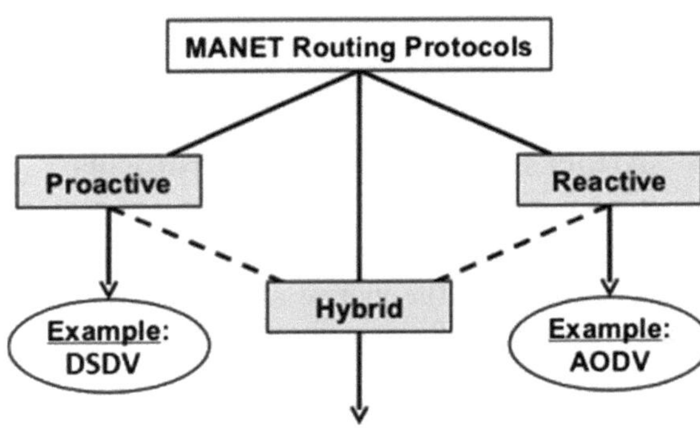

Fig. 1 Types of routing protocols

Proactive routing protocol is a table driven routing protocol, in this protocol each node maintains a routing table which contains information of every other node in the mesh and their respective paths. There are two types of messages that are sent by the nodes in proactive routing protocol and they are: Hello messages and Topology Control (TC) messages. When a new node enters the mesh, it will broadcast a hello message to mark its presence to every other node that is prevailing in the network before any routing topology messages are mutually exchanged. Dynamic destination sequenced distance vector (DSDV) routing protocol also known as Global state routing and Zone based hierarchical link state routing protocol uses the table driven or proactive routing method. There are few shortcomings of this routing mechanism. Due to the dynamic topology of the network the frequent broadcast of hello messages and topology messages consume the limited bandwidth. Each node individually requires a memory block to store and maintain the information of all the other nodes in the network which increases the overhead. Thus the mechanism of reactive routing is widely used.

In Reactive routing protocols, the nodes discover the route to the destination only when data has to be transmitted. There are three types of packets [2] (RREQ, RREP and RERR) which are forwarded by nodes. These packets assist nodes in route discovery. Each node individually updates two counters (source sequence number and hop count), and the address of source node and destination node into their routing tables. Sequence Number assures that all paths discovered are loop free and helps to find the shortest path to the destination node. RREQ (Route Request) packet is used for route discovery; RREP (Route Reply) packet is the acknowledgement which the destination node sends to the source node when an authentic and smallest path is discovered. RERR (Route error) packet is generated when there is some error or vulnerability in the already established source to destination route. Ad Hoc on Demand Distance Vector routing (AODV), Dynamic Source Routing Protocol (DSR), Temporally Ordered Routing Algorithm (TORA) and Cluster Based Routing Protocol are invoked on Reactive Routing. Hybrid routing protocols retain the address of its adjacent nodes and for transmission beyond adjacent nodes; it uses reactive routing protocols for route discovery. In all the present day research on MANET, this is the basic conviction that Network is healthy and reliable. But many operation of MANET are in untrustworthy condition. Consequently, most MANET protocols are vulnerable to various types of attacks.

3 Attacks on Routing and Protocols

Node Failure or Link Failure is two types of failure which can occur during any attack on routing. The motive of any attack is to disturb the flow of data or to modify the packet. Reasons for attacks in MANET are [1]:

- Dynamic Topology: In Ad Hoc network, any node can join or leave the mesh at any point of time. As node can join at any time it very difficult to determine, which node is malicious or not [3].
- Wireless Interface: All nodes in Ad Hoc Network are connected with wireless network. Wireless network have lesser bandwidth then that of wired or physical network. So it is easier for attacker to target an address [3].
- Lack of any Standard Defense mechanism: There is no hierarchy of defense protocol. So attack can come from anywhere may be externally or internally [3]. We can categorize the attacks as follow [4, 5].

3.1 Based on Secrecy and Authentication

(A) Eaves Dropping Attack: Eaves Dropping is an unauthorized real time forestall of a hidden communication [6]. This network layer attack emphasis on to put cuffs on small packets and accessing the data in search of any information. Mostly occur when there is chance of Encryption. In eaves dropping attacker is in search of any sensitive information like password, session token or any intimate information. The mechanism of network access control and strong encryption can avoid Eaves dropping attack.

(B) Packet Replay: In course of routing when an attacker utterly replays the wiretap frames as they were or with some manipulation in the packet, this type of attack is called as packet replay [7]. Attacker continuously transmits the same packet to generate illegitimate traffic. This causes increase in level of traffic and interference in the mesh [6]. Packet replay attack is more effective when packet is broadcasted over range of one hop only. Even one attacker can declass route throughput by 61%. Packet replay can be prevented by use of COPS, Digital Signatures and Bloom filters.

(C) Spoofing of Packets: Spoofing is a technique used to acquire illegitimate access to the packets by attacker via sending message to the node with a forging address revealing like a trusted node [8]. Attacker use trusted relationship exhibited by systems that are appraised as internal or trusted nodes. Spoofing generally occur when attacker use source address of a within range trusted internal node or it uses an authentic external node. Attacker may inject malicious data into the system [6]. It may modify the routing table to node for receive all the packets and replying like an authentic address. Spoofing attack is easier because authentication is based on only source address and router only requires destination address. For discovering spoofing on a system, we compare process accounting chronicle between systems on the current mesh. With the use of Source Address Validation and by installing a filtering router spoofing can be avoided.

3.2 Based on Integrity

(A) Sybil Attack: Attacker uses numerous identities and uses them to attain a large ascendancy. Seeing that all Sybil identities are clout by the attacker, he can maliciously inject wide ranging activities in the system [6]. Sybil attacks degrade the trust mechanism in communication of particular system [9]. Sybil attack is basically to degrade the reputation of the sensor and Ad Hoc system. By reducing the token issuing rate, we can stop steady accumulating of Sybil nodes. Defenses against Sybil attack are verification of key sets for random key redistribution, registration and radio resource testing.

(B) Black Hole: In black hole attack, attacker uses the malicious node to reveal itself having the shortest path for reaching destination whose packets it want to head off. In flooding based routing, if the RREP packet of malicious nodes reaches before the real node having authentic packet, a counterfeit route has been created [2, 6] as shown in Fig. 2. Then it depends upon the attacker whether it drops the packet or it performs denial of service attack [3].

The remedies for black hole attack are to search more than one path redundant routes. The source node will broadcast a ping packet for destination via all three discovered paths with different sequence number. The malicious node will drop the ping packet, and route will be discarded. Packets will be routed through the other discovered path. APRAODV protocol is also a bonafide solution for preventing black hole attack as it increases packet delivery ratio with minimum increase in mean node to node delay.

(C) Sink Hole: In sinkhole attack, Attacker creates a malicious node and tempts the network traffic by its fake routing table. Attacker may amend packet or drop the packet mutely [9]. Sinkhole upturns network overhanging and cutback network's lifetime by boosting energy utilization. Sink hole is used to commence selective forwarding attack or spoofing [5, 6]. It is very difficult to negate sink hole because imparted routing information is difficult to validate [10]. Counter measures for sink hole attack is Data Consistency and Network Flow Information Approach, Hop-Count Monitoring Scheme, Monitoring node's CPU usage and Mobile AgentBased Approach.

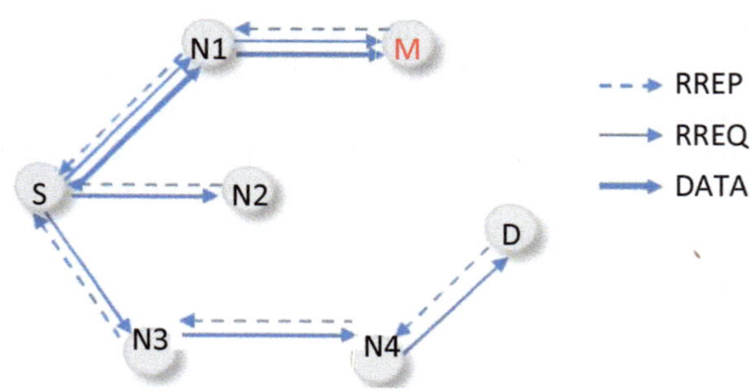

Fig. 2 Black hole attack by malicious node 'M', while routing from node 'S' to 'D'

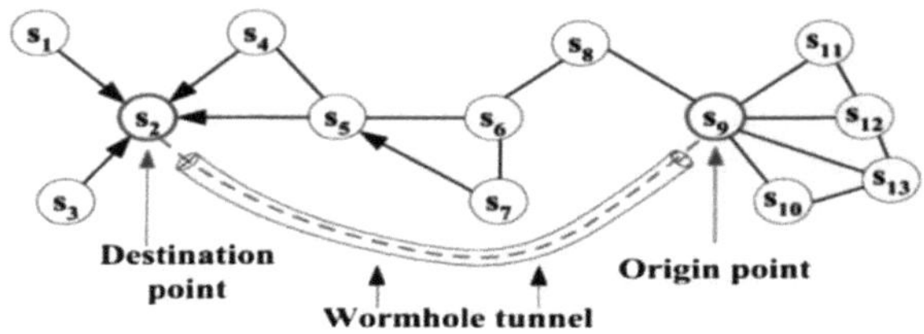

Fig. 3 Wormhole attack, via node 'S_2' and 'S_9' attacker is tunneling the packets during the course of routing

(D) Worm Hole: In worm hole attack a malicious node abduct packets from a node and transmit them to a far location. In network, attacker cuts the edge of route by which two nodes are communicating and ensnare the packet to packet to a divergent location. It draws out packet through a tunnel or a fabricated path [5]. The tunnel is either a wired link or high frequency link. It may drop the packet or roll to some other node to avoid detection. In wormhole attacker do not require any key for encryption [6]. It only needs source and destination node and a channel for tunneling packets as in Fig. 3.

Worm holes attack is done in two modes as hidden mode and participation mode. Packet encapsulation and packet replay eventuate hidden mode Wormhole attack. High power and out of band transmission leads to wormhole attack in participation mode [3]. Worm hole attack leads to bypassing of enormous network traffic and shatter routing. Existing solutions for wormhole attack are transmission time based mechanism, and digital signature based approach. Time of flight is one of the best mechanisms against wormhole attack, as it is based on time of flight of particular packets. As suggested by Capkun et al. [8], round trip travel time of message (RTT). RTT time is calculated by taking difference of time at which acknowledgement received and time at which packet is delivered. Consider 'd' distance between nodes in maximum range of transmission 'R' and the RTT time 'α' and presuming wireless signal travels with speed of light 'c'. We can write time α as;

$$d = \alpha \times \frac{c}{2} \quad \cdots \quad (a); \quad \alpha = \frac{2d}{c} \quad \cdots \quad (b)$$

We can say because the maximum verified range of communication is R.

$R > d$ (d is within the range of transmitting node);

$$R > \alpha \times \frac{c}{2} \quad \cdots \quad (c); \quad \alpha < \frac{2R}{c} \quad \cdots \quad (d)$$

The standard protocol for MANET is wireless Ad Hoc network 802.11 Medium Access Control (MAC), which have a predefined short inter frame transmission range (S) of 10 μs. Let's take an example, maximum transmission range (R) is 600 meters is assumed and speed of light is 3×10^8 m/s. So the order of round trip travel time is much smaller than the magnitude of delay required by the 802.11 protocol (see Eq. d). In RTT based mechanism, one address sends a packet to another and acknowledgement is received with in the time as wireless signal travel between nodes only. Thus if there is some malicious Worm hole Node it will not travel further, cannot be received in time calculated in Eq. (e).

$$\alpha = \frac{(2\times 600)}{3\times 10^8} = 0.000004\,\mu\text{s};\quad \alpha < \frac{2d}{c} + S \quad \cdots \quad \text{(e)}$$

3.3 Based on Availability: Denial of Service

Attackers deplete resource in network with malicious nodes to decline the service. Attacker affects the network between server and router CPU cycle, server interrupt processing capacity and particular protocol data structure. The aim is to flood no of request on router or at any node affecting the accessibility for the routing. Transmitting numerous ICMP packets from multiple addresses to unable system for processing traffic. If system is unable to process traffic, ultimately it will be unable to provide service to the network and the whole system goes down [9]. If attacker attempt to deny Bandwidth to TCP flow while transmitting at low throughput, this category of DOS attack is known as low rate denial of service attack or shrew attack. When attacker uses multiple addresses to initiate a amalgamate Denial of service attack averse to a single or multiple node, using client/server technology. We can avoid DOS attack by using good firewall mechanism, and by dumping the first SYN [8] packet [3]. DOS can be avoided by automatic dynamic changing of IP Address of the system which will create problem for the attacker for updating address of the nodes being attacked.

4 Comparison of Attacks

Active attacks are frequently carried out attack in which the flow of message transmission breaks off. Passive attack does not alter the data transmission between nodes but indulge non-permitted listening of messages. Eavesdropping is the best example of passive attack on MANET. Attacks can occur in distinguished layers of Network protocol stack. Physical layer is hardware deployed and also needs the hardware infrastructure to come into the operation. So, attack on the physical layer is easier for the attacker than the attack on other layer because it do not require much knowledge of technology. Eavesdropping is identified as an attack on physical layer. Active attacks can be classified as Network layer attack, Transport layer attack and Application layer attack. Network layer provide prerequisite connection between nodes through hop to hop mechanism. In Ad Hoc network, each node forward packets whether they are RREQ, RREP or Ping Packets through the path having the highest priority in routing table. Malicious node implant itself in

Table 1 Classification of Attacks

Type of Attack	Active or passive	Internal or external	Dropping of packets	Modification of packets	Path fabrication	Affect traffic	Layer
Eaves dropping	Passive	Internal	No	No	No	No	Physical
Packet replay	Active	Both	No	No	No	Yes	Network
Spoofing of packets	Active	Both	Yes	Yes	No	No	Network
Sybil	Active	External	No	Yes	No	Yes	Network
Black hole	Active	External	Yes	No	No	No	Network
Sink hole	Active	External	Yes	Yes	No	Yes	Network
Worm hole	Active	Internal	Yes	No	Yes	No	Application or network
Denial of service	Active	Both	No	No	No	Yes	Network or DLL

Table 2 Prevention techniques for preventing attacks

Attack	Prevention technique	
Eavesdropping	Strong network access control	Robust encryption
Packet replay	Use of COPS, bloom filter	Digital signature
Spoofing of packets	Use of source of address validation	Installation of filtering router
Sybil	Use of radio resource testing	Random key redistribution
Black hole	Use of APRADOV protocol	SAODV protocol
Sink hole	Network flow information approach	Hop count monitoring system
Worm hole	Transmission time based mechanism	Digital signature based approach
Denial of service	Dynamic change of IP addresses	Resilient firewall mechanism

active path or to absorb network traffic, this is the basic idea of Network layer attack. Attacker creates routing loops to inject blockage or obstruction in the network. Black hole attack, Sybil attack, Sinkhole attack, Replay attack and Worm hole attack are Network layer attack. Every packet is accessed in Application layer. Application layer have HTTP, SMPT and FTP protocols, which can provide the access point to the attacker. So, this gives indebtedness to the attacker for accessing the network. Worm Hole, DOS attack and malicious code attack are application layer attack. It may attack both the operating system and the user (Tables 1 and 2).

5 Study of Simulations

For studying simulation outcomes for this paper, a MANET network is established over NS2 platform. The protocol used is AODV, and Blackhole and Wormhole nodes are configured. The implementations properties of the Blackhole and

Wormhole are provided by manipulating the code present in protocol implementation files dsr.cc and dsr.h files. Blackhole and Sinkhole node and dropper agents are generated and these are combined to the nodes which have to work as a malicious node. It is simulated using the front end TCL language used in NS2. The method for creating the bogus RREQ request_fake_send() is written in dsr.cc and called as agent by the malicious node in the mesh. To analyses the attacks parameters like send packets, received packets, total packets forwarded, routing packets, throughput, hop count and total energy were used.

The simulation result of Blackhole and Wormhole attack for different network size is shown in Figs. 4, 5, and 6. While comparing the parameters in normal routing with same during attack, there is a remarkable difference mathematically. This difference points toward the degradation in transmission of packets caused due to attack. The efficiency or Packet delivery ratio of the system for 20 nodes is 62% but while attack it is downturn to 7% and for 40 nodes it reduces to 8% from 53%. These figures are enough to represent the deterioration caused by black hole attack by dropping packet via its malicious node as described in Figs. 4 and 5.

In Wormhole attack as shown in Fig. 6, 417 packets are received while attack but only 396 packets are received in normal operation for mesh of 20 nodes and for 40 nodes, 1920 packets are received during attack while 1602 packets are received in conventional way. Packet delivery ratio is increased to 65% from 57% hop count

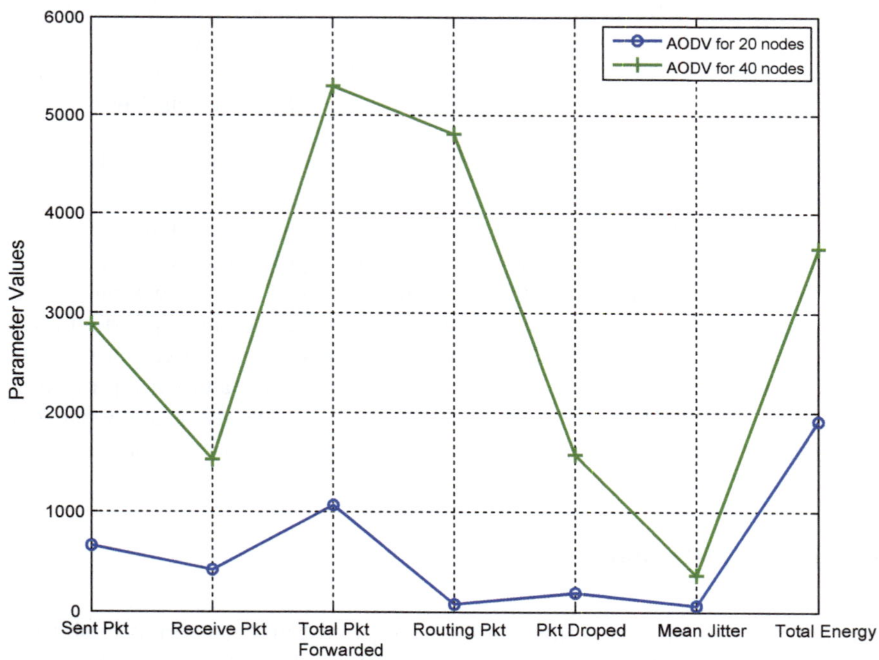

Fig. 4 Performance graph on 20 and 40 nodes operating in conventional conditions

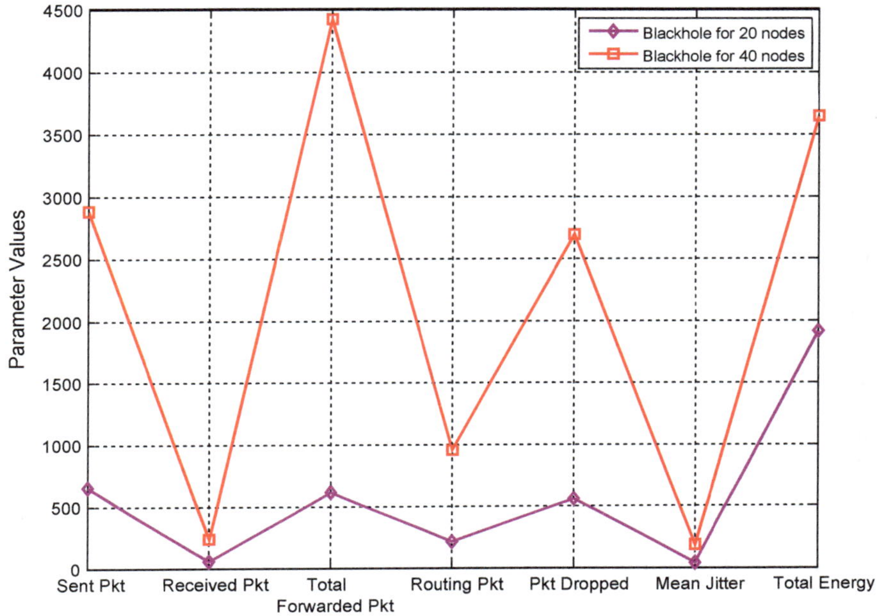

Fig. 5 Performance graph on 20 and 40 nodes during Blackhole attack

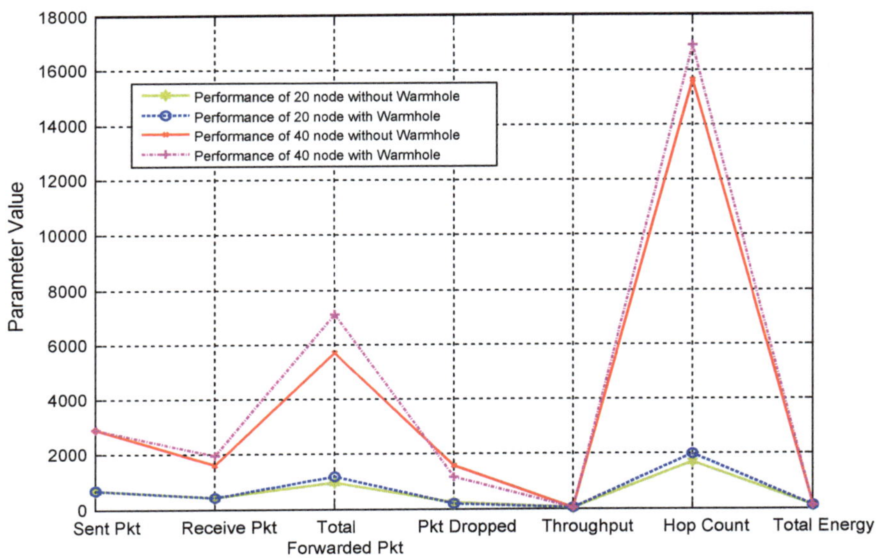

Fig. 6 Performance graph on 20 and 40 nodes during Wormhole attack

is upturn to 16,880 from 15,596. This indicates the way by which Wormhole attacker is inducing junk packets for consuming the bandwidth and degrading the system.

6 Conclusion

All attacks on Ad Hoc network outstandingly degrade network potential and are dangerous to the Network's security. In this paper the possible attacks on MANET with their respective detection mechanism and Prevention has been discussed. Attacks have been classified on the basis of their threat to availability, integrity, Secrecy and Authentication. In all MANET protocol whether it is AODV, DSDV or DSR there is no security or defense mechanism against the malicious node or attacker. Achieving security in Ad hoc network is not a simple task as there is need of having a new protocol comprising a line of defense mechanism against all risk and threats, whether they are from internal node or external node. Use of an active intrusion detection system along with authorization and encoding system can accustom all the possible failures.

References

1. Ning, P., & Sun, K. (2005). How to misuse AODV: a case study of insider attacks against mobile ad-hoc routing protocols. *Ad Hoc Networks*, *3*(6), 795–819.
2. Noureldien A. Noureldien, "A novel taxonomy of MANET attacks", Electrical and Information Technologies (ICEIT), 2015, Electrical and Information Technologies, Department of Computer Science, University of Science and Technology, Omdurman, SUDAN, Pages: 109–113.
3. Gagandeep, Aashima, Pawan Kumar, "Analysis of Different Security Attacks in MANETs on Protocol Stack A-Review", International Journal of Engineering and Advanced Technology (IJEAT), ISSN: 2249 – 8958, Volume-1, Issue-5, June 2012.
4. Sanzgiri, K., Dahill, B., Levine, B. N., Shields, C., & Royer, E. M. B. (2002, November). A secure routing protocol for ad hoc networks. In *Network Protocols, 2002. Proceedings. 10th IEEE International Conference on* (pp. 78–87). IEEE.
5. H. Chris Tseng and B. Jack Culpepper "Sinkhole intrusion in mobile ad hoc networks: The problem and some detection indicators", In Computers & Security vol. 24, pages 561–570, 2005.
6. V. M. Zapata and N. Asokan. Securing ad hoc routing protocols. In D. Maughan and N. Vaidya, editors, Proceedings of the 3rd ACM workshop on Wireless Security (WiSe '02), Atlanta, Georgia, US, September 28, 2002, pages 1–10.
7. ZiFeng, Jianxia Ning, Ioannis Broustis, Konstantinos Pelechrinis, Srikanth V. Krishnamurthy and Michalis Faloutsos, "Coping with Packet Replay Attacks in Wireless Networks", 2011 8th Annual IEEE Communications Society Conference on Sensor, Mesh and Ad Hoc Communications and Networks, Pages 245–257
8. Templeton, S. J., & Levitt, K. E. (2003, April). Detecting spoofed packets. In *DARPA Information Survivability Conference and Exposition, 2003. Proceedings* (Vol. 1, pp. 164–175). IEEE.

9. Bing Wu, Jianmin Chen, Jie Wu, Mihaela Cardei, A Survey on Attacks and Countermeasures in Mobile Ad Hoc Networks, Springer, 2006. Pages 107–139.
10. Zhenzhen, Y. and H. Yingbo, "Stability of wireless relays in mobile ad hoc networks", in Proceedings of Acoustics, Speech, and Signal Processing, (TCASSP '05). IEEE International Conference on (Volume: 3), 2005, Vol 3, pages 465–468.

EKMT-k-Means Clustering Algorithmic Solution for Low Energy Consumption for Wireless Sensor Networks Based on Minimum Mean Distance from Base Station

Bindiya Jain, Gursewak Brar and Jyoteesh Malhotra

Abstract EKMT-k-means clustering algorithmic solution is one of the well known methods among all the partition based algorithms to partition a data set into group of patterns. This paper presents an energy efficient k-means clustering algorithm named EKMT which is based on concept of finding the cluster head minimizing the sum of squared distances between the closest cluster centers and member nodes and also considers the minimum distance between cluster heads and base station. In the proposed protocol the effort was made to improve the energy efficiency of the protocol by re-selecting the cluster head among the most possible cluster heads on the basis of the least distance between new selected cluster head and the base station thus improves throughput and delay.

Keywords Wireless sensor networks · Clustering · k-means
Cluster centroid · Residual energy

1 Introduction

In a Wireless sensor network, a huge number of sensor nodes are deployed in the monitored area and the monitored data is collected by external base station. In the network when battery of node dies a node is no longer useful. This is the reason

B. Jain (✉)
DAV Institute of Engineering and Technology, Jalandhar, India
e-mail: bindiyajain29@gmail.com

G. Brar
BBSBEC, Fatehgarh Sahib, India
e-mail: brargs77@rediffmail.com

J. Malhotra
Guru Nanak Dev University, Jalandhar, India
e-mail: jyoteesh@gmail.com

G.M. Perez et al. (eds.), *Networking Communication and Data Knowledge Engineering*, Lecture Notes on Data Engineering and Communications Technologies 3, https://doi.org/10.1007/978-981-10-4585-1_10

why researchers are now days developing new routing algorithms for wireless sensor networks to save energy and improve QoS parameters like delay, throughput and jitter. Heinzelman et al. [1] developed a hierarchical adaptive clustering routing protocol LEACH to improve energy efficiency and reliable transmission in WSN. New clustering routing algorithms provides energy efficient approaches in data redundancy by reducing the communication distance for wireless sensor networks by data aggregation [2]. In LEACH the run time of network is broken into several rounds by dividing the sensor nodes in the network into several clusters. In this paper, we are introducing an improvement in the algorithms for routing to control traffic and manage energy consumptions. There are various algorithms available for routing. Another efficient method for conserving the energy of sensor networks is data gathering. To remove the redundant data and save the transmission energy is the main purpose of data gathering [3, 4]. LEACH protocol leads to non uniform energy consumption in the network In earlier leach based protocol, cluster head selection criteria is not considering the distance between base station and nearest cluster head having minimum energy [5]. The proposed protocol called EKMT-K means will optimize the energy of the WSN and increase the sensor lifetime as compared to earlier k-means protocols available. Although some works have been done on the performance and improvement of LEACH protocol [6–8], there is little work on energy efficient k-means algorithm for efficient cluster head selection method to maximize the energy efficiency of wireless sensor network. It is based on the k-means algorithm model for finding the cluster head having maximum energy among all the nodes and minimizing the sum of Euclidean distances between the cluster head and member nodes. This paper is focussed on this issue and discusses how to reduce energy consumption and improve QoS parameters like increased throughput and lesser end to end delay and low jitter of cluster-based wireless sensor network by proposing an algorithm named, EKMT-k means i.e. Energy Efficient k-means clustering in which the k-means clustering technique is applied for choosing cluster heads and also the distance from the base station to cluster head is considered in our new technique which plays an important role in their energy consumption. This algorithm is completely tested on simulation in MATLAB.

2 Related Work

The limitation in the energy resources of nodes are the main challenging issue in the process of developing routing protocols for the Wireless Sensor Networks. The node shall be the cluster head next time that is having higher energy in comparison to other nodes in the cluster [9]. ELEACH is the Enhanced LEACH. In ELEACH there are same rounds as there in LEACH algorithm. But the difference is there are two more terms added these are residual energy and consumed energy. In ELEACH on the basis of the residual energy the cluster head is selected after each cycle, means the node having higher residual energy will be the cluster head in the next round [10]. With this all nodes remain alive for long time. The other distinctive clustering routings protocols in WSNs include Hybrid Energy-Efficient Distributed clustering (HEED) [11], Leach (CKM), EBK means clustering, Unequal Clustering Size (UCS), Base-Station Controlled Dynamic Clustering Protocol (BCDCP), Energy Efficient Clustering Scheme (EECS), Energy-Efficient Uneven Clustering (EEUC) model. k-Means algorithm [12] uses Euclidean distance metric on the basis of each node's coordinates (X and Y) to group similar sensors and energies for choosing cluster head, the information about the node's positions and energies is obtained by every node by exchanging messages among themselves. K-means algorithm is very popular clustering method proposed by Geon [13] for cluster head selection. Yanlin Gong [14] proposed a clustering protocol based on balanced serial k-means algorithm which computes the distance formula and reduces the total sum of squared distances between all the non cluster head nodes and closest cluster centers and make the system efficiency maximum. Shalli [15] proposed multi-hop data aggregation by forming co-ordination in hierarchal clustering. LEACH [16–18] forms clusters by using distributed algorithm. Cross layer protocol [19] using optimal cluster head selection to optimize the energy efficiency by balancing the load on the clusters. Yang [20] proposed that any intruder trajectory line cutting across a region of interest will be detected by at least K sensors. For energy efficiency, an efficient sensor movement to satisfy the line K-coverage while minimizing the total sensor movements, which is named as LK-Min Movs problem is scheduled. Nidhi [21] proposed a mobile cluster head data collection model for wireless sensor heterogeneous networks. The proposed MCHDC model performs much better than random mobility model and horizontal line mobility model.

3 Proposed k-Means Clustering Based Energy Efficient Protocol

Step 1: Initial Cluster formation phase: 100 nodes are generated randomly in 400 * 400 m^2 area. Each sensor node has given unique identity number. In this phase we calculate the center associated to each cluster. Cluster head is chosen on the bases of the distance from the centroid. Assume that WSN of nodes noted by x is divided into k clusters $\{C_1, C_2, \ldots C_k\}$ such that $C_1 \cup C_2 \cup \ldots C_k = \{x_1, x_2 \ldots x_n\}$ and $C_i \cap C_j = \phi$ Let c_k be the mean or center of cluster groups C_k.

$$c_k = \frac{1}{N} \sum_{x_i \in C_k} x_i \tag{1}$$

The objective of k-means is to minimize the total Euclidean distance between cluster head and cluster members. K-means uses minimum distance formulae algorithm that minimizes the sum of distances from each sensor node to its cluster centroid. It moves sensor node between clusters until the sum cannot be decreased further. The result is a set of clusters that are as compact and well-separated as possible.

$$I\{C_1, C_2 \ldots C_k\} = \sum_{k=1}^{n} \sum_{x_i \in C_k} \|x_i - c_k\| \tag{2}$$

where $\|x_i - c_k\|$ is the Euclidean distance between x_i and c_k. From the Eq. (2) the sensor node which is at least distance from centroid will be assigned as cluster head. The BS sets up the TDMA schedule for every cluster, the cluster information and TDMA schedules is then broadcasted back to all the sensor nodes. The time slot is assigned to each non-cluster head node which starts to transmit data to CH. The CHs can perform the data aggregation function when all the data are received and then send the collected data to the BS. After all the data transmission from the CHs to the BS is finished, the next round begins from step 2. Step 2: Cluster head determination: Cluster head is selected on rotation basis when the residual energy $\overline{E(r)}$ of the cluster head falls below the threshold E_{th} energy level. The new node with highest energy will be selected as cluster head. The cluster-head is decided on the basis of the residual energy and the distance from node to cluster-center.

$\overline{E(r)}$ is the residual energy

$$\overline{E(r)} = \frac{\sum Energy\,(N)}{Total\,number\,of\,alive\,nodes}$$

where

$$\sum Energy\,(N) = total\,energy\,of\,n\,nodes$$

Highest current energy $E_{initial} = \frac{\overline{E(r)}}{d_i}$ where $d_i = \sqrt{(x_{ch} - x_i)^2 + (y_{ch} - y_i)^2}$ is the distance. The node with highest current energy will be elected as cluster head.

The residual energy of the rth round is calculated to find the network load

$$E(r) = E_{initial} - (r-1)E_{round} \tag{3}$$

d_i is the distance between current cluster head and node xi. If (x_{ch}, y_{ch}) and (x_i, y_i) are the location co-ordinates of current cluster head and node x_i. The function $F(d_i(x_i), \omega_i)$ can be used to decide node x_i in its cluster to be cluster head. As shown in Fig. 1 node with unique ID 1 has minimum distance from the centroid so it is chosen as cluster head.

$$F(d_i(x_i), \omega_i) = d_i(x_i) * \exp(1/\omega_i) \tag{4}$$

where $\omega_i = \frac{E_i(r)}{\overline{E(r)}}$ where r is the current round, $d_i(x_i)$ is the distance from node x_i to its cluster centre.

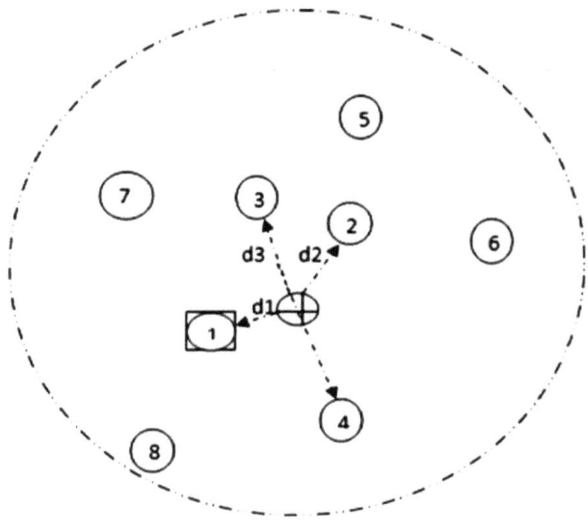

Fig. 1 Cluster head selection by k-means algorithm

Step 3 Re-selection of cluster heads: Cluster heads are reselected on the bases of residual energy and k-means i.e. the cluster head is shifted if the energy of the head is below threshold and the node which is second best according to k-means and has energy greater than threshold is selected as the new cluster head. Up to this point the process is same as KMT, in the k-means clustering technique is also applied on the cluster the k-means also having some distance from the base station. EKMT clustering technique is also applied on the cluster heads as they are from the base station. Firstly the cluster head with minimum distance i.e. d1 is made the cluster head. During re-selection when the energy of the cluster head with d1 distance is reduced and goes below the threshold the new cluster head has to be selected. For selecting new cluster head the second best candidate with distance d2 should be made the cluster head in accordance with KMT but in EKMT the distance from the base station is also considered. An energy coefficient is calculated on the bases of the fact that even though the node with distance d2 has less distance from the centroid as compared to the third best node with distance d3, but the distance of second best node is greater than the third best node from the base station i.e. D2 > D3. Hence as the distance difference in D3 − D2 becomes more important than d2 − d3, third best node 3 is selected as the cluster head shown in Fig. 2.

$$E_{c(i,j)} = \frac{D_i - D_j}{d_i - d_j} \tag{5}$$

where $E_{c(i,j)}$ represents energy constant of respective nodes from base station, D_i, D_j represents distance of nodes from base station and d_i, d_j represents distance of nodes from centroid.

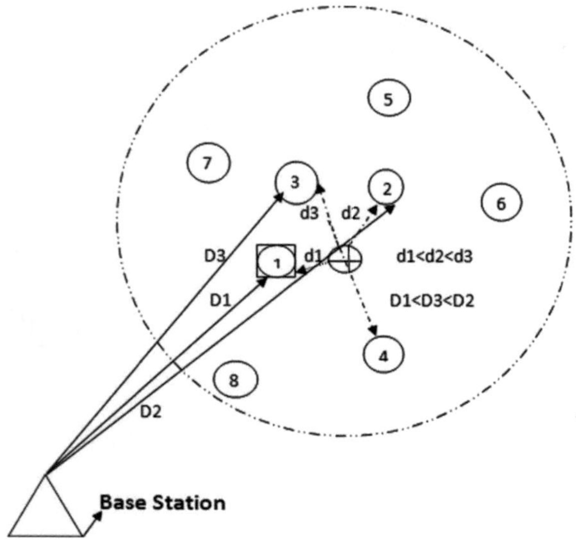

Fig. 2 Re-orientation of the cluster in view of new CH selection by using minimum mean distance algorithm

Residual Energy calculation in round r

$$E(r) = (r-1)E - (E_c * d(r)) \tag{6}$$

The most preferred cluster head now will be the node having maximum residual energy according to equation.

A simple radio model as in [1] is used as shown in Fig. 3. Transmission equation (7) and reception equation (8) are used to calculate energy consumed. It is also assumed that depending upon distance between transmitter and receiver d^2 and d^4 power loss law [22] for free space and multipath fading channel respectively. If the distance is less than a threshold, the free space (d^2 power loss model) is used; otherwise, the multi path (d^4 power loss) model is used. Thus, to transmit an l-bit message at distance d, the radio expands

$$E_{Tx}(l,d) = E_{elec} * l + \varepsilon_{fs} * l * d^2 \; d < d_{thres} \tag{7}$$

$$E_{Tx}(l,d) = E_{elec} * l + \varepsilon_{mp} * l * d^4 \; d > d_{thres}$$

where the threshold is $d_{thres} = \sqrt{\frac{\varepsilon_{fs}}{\varepsilon_{mp}}}$

To receive the l-bit message, the radio expands

$$E_{Rx}(l) = l * E_{elec} \tag{8}$$

$E_{RX}(l)$ is the energy spent in reception of a l-bit message, E_{elec} is the base energy required to run the transmitter or receiver circuitry.

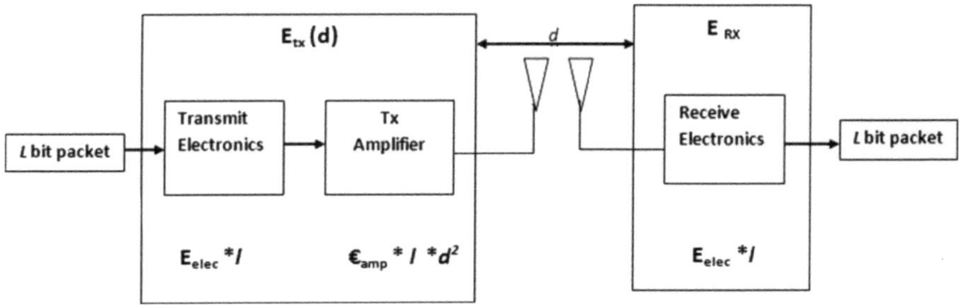

Fig. 3 Radio energy dissipation model

4 Simulation Results

The proposed EKMT-k-means clustering algorithm based low energy consumption protocol for Wireless sensor networks is tested on MATLAB. Then simulate the new proposed EKMT algorithm based low energy consumption protocol, whose results are compared with the results of already existing k-means algorithm protocol KMT. As energy depletion issue is major in wireless sensor networks, so we evaluate EKMT-k means by investigating the residual energy in the network.

The following Fig. 4 shows that the in proposed EKMT protocol total energy get consumed in 990 rounds whereas in KMT total energy get consumed in 780 rounds only. So an improvement in residual energy is more than 20% is observed in proposed protocol. Throughput of a network can be represented by number of data units successfully received through a network in a unit of time. Figure 5 shows the throughput analysis of proposed EKMT protocol with the existing KMT protocol. As shown in Fig. 5, EKMT protocol number of packets received at base station is 2500 in 100 rounds but in KMT only 1900 packets are received at base station. It is investigated that the proposed EKMT can successfully do the data aggregation from the source to base station

Jitter is the variation in packet arrival time. If the different packets at destination reach at receiver end with different delays than jitter will be more and it will effect the sensor network. So jitter should be minimum. Figure 6 shows the jitter for two protocols. It is visually proven that jitter for EKMT is very less as compared to already existing KMT protocol up to 90 rounds.

Delay is the total amount of time the system takes to send the data from source to base station. Figure 7 shows that the proposed protocol EKMT takes lesser time to aggregate and forward data to base station than existing KMT protocol. From the experimental results, it obviously proves that proposed system can perform better than the existing protocol. Hence the objective of EKMT-k means is achieved.

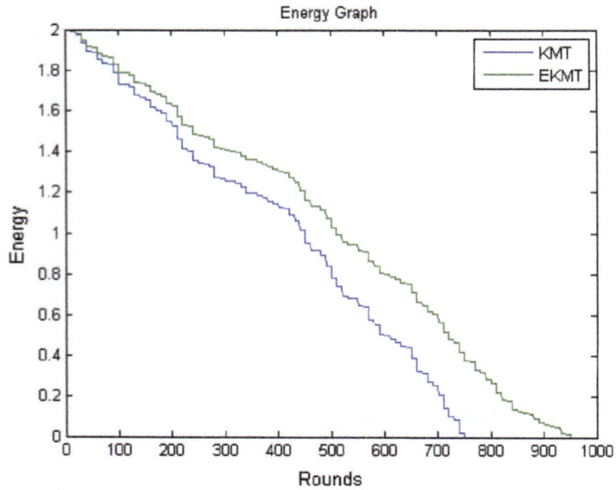

Fig. 4 Residual energy analysis with varying rounds for proposed EKMT protocol and KMT protocol

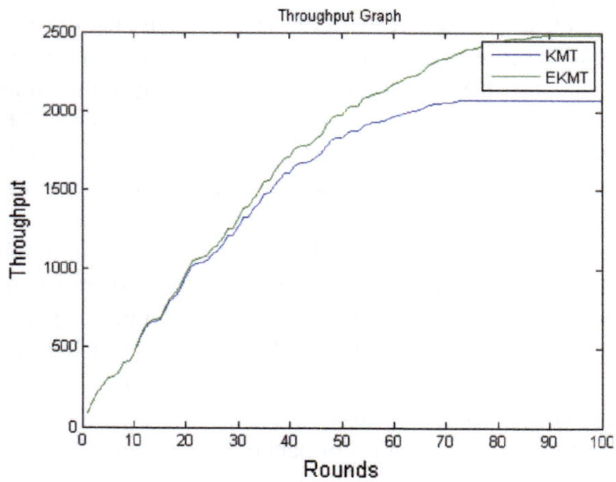

Fig. 5 Throughput analysis with varying rounds for proposed EKMT protocol and KMT protocol

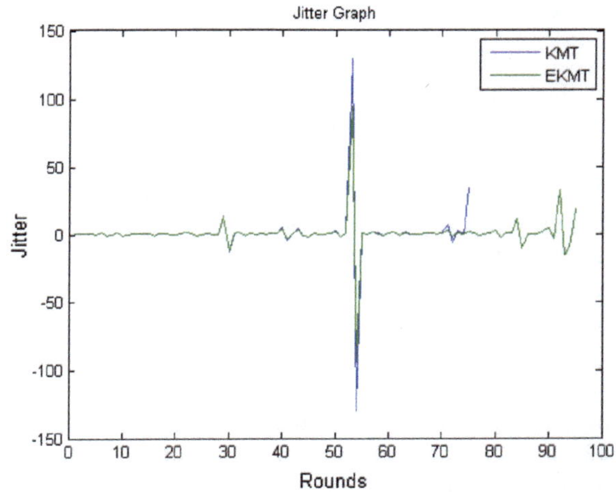

Fig. 6 Jitter with varying rounds

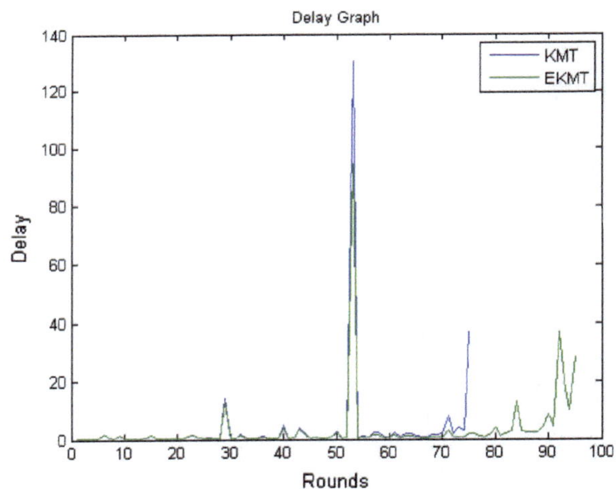

Fig. 7 Delay analysis with varying rounds

5 Conclusion

In this paper EKMT k-means clustering algorithmic solution based low energy consumption protocol for wireless sensor networks is proposed. In earlier the energy efficiency was improved by other means other than considering the distance from base station. Our new protocol called EKMT–k mean algorithm improves energy efficiency by considering the least distance between new selected cluster head and the base station. The proposed protocol can be realized by using k-means technique when the residual energy of cluster head is below threshold then re-selecting the cluster head using minimum distance between base station and new cluster head. The results shows that our protocol significantly improves the energy consumption in the network and much better throughput than earlier protocols. The performance metrics like jitter and delay are improved in our approach. The factors affecting cluster formation, selection of cluster head and CH communication with base station are open issues for future research. This investigation will be highly useful for energy efficient wireless sensor network.

References

1. W. Heinzelman, A. Chandrakasan, and H. Balakrishnan, "Energy-Efficient Communication Protocols for Wireless Micro sensor Networks (LEACH)" in HICSS, vol. 8, Maui, Hawaii, January 2003, pp. 1–10.
2. V. Pal, G. Singh and R. P. Yadav, "SCHS: Smart Cluster Head Selection Scheme for Clustering Algorithms in wireless Sensor Networks" Wireless Sensor Network, 4, 2012, pp. 273–280.
3. Y. Zhu, W W-d, J. Pan and T y-P, "An energy efficient data gathering algorithm to prolong lifetime of wireless sensor networks," Comput. Commun. 33, 2010 pp. 639–647.
4. S. Gao, H. Zhang and S. Dass, "Efficient data collection in wireless sensor networks with path – constrained mobile sinks," IEEE TRANS. Mob. comput. 10, 2011, pp. 592–608.
5. P. Bawa, B. Singh and V. Shokeen "Enhancement in Leach Protocol for Wireless Sensor Networks." International Journal of Computer Applications (0975 – 8887), July 2013, Volume 74– No. 21.
6. J. Hu, Y. Jin, and L. Dou, "A Time-based Cluster-Head Selection Algorithm for LEACH," IEEE Symposium on Computers and Communications, 2008, pp. 1172–1176.
7. H. Yang and B. Sikdar, "Optimal Cluster Head Selection in the LEACH Architecture", IEEE International Conference on performance, Computing, and Communications, 2007, pp 93–100.
8. TejalIrkhede and P. Jaini, "Cluster and Traffic Distribution Protocol for Energy consumption in Wireless Sensor Network", 2013 IEEE.
9. H. Yang and B. Sikdar, "Optimal Cluster Head Selection in the LEACH Architecture", IEEE International Conference on Performance, Computing, and Communications, pp 93–100, 2007.
10. M. Usha and N. Sankarram, "A Survey on Energy Efficient Hierarchical (Leach) Clustering Algorithms in Wireless Sensor Network", International Journal of Innovative Research in Computer and Communication Engineering ISO 3297: 2007.
11. O. Younis, S. Fahmy, "HEED: A hybrid, energy-efficient, distributed clustering approach for adhoc sensor networks." IEEE Trans. Mobile Comput., Vol. 3, 2004, pp 366–379,.

12. P. A. Forero, A. Cano, and G. B. Giannakis, "Distributed Clustering using Wireless sensor networks", IEEE Journal of selected topics in signal processing, August 2011, Vol 5, No. 4.
13. G. Y. Park, H. Kim, H. W. Jeong, and H. Y Youn, "A Novel Cluster Head Selection Method based on K-Means Algorithm for Energy Efficient Wireless Sensor Network", 27th International Conference on Advanced Information Networking and Applications Workshops, 2013.
14. Y. Gong, G. Chen and L. Tan. "A Balanced serial K-means based clustering protocol for wireless sensor networks", Oct. 2008, IEEE. pp 1–6.
15. S. Rani, J. Malhotra, R. Talwar, "Energy Efficient Chain based co-operative routing protocol for wireless sensor networks" Elsevier, Journal Applied Soft computing, Oct. 2015, pp. 1568–4946.
16. J. N. AL-Karaki and A. E. Kamal, "Routing techniques in wireless sensor networks: survey," IEEE Wireless Communication, 2004, vol. 1, pp. 6–28.
17. W. B. Heinzelman, A. P. Chandrakasan, and H. Balakrishnan, "An application- specific protocol architecture for wireless microsensor networks," IEEE Transactions on Wireless Networking, vol. 1, 2002, pp. 660–670.
18. A. Abbasiand, M. Younis: "A survey on clustering algorithms for wireless sensor networks," Computer Communication 2007, vol. 30, pp. 2826–2841.
19. S. Rani, J. Malhotra, R. Talwar, "On the Development of Realistic Cross Layer Communication Protocol for Wireless Sensor Networks," Sensor Network, 2014, 6, 57–66.
20. W. Yang, W. Shuan, X. Gfeng, W. Fan, G. Chen "Minimizing mobile sensor movements to form a line K-coverage," Peer-to-Peer Netw. Appl., published online 2015, doi:10.1007/s12083-016-0469-9.
21. G Nidhi, S Sanjeev, V Renu, "Energy-efficient mobile cluster-head data collection model for wireless sensor networks." Turk J Elec Eng & Comp Sci 2016 24: 3448–3458 doi:10.3906/elk-1406-155.
22. A. Paulraj, R. Nabar, and D. Gore, "Introduction to Space-Time Wireless Communications", Cambridge, U.K., Cambridge Univ. Press, 2003.

A Review of FIR Filter Designs

Bharat Naresh Bansal, Amanpreet Singh and Jaskarn Singh Bhullar

Abstract Digital filters are commonly used as an essential element of Digital Signal Processing (DSP) System. Digital filter can be used for developing many designs, which are impractical or impossible in Analog filter. Digital filters may be more expensive than an equivalent analog filter due to their increased complexity. Finite Impulse Response (FIR) and Infinite Impulse Response (IIR) are the two types of digital filters; FIR filters are preferred over IIR filters due their properties like inherent stability and linear phase. In this paper various techniques for FIR filter has been described and analyzed various factors which effects the performance of FIR filter in communication system especially in multi-standard communication.

Keywords FIR filter · Computational complexity · Coefficient Power consumption · Reconfigurable

1 Introduction

In the present scenario, Digital Signal Processing (DSP) is widely used in portable battery powered products based on wireless communication systems and various electronic devices. In DSP, finite impulse response filters (FIR) and filter banks (FBs) have attractive properties as the stability can be guaranteed and linear phase can be easily achieved. Therefore, these filters are popular in numerous applications such as communication systems, signal processing, biomedical instruments etc.

B.N. Bansal (✉) · J.S. Bhullar
MIMIT, Malout, India
e-mail: bharat.mimit@gmail.com

J.S. Bhullar
e-mail: bhullarjaskarn@gmail.com

A. Singh
IKGPTU, Kapurthala, India
e-mail: amanpreet1970@gmail.com

G.M. Perez et al. (eds.), *Networking Communication and Data Knowledge Engineering*, Lecture Notes on Data Engineering and Communications Technologies 3, https://doi.org/10.1007/978-981-10-4585-1_11

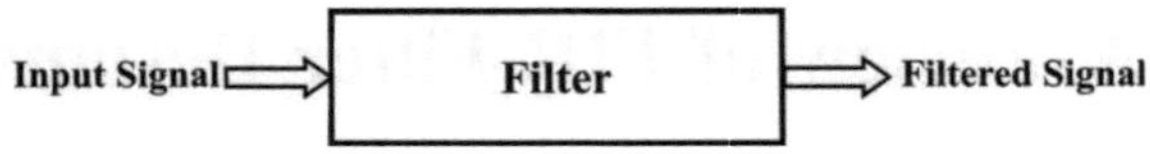

Fig. 1 Basic block diagram of filter

1.1 Filter

Filter [1, 2] is a device used to extract useful information from the input by removing unwanted component like noise. The basic block diagram of filter is shown in Fig. 1. In this figure the input signal is made up of information and noise signal. The filter removes the noise signal from the input signal and gives output as filtered signal i.e. information signal.

The primary functions of filters are:

- To confine a signal into a prescribed frequency band as in lowpass, highpass, and bandpass filters.
- To modify the frequency spectrum of a signal as in channel equalization and audio graphic equalizers.
- To decompose a signal into two or more sub-bands as in FBs, frequency multiplexers and sub-band coders.
- To model the input output relationship of a system such as telecommunication channels, human vocal tract, and music synthesizers.

The three key components of a digital filter are adder, delay and multiplier. Cost of each of these components relies very much upon the implementation approach.

1.1.1 Need of Filtering

Filtering [1, 2] is the most widely used complex signal processing operation and its main goal is to change the frequency spectrum according to the given specifications. The classical filtering is time invariant by keeping sampling frequency and filter order unique. Filters are used in various communication applications such as channel equalization, noise reduction and symbol detection. In signal processing, the role of a filter is to remove unwanted signal or to extract useful component in the frequency range of the signal.

1.1.2 Digital and Analog Filtering

Digital filtering [1, 2], has widespread applications in wireless communication systems to perform various functions like matched filtering, channel equalization, channelization, audio and video processing. Digital filters are divided in two categories namely recursive and non-recursive filter, which are also known as Infinite Impulse Response (IIR) filter and Finite Impulse Response (FIR) filter. For a given filter order, IIR filter has much sharper transition characteristic and has nonlinear phase characteristics. It is difficult to use IIR filter in adaptive filter applications. FIR filter has a linear phase response when filter's coefficients are symmetric. FIR filters are always stable but IIR filter has stability problem.

The digital communication is attractive due to constant improvement in digital circuit technology. Digital circuits are preferred over analog circuits because of its insensitivity to temperature and supply voltage. But for the successful implementation of digital communication, there is need to reduce significant power consumption in digital filter.

1.1.3 Finite Impulse Response (FIR) Filter

FIR filter [1, 2] is one of the most common components of DSP communication. FIR filtering operation is implemented by convolving the input data samples with the desired unit impulse response of the filter. To implement FIR filter, variety of implementation methods are developed to reduce the extra computational complexity.

Various tasks like spectral shaping, interference cancellation, matched filtering and channel equalization are performed with FIR filters.

Output Y_n, of an N-tap FIR filter is given by the weighted sum of latest N input data samples.

$$Y_j = \sum_{0 \le k \le N} C_k X_{j-k} \quad (1)$$

The C_k's are called the coefficients of the filter, X_j and Y_j are the jth terms of the input and output sequences, respectively. Figure 2 depicts the basic functional block diagram of FIR filter.

The difference equation from Fig. 2 which defines the output in term of input as:

$$y[n] = a_0 x[n] + a_1 x[n-1] + \ldots\ldots\ldots + a_N x[n-N] \quad (2)$$

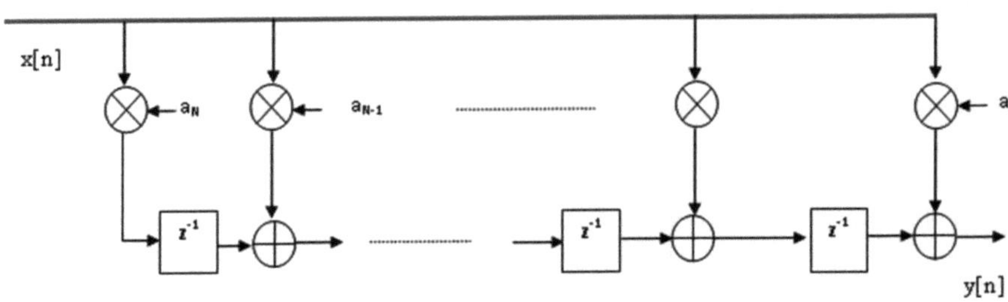

Fig. 2 Basic functional diagram of FIR filter

where y[n] is the output signal, x[n] is the input signal, a_i are the filter coefficients, also known as tap weights, and *N* is the filter order and N^{th} order filter has (N + 1) terms on the right hand side. These are commonly referred to as taps (the number of inputs), and one may speak of "N^{th} order/N + 1 tap filter".

1.1.4 Design Parameters and Constraints

The design of linear phase FIR filters with discrete valued coefficients has attracted great attention because of features like, attractiveness in a various applications as well as the difficulty of the problem due to the possibility of large number of filter coefficients. Approaches having low computational complexity as well as energy consumption are preferred to implement FIR filter. There are two major challenges with FIR filters:

(a) The deviation from the infinite precision filters resulted in degradation of filter performance which can be compensated by increasing the level of precision.
(b) FIR filters are dominated by a large number of multiplications, resulting into increase area and power, which can be reduced by replacing such operator into addition, subtraction and shifting operation.

2 FIR Filter Design

This section describes the work carried out by the researchers so far in the field of FIR filter design on basis of approximation error, SNR, computational complexity, reconfigurability, power consumption or energy efficient.

Benvenuto, Franks and Hill (1984) [3] worked on design for FIR filters with powers-of-two coefficients. Authors used signal to noise ratio (SNR) criterion to select the value of the filter coefficients and also explored the relationship between the size of the coefficient alphabet, filter sampling rate and the number of coefficients. The designed filter gave 2–3 dB improvement in terms of SNR. There was tradeoff between the size of the coefficient alphabet and the filter sampling rate.

Lim (1986) [4] proposed a frequency masking technique to design digital filter having linear phase. If each delay element of a linear phase low-pass digital filter was replaced by M delay elements, an (M + 1) band filter was resulted. When the output of this filter subtracted from delay version of the input, a complimentary filter was produced. A narrow transition band filter with very sharp transition was obtained initially by masking frequency response of complimentary filter, followed by combining resulting response to frequency response of original filter. The computational complexity of the filter was less because of less number of arithmetic operations. The complexity of the filter could be further reduced by using the power-of-two design technique [5–7].

Yang, Liu and Lim (1988) [8] designed a FIR filter based on frequency response masking technique which results a sharp transition having a linear phase. The design is a generalization of the frequency response masking technique [9] and achieved 20% more saving than technique developed by Lim [9]. The method could be used for bandpass filters as well as high pass filters.

Saramaki, Neuvo and Mitra (1988) [10] presented interpolated FIR filter with less number of computations. Authors reduced the required number of multiplier by optimizing the overall interpolated FIR (IFIR) structure which included the interpolator, shaping filter and interpolator factor. The method was close to the best design, as in addition to a small number of arithmetic operations, the number of delays was only 10–20% more than the conventional design. The results derived in the method had significant improvements over canonical multiplier efficient FIR filter designs. This method could be developed for medium wide bandwidths by using the methods of Lim [4] and Jing and Fam [11].

Samueli (1989) [12] presented a design of multiplier-less FIR filter with powers of two coefficients using an improved algorithm. The larger coefficients were allocated with an extra nonzero digit in the canonic signed digit (CSD) code to compensate for non-uniform nature of the CSD coefficient distribution. Two stages were combined in this algorithm: (a) Search for an optimum scale factor, and (b) A local bivariate search in the neighborhood of the scaled and rounded coefficients. There was an increase in filter complexity due to additional CSD digits, but it was minimal. Moreover, an improvement in the frequency response was also obtained. This design could further be used in combination with other techniques for reducing FIR filter complexity such as the interpolated FIR filters [13] or the prefilter structures [14].

Zarour and Fahmy (1989) [15] worked on variation of filter coefficients to change the frequency characteristic of a filter during signal processing. This was achieved using "variable filters" in which coefficients were directly obtained from specified spectral parameters. The coefficients of filter were expressed as analytical functions of the frequency specifications by method of curve fitting technique. Authors also proposed an approach for designing filters which optimizes several fixed filters which have different spectral parameters value. The main limitation of the method was requirement of large design time. Moreover, it is based on direct form realization which is highly sensitive to quantization errors.

Lim and Lian (1994) [9] introduced a frequency response masking approach for digital filter design. Authors also worked on a technique for complexity reduction using masking filter factorization and also presented three methods to reduce the complexity of a filter. These methods always have different group delay and round off noise performances and therefore, the analysis of their performance measuring parameters is always tough task.

Mehendale, Sherlekar and Venkatesh (1995) [16] proposed an algorithm to reduce power dissipation in FIR filter when it was implemented on a digital signal processor by optimizing coefficients of the filter. Authors designed an algorithm to minimize the Hamming distance for reducing power dissipation by coefficient scaling and perturbation. Experimental results show that 36% reduction in the total

Hamming distance; therefore, significant reduction in power dissipation of coefficient memory data bus and multiplier was achieved. Hamming distance minimization and minimizing number of 1's was a conflicting requirement in this algorithm; hence there was need to develop a unified cost model that assigns appropriate weights to these two factors.

Lian (1995) [17] introduced a very sharp half-band FIR filter using the frequency response masking technique. The method was generalized to reduce the complexity. Lian derived an expression for the impulse response up-sampling ratio which produces a minimum complexity design and also considered a multi-stage masking structure to achieve further savings in the number of multiplier. It minimizes the complexity of a half-band filter design using frequency response masking technique. The method had reduced realization complexity of the filter but its group delay was larger than direct form FIR filter with same approximation accuracy. This property was not desirable in many applications.

Xu, Wang and Parhi (1996) [18] implemented a power efficient, order configurable FIR filter with programmable coefficient. Configurable processor array (CPA) and phase locked loop (PLL) were the two main parts of the design. This design was implemented using Mentor Graphics tools. Efficient power consumption was one of the important advantages of the architecture. The 8 bit parallel multipliers which determined the critical path of the system was a key part of the CPA. PLL circuit was used that automatically controls the internal supply voltage. The design example proved that PLL provided the lowest voltage allowable while achieving the required output. There is increase in power consumption of voltage level shifter with increase in level of voltage shift. Hence, there is a tradeoff between level of voltage shift and power consumption.

Hartley (1996) [19] presented a design for optimizing CSD multipliers by sharing subexpressions. It accommodates different requirements in hardware and software contexts. In a software context, the goal was to minimize the number of operations required for the evaluation of an expression. In hardware, the determined gains made by subexpressions sharing were more complicated because there was difficulty to change the order of expression's evaluation. Moreover, the method proved that for pipelined circuits, it was important to consider both the number of adders and pipeline latches used, which was more difficult to determine. The described design based on finding common subexpressions in the CSD representation of the filter coefficients, resulted in decrease the number of operators by about 50%. Further, 33% saving of the number of additions were achieved by sharing subexpressions in pair. The subexpression sharing technique reduces the computational complexity in the implementation of FIR filters. The design was not applicable to receive data in real time since the modification of filter coefficients was difficult to accomplish in real time.

Sankarayya, Roy and Bhattacharya (1997) [20] implemented an algorithm based on differential coefficients for realizing low power FIR filter. A low power FIR filter was realized using this algorithm in which different orders of differences between coefficients were convolute with input data rather than directly with the coefficients. This algorithm resulted in a less number of computations per convolution as compared to directly using the coefficients, hence there was less energy dissipation and improvement in speed. The algorithm requires more memory storage and its access because stored computations results were reused.

Chang and Jen (1998) [21] presented a low power FIR filter design using various order of difference for coefficient and first order difference of the input. Authors also formulated the FIR operations with the differences in coefficient as well as input value. Distributed Arithmetic (DA) approach together with differential coefficients/input method (DCIM) was used to utilize the probability distribution such that energy consumption to be further reduced. The design example proved that less power is dissipated as compare to the previous design with the appropriate performance. The approach has poor performance in linear and square root memory model due to the large requirement of large memory.

Ramanathan, Visvanathan and S.K. Nandy (1999) [22] introduced a computational engine for multirate FIR filter design. It was programmable and comprises of both data and control path. It was generic to support FIR filters, decimation FIR filters and interpolation FIR filters. The work was derived from the base folded approach of multirate FIR filtering algorithms. This approach met the performance requirements as well as minimize the power dissipation and system cost. Authors also discussed the mapping procedure of algorithms with the computational engine to achieve the desired performance constraints of throughput and power consumed.

Hong and Stark (1999) [23] worked on energy efficient, low power FIR filter which was designed to achieve high throughput for communication systems. Critical delay and switching activity were reduced which optimizes the power consumed by the digital matched filter. The filter coefficients were designed in such a way that the spectral concentration in the transmission bandwidth of the channel with zero ISI was optimized while minimizing the heat dissipation. Authors also designed the filter by restricting its coefficients to powers of two and use finite non-zero coefficient bits to reduce power dissipation and increase throughput. There was always a possibility that the method did not converge by restricting the coefficient to a discrete power of two coefficient space. Moreover, there was no possible way to decrease condition variables simultaneously.

Coleman (2001) [24] proposed a method in which the coefficients were expressed in a higher radix number system that reduced a large linear combination problem to a smaller one. A well-chosen sequence of such number systems then leads to a cascade of these problem-reducing networks that jointly solve the original problem with significant overall computational efficiency. The method was based on MATLAB programming and significant solutions were achieved for higher order filters. Simulation results proved that FIR filter was designed with −95dBrms approximation error using less than two subtract or add operations per coefficient. It reduces approximately 60% computational complexity relative to the normal CSD method.

Lian, Zhang and Ko (2001) [25] proposed an improved Frequency Response Masking (FRM) approach to design a sharp FIR filters which gives a FIR filter having less number of multipliers. It helps to reduce the computational complexity. One of the sub-filters in FRM approach was replaced by IFIR filter which results

21% saving of multipliers as compared to simple FRM technique. A multi-stage frequency response masking design was achieved with less number of multipliers at a cost of longer group delay. Further savings in the number of multiplier could be achieved by combining the above method with other masking filter complexity reduction techniques [16, 22].

Chan, Liu and Ho (2001) [26] worked on multiplier-less reconstruction modulated filter banks using Sum of Power of Two (SOPOT). The method was focused on a modified factorization of the discrete cosine transform (DCT-IV) matrix and the lattice configuration of the prototype filter had no loss. It represents the coefficients in SOPOT form without affecting the perfect reconstruction condition using Genetic Algorithm (GA) as a search engine. The FIR filter with better frequency characteristic was designed with a very low implementation complexity. However, the methods based on SOPOT coefficient multiplications could be implemented effectively with simple shifts and additions.

Tang, Zhang and Min (2002) [27] presented a programmable high speed FIR filter having CSD encoding coefficients which results a multiplier-less filter. In order to reduce the complexity and enhance the compatibility of FIR, a new programmable CSD encoding structure was employed. The work exhibits improvement of speed in filter and reduction in area as compared to previous FIR filter design. The filter was designed which operated at maximum frequency of 100 MHz.

Park, Jeong, Meimand, Wang, Choo and Kaushik Roy (2004) [28] designed a digital FIR filter for low power and high performance applications using programmable Computation Sharing Multiplier (CSHM) technique. The multiplications in filtering process were represented by combination of shift and add operations in CSHM technique. The common computations were shared without additional memory area. The design achieved low power and high performance for FIR filter implementation due the sharing property of CSHM technique. Further, improvement in power and performance was achieved using carry select adder and conditional capture flip-flop which was a circuit level efficient technique. Using this technique a 10 tap FIR filter was designed and fabricated in Complementary Metal Oxide Semiconductor (CMOS) 0.25 m technology.

Vinod and Lai (2005) [29] proposed a multiplication algorithm for realizing low complexity FIR and IIR filters which have minimum number of full adders and improved speed. Authors also described coefficient partitioning, in which each coefficient partitioned into two subcomponents. This reduces the shifts of the operands of the adders. Furthermore, there was reduction in adder of multiplier by combining three approaches; coefficient partitioning algorithm, PFP representation and Common Subexpression Elimination (CSE) method. The algorithm achieved an average full adder reduction of 30% for FIR filters and 20% IIR filters over CSE methods. The algorithm did not base on the statistical distribution of shifts between the 2-bit Common Subexpression (CS) because it did not employ Super Subexpressions (SSs).

Kidambi (2006) [30] designed a linear phase FIR filter with variable bandwidth based on closed form expressions. The filter was designed as a combination of filters having fixed coefficients in a least square sense. It formulates an error function which reflected the difference between the practical filter and desired variable bandwidth filter. Significant savings in computational complexity was achieved in obtaining the entries of the block symmetric matrix. The filter did not have salient features such as flexibility and reconfigurability.

Chen and Chiueh (2006) [31] designed a low power digit reconfigurable FIR filter which was compact and had a wide range of precision. To implement filter with finest granularity was adopted and resulted in an architecture used for reconfigurable FIR filter. The filter implemented with this approach was easily configured as a pulse shaping and matched filters. The FIR architecture also has scalability, modularity and cascade-ability which justify its utility to Very Large Scale Integration (VLSI) implementation. It has incurred large overhead to support reconfigurable schemes due to arbitrary nonzero digit assignment. This design had high hardware resource utilization which made it impractical for resource constrained SDR handsets.

Eshtawie and Othman (2007) [32] designed an algorithm for representation of FIR filter response by varying number of non-zero coefficients and their values. Authors also represented phase and frequency response of FIR filter using minimum number of non-zero coefficients. Hence, the algorithm reduces the number of adders as well as addition operation required to get the filter output, which further reduced the computational complexity, area usage, processing time and power consumption. The algorithm implemented the FIR filters, require half of the number of additions as compared to simple filter which provides less computational complexity with high speed. Verilog Hardware Description Language (VHDL) was used to design the filter. The algorithm could be more beneficial if used with multiplierless approaches like DA.

Wang and He (2008) [33] presented a FRM based FIR filters which uses back-propagation neural network algorithm. In this sparse coefficients were used for implementation of filter. The resulting filter has less computational complexity as the implementation required less number of multipliers and adder as compared to conventional design methods. In this design, coefficients of overall sub-filters were optimized simultaneously, as a result the joint optimization leads to improve design performance. Authors failed to address the issue such as convergence rate. Moreover, the filter was not flexible and reconfigurable.

Mahesh and Vinod (2008) [34] worked on coefficient decimation approach for realizing reconfigurable FIR filters. To achieve multiband frequency response, the coefficients of an FIR filter were decimated by order M. Authors also investigated the possibility of obtaining a decimated frequency response and filter-bank. The approximation error could further be improved using other soft computing tools and linear programming.

Meher, Chandrasekaran and Amira (2008) [35] designed area, power and delay efficient FIR filter by systolic decomposition of DA based inner-product computation. The FPGA realization possessed less area-delay complexity compared with conventional DA based design of FIR filter. Energy per operation, energy density and energy throughput were three key parameters for energy estimation which were analyzed for the FIR filter realization. The FPGA realization was fully realizable, modular and scalable; therefore, it could be easily used as an IP core in a number of environments.

Mahesh and Vinod (2008) [36] proposed CSE based algorithm for the design of higher order FIR filters with a lesser number of adders. Binary representation was used for coefficients in filter design. The number of unpaired bits was considerably less in case of binary coefficients in comparison to CSD coefficients. The reduction in number of adders was achieved without much increase in logic depth. The algorithm reduced the computational complexity by combining three techniques: (a) Binary Horizontal Subexpression Elimination (BHSE), (b) Binary Vertical Subexpression Elimination (BVSE) and (c) Binary Super Subexpression Elimination (BSSE). The reduction of logical operators is achieved using this algorithm which was slightly inferior for lower order filters.

Aktan, Yurdakul and Dundar (2008) [37] proposed an algorithm for the design of hardware efficient and lower power linear phase FIR filters. The algorithm found filter coefficients with reduced number of SOPOT terms. The value of coefficient was fixed using linear programming because it was a branch and bound based algorithm. The design comprises fewer number of SOPOT terms and shorter in wordlength. The proposed design consumed 20% less power and up to 30% reduction in number of SOPOT terms on an average than unoptimized coefficients. The discrete coefficients in an optimized filter were observed in reasonable time. The proposed design was superior in terms of SOPOT, power performance, design time, hardware complexity, and term count for high order filters.

Choi, Banerjee and Roy (2009) [38] presented a technique to design a FIR filter without using the entire coefficients. Using a level constrained CSE algorithm to calculate each of the coefficient outputs in a filter. The presented design technique caters to low energy requirements and give reasonably accurate response with tolerance to large process variations. As compared to conventional CSE based filter implementation this technique required less number of computation steps. This filter operates at low voltages to reduce power consumption under minor quality degradation in the output response.

Baran, Wei and Oppenheim (2009) [39] designed a sparse FIR filter using linear programming algorithms. Array design and acoustic equalization were the key factors to evaluate the algorithms. Authors also discussed several estimated polynomial time approaches in which linear programming was used to design sparse filters. The sparse filter means filter have a small number of nonzero coefficients. This type of filter eliminates the arithmetic operation corresponding to zero value coefficients and hence the computational complexity of filter reduced.

Smitha and Vinod (2009) [40] proposed a low power reconfigurable filter architecture based on decimation, frequency masking and interpolation for dynamic channels adaption in cognitive radio handsets. The architecture provided area-time complexity reduction about 10.7% as compared to the conventional reconfigurable FIR filter. The results showed that an average gate count reduction of 10.7% and power reduction of 5.9% over per channel approach was achieved when implemented on Xilinx Virtex 2 FPGA.

Lin, Vinod and Samson See (2011) [41] proposed a new flexible multistage coefficient decimation filter bank (MS-CDFB) which comprises a low complexity and wideband spectrum sensing in cognitive radios. The complexity reduction of design over DFTFB is 30%. Moreover, the complexity further reduced up to 85%, if identical sensing was produced by both the spectrum sensors. They also showed that given design provides superior accuracy of the estimated power spectrum than discrete fourier transform filter bank (DFTFB) whenever channel separation was narrow. MS-CDFB obtained a dynamically reconfigurable multiband frequency response using fixed-coefficient FIR filter.

Sheikh and Gustafsson (2012) [42] designed a coefficient decimation FIR filter using linear programming taking all configuration mode. The designed filter has a property of low complexity and re-configurability. Authors also obtained the decimated impulse response in two possible ways for even decimation factors and effect of this was numerically found. The approximation error obtained by this technique is better than Mahesh and Vinod [34]. The approximation error could further be improved implementing a filter in MATLAB having D^{th} coefficient same and either removing or replacing in between coefficient to zero.

Rusu and Dumitrescu (2012) [43] implemented a FIR sparse filter which has lower complexity as compared to full filters by maintaining a good performance level. The zero value coefficients were introduced in bulk by reweighting of minimization stages. Extra coefficients were eliminated due to a small number of greedy iterations. Switching factor and convergence rate were the major constraints in work carried out by these authors.

Bhattacharjee, Sil and Chakrabarti (2013) [44] work was focused on the design of FIR filter which requires a low power in DSP applications based on FPGA. Authors also implemented the various forms of FIR filter on FPGA and compared their performances in terms of delay, frequency of operation, resource and power utilization. The work was focused on custom design and their implementation, but failed to provide a thorough analysis of various FIR circuit in terms of FPGA design metrics of resource, power utilization and delay.

Ambede, Smitha and Vinod (2013) [45] discussed an improved coefficient decimation approach to design low complexity FIR FBs both non-uniform and uniform for different communication channelizers. They also obtained the sub-bands of desired bandwidths by subtraction of the highpass and lowpass responses, which were coefficient decimated frequency responses of the simple filter. The designed FB supported multistandard channelization. The approach achieved high degree of reconfigurability and less complexity than coefficient decimation method (CDM) based approaches. Further, as compared to progressive decimation filter bank (PDFB) technique, this approach provided superior stopband and transition band characteristics. Moreover, this approach reduced 74% of computational complexity as compared with PDFB for non-uniform fixed frequency multi-standard channelization. Furthermore, this approach reduced 95.67% of computational complexity as compared with PDFB using the method for multistandard channelization with variable subband locations.

Qaisar, Fesquet and Renaudin (2014) [46] presented a computationally efficient signal processing approach based upon adaptive rate filtering. The local variations in the input signal decided the change in the sampling frequency as well as in the filter order. The computational complexities and output qualities of the technique were compared to the conventional design for a speech signal. A drastic decrease in computational and a significant processing power reduction was observed in simulation results as compared to other approaches. An approach to compute the filtering error was also devised. It was shown that errors made by the approach were minor and a higher precision could be achieved by increasing the Level Crossing Sampling Scheme based analog to digital converter (LCADC) resolution or the interpolation order. Major drawback of this design is that the significant signal parts should be sampled locally at higher rates.

Ambede, Shreejith, Vinod, and Fahmy (2015) [47] designed a variable digital filter for SRR channelizer based on the low complexity improved coefficient decimation method. Different types of responses were obtained using a single set of prototype filter coefficients. The design had a limitation of requirement of high order prototype filter when sharp transition BWs were desired, which may lead to high computational complexity.

2.1 Observations for FIR Filter

From the Sect. 2 the following observations have been drawn:

(a) FIR filters are preferred over IIR filters in most of the communication systems due to its properties such as linear-phase, stability and low sensitivity to coefficient.
(b) The transition width of a very sharp FIR filter is inversely proportional to the filter length. Hence, these FIR filters are high order filters and leads to implementation problems.
(c) FIR filter as compared to IIR filter require large number of arithmetic operations, which requires high power and reduces its speed and in turn increases its computational complexity.
(d) A filter design procedure involves suitable values of the design parameters which are based on given design requirements and specifications. It is almost impossible to attain reconfigurability without increasing complexity in terms of hardware as the filter coefficient has wide range in precision for different applications.
(e) Reduce computational complexity is accomplished by eliminating programmability of the coefficients in many application, which allows optimization of the hardware corresponding to particular fixed coefficient set and vice versa.
(f) FIR filters either in linear phase or nonlinear require multipliers which are almost equal to or half the length of the FIR filter. Hence, FIR filters require

more power consumption and area as they utilize huge number of multipliers. Therefore, the design problem of low power and small area FIR filters require more attention. The complexity of the FIR multiplier is dominated by the number of adders (subtractors) employed in the coefficient multipliers. The number of adders needed in the multipliers is proportional to the coefficient wordlength.

(g) Coefficient decimation approach in FIR filter produces different frequency responses without changing the coefficient values. Different types of filter such as lowpass, highpass and bandstop are obtained from these frequency responses using FRM technique.

(h) FIR filters are implemented with property of reconfigurability using coefficient decimation approach which is based upon interpolating or/and decimating a fixed set of coefficients. There is no need to store different filter coefficient sets. As the shift and add approaches are directly applied which simplify the multipliers.

(i) DSP based programmable filters have advantages of high sample rate and flexibility but not much considerable in mobile communication applications where low power consumption and high output are required.

(j) Many approaches are used to reduce the power consumed by the digital filters, which includes techniques such as SOPOT, CSD, CSE and PFP for transforming the binary representation of the filter coefficients. These approaches reduce the arithmetic computation, which are required to get filter output without degradation in the performance.

3 Conclusions

The need and background of research work are clearly found by carrying out literature review. It gives related solution regarding improvements in the study already done and allows unsolved problems to come out and hence define all boundaries related to development of the research work. A lot of literature has been reviewed in connection with optimization systems in digital filter design focusing on multi-standard communication system.

Literature survey has been carried out for FIR filter design on basis of energy efficiency, less computational complexity, reconfigurability and its importance in communication system. The performance of the FIR filter design technique in the reference to approximation errors, computational complexity and reconfigurability has been evaluated which showed that there is need to decrease computational complexity as well as power consumption of a fir filter. Moreover, filter should have property of reconfigurability and flexibility for the effective multi-standard communication.

References

1. Mitra, S. K. (2006), Digital Signal Processing, New Delhi: Tata Mc-Graw-Hill.
2. Proakis, J. G. and Manolakis D. G. (2007), Digital Signal Processing: Principles, Algorithms and Applications, New Delhi: Prentice Hall of India.
3. Benvenuto, N., Franks L.E. and Hill F.S. (1984), "O*n the Design of FIR Filters with Powers-of-Two Coefficient*", IEEE Transactions on Communications, **32**(12), pp. 1299–1307.
4. Lim Y. C. (1986), "*Frequency Response Masking Approach for the Synthesis of Sharp Linear Phase Digital Filters*", IEEE Transactions on Circuits and Systems, **33**(4), pp. 357–364.
5. Lim, Y. C. and Parker, S. R. (1983), "D*iscrete Coefficient FIR Digital Filter Design Based Upon an LMS Criteria*," IEEE Transactions on Circuits and Systems, **CAS-30**, pp. 723–739.
6. Lim, Y. C., Parker, S. R. and Constantinides, A. G. (1982), "Finite Wordlength FIR Filter Design Using Integer Programming Over a Discrete Coefficient Space," IEEE Transactions Acoustic, Speech and Signal Processing, **ASSP-30**, pp. 661–664.
7. Lim, Y. C. and Parker, S. R. (1983), "*FIR Filter Design Over a Discrete Powers-Of-Two Coefficient Space*," IEEE Transactions Acoustic, Speech and Signal Processing, **ASSP-31**, pp. 583–591.
8. Yang, R., Liu, B. and Lim, Y. C. (1988), "*A New Structure of Sharp Transition FIR Filters using Frequency Response Masking*", IEEE Transactions on Circuits and Systems II: Express Briefs, **35**(8), pp. 955–966.
9. Lim, Y. C. and Lian, Y. (1994), "*Frequency-Response Masking Approach for Digital Filter Design: Complexity Reduction via Masking Filter Factorization*", IEEE Transactions on Circuits and Systems II: Analog and Digital Signal Processing, **41**(8), pp. 518–525.
10. Saramaki, T. and Mitra, S. K. (1988), "*Design of Computationally Efficient Interpolated FIR Filters*", IEEE Transactions on Circuits and Systems, **35**(1), pp. 70–88.
11. Jing, Z. and Fam, A. T. (1984), "*A New Structure for Narrow Transition Band, Lowpass Digital Filter Design*" IEEE Transactions Acoustic, Speech and Signal Processing, **ASSP-32**, pp, 362–370.
12. Samueli, H. (1989), "*An Improved Search Algorithm for the Design of Multiplier-less FIR Filters with Powers-of-Two Coefficients*", IEEE Transactions on Circuits and Systems, **36**(7), pp. 1044–1047.
13. Saramaki, T., Neuvo, Y. and Mitra, S. K. (1988), "*Design of Computationally Efficient Interpolated FIR Filters*," IEEE Transactions Acoustic, Speech and Signal Processing, **35**, pp. 70–88.
14. Adams, J. W. and Willson, A. N. (1984), "*Some Efficient Digital Prefilter Structures,*" IEEE Transactions on Circuits and Systems, **CAS-31**, pp. 260–265.
15. Zarour, R. and Fahmy, M. M. (1989), "*A Design Technique for Variable Digital Filters*", IEEE Transactions on Circuits and Systems, **36**(11), pp. 1473–1478.
16. Mehendale M., Sherlekar, S.D. and Venkatesh, G. (1995), "*Coefficient Optimization for Low Power Realization of FIR Filters*", IEEE Workshop on VLSI Signal Processing, Japan, pp. 352–361.
17. Lian Y. (1995), "*The Optimum Design of Half-Band Filter using Multi-Stage Frequency-Response Masking Technique*", Signal Processing, **44**(3), pp. 369–372.
18. Xu, C., Wang, C. Y. and Parhi, K. K. (2011), "*Order-Configurable Programmable Power-Efficient FIR Filters*", International Conference on High Performance Computing, Trivandrum, pp. 357–361.
19. Hartley, R. I. (1996), "*Subexpression Sharing in Filters Using Canonic Signed Digit Multipliers,*" IEEE Transactions on Circuits and Systems-II: Analog and Digital Signal Processing, **43**(10), pp. 677–688.
20. Sankarayya N., Roy, K. and Bhattacharya, D. (1997), "*Algorithms for Low Power Fir Filter Realization Using Differential Coefficients,*" IEEE Transactions on Circuits and Systems—II: Analog and Digital Signal Processing, **44**(6), pp. 488–497.

21. Chang, T. S. and Jen, C. W. (1998), "*Low Power FIR Filter Realization with Differential Coefficients and Input*", IEEE International Conference on Acoustics, Speech and Signal Processing, Seattle, WA, pp. 3009–3012.
22. Ramanathan, S., Visvanathan, V. and Nandy, S. K. (1999), "*A Computational Engine for Multirate FIR Digital Filtering*", Signal Processing, **79**(2), pp. 213–222.
23. Hong, S. and Stark, W. E. (1999), "*Performance Driven Coefficient Optimization for High throughput Energy Efficient Digital Matched Filter Design*", IEEE Pacific Rim Conference on Communications, Computers and Signal Processing, Victoria, BC, pp. 321–324.
24. Coleman, J. O. (2001), "*Cascaded Coefficient Number Systems Lead to FIR Filters of Striking Computational Efficiency*," IEEE International Conference on Electronics, Circuits and Systems, Washington, DC, USA, pp. 513–516.
25. Lian Y., Zhang, L. and Ko, C. C. (2001), "*An Improved Frequency Response Masking Approach for Designing Sharp FIR Filters*", Signal Processing, **81**(12), pp. 2573–2581.
26. Chan, S. C., Liu, W. and Ho, K. L. (2001), "*Multiplierless Perfect Reconstruction Modulated Filter Banks with Sum-of-Powers-of-Two Coefficients*", IEEE Signal Processing Letters, **8**(6), pp. 163–166.
27. Tang, Z., Zhang, J. and Min, H. (2002), "*A High-Speed Programmable CSD Coefficient FIR Filter*", IEEE Transactions on Consumer Electronics, **48**(4), pp. 834–837.
28. Park, J., Jeong, W., Meimand, M. H. Yongtao Wang, Hunsoo Choo and Kaushik Roy (2004), "*Computation Sharing Programmable FIR Filter for Low-Power and High-Performance Applications*" IEEE Journal of Solid-State Circuits, **39**(2), pp. 348–357.
29. Vinod, A. P. and Lai, E. M. K. (2005), "*An Efficient Coefficient-Partitioning Algorithm for Realizing Low-Complexity Digital Filters*," IEEE Transactions on Computer-Aided Design of Integrated Circuits and Systems, **24**(12), pp. 1936–1946.
30. Kidambi, S. S. (2005), "*An Efficient Closed-Form Approach to the Design of Linear-Phase FIR Digital Filters with Variable Bandwidth Characteristics*", Signal Processing, **86**(7), pp. 1656–1669.
31. Chen, K. H. and Chiueh, T. D. (2006), "*A Low Power Digit Based Reconfigurable FIR Filter*", IEEE Transaction on Circuits Systems II, **53**(8), pp. 617–621.
32. Eshtawie, M. A. M. and Othman, M. B. (2007), "*An Algorithm Proposed for FIR Filter Coefficients Representation*" International Journal of Electrical, Computer, Energetic, Electronic and Communication Engineering, **1**(2), pp. 126–132.
33. Wang, X. H. and He, Y. G. (2008), "*A Neural Network Approach to FIR Filter Design Using Frequency-Response Masking Technique*", Signal Processing, **88**(12), pp. 1695–1704.
34. Mahesh, R. and Vinod, A. P. (2008), "*Coefficient Decimation Approach for Realizing Reconfigurable Finite Impulse Response Filters*", IEEE International Symposium on Circuits System, Seattle, WA, pp. 81–84.
35. Meher, P. K., Chandrasekaran, S. and Amira, A. (2008), "*FPGA Realization of FIR Filters by Efficient and Flexible Systolization using Distributed Arithmetic*", IEEE Transactions on Signal Processing, **56**(7), pp. 3009–3017.
36. Mahesh, R. and Vinod A. P. (2008), "*A New Common Subexpression Elimination Algorithm for Realizing Low-Complexity Higher Order Digital Filters*", IEEE Transactions on Computer-Aided Design of Integrated Circuits and Systems, **27**(2), pp. 121–134.
37. Aktan, M. Yurdakul, A. and Dundar, G. (2008), "*An Algorithm for the Design of Low-Power Hardware-Efficient FIR Filters*", IEEE Transactions On Circuits And Systems—I: Regular Papers, **55**(6), pp. 1536–1545.
38. Choi, J. H., Banerjee, N. and Roy, K. (2009), "*Variation-Aware Low-Power Synthesis Methodology for Fixed-Point FIR Filters*", IEEE Transactions on Computer-Aided Design of Integrated Circuits and Systems, **28**(1), pp. 87–97.
39. Baran, T., Wei, D. and Oppenheim, A. V. (2010), "*Linear Programming Algorithms for Sparse Filter Design*", IEEE Transactions on Signal Processing, **58**(3), pp. 1605–1617.
40. Smitha, K. G. and Vinod, A. P. (2009), "*A New Low Power Reconfigurable Decimation Interpolation And Masking Based Filter Architecture For Channel Adaptation In Cognitive Radio Handsets*," Physical Communication, **2**(1), pp. 47–57.

41. Lin, M., Vinod, A.P. and Samson See, C. M. (2011), "*A New Flexible Filter Bank for Low Complexity Spectrum Sensing in Cognitive Radios*", Journal of Signal Processing Systems for Signal, Image and Video Technology, **62**(2), pp. 205–215.
42. Sheikh, Z. U. and Gustafsson, O. (2012), "*Linear Programming Design of Coefficient Decimation FIR Filters*", IEEE Transaction Circuits System II, **59**(1), pp. 60–64.
43. Rusu, C. and Dumitrescu, B. (2011), "*Iterative Reweighted Design of Sparse FIR Filters*", *Signal Processing*, **92**(4), pp. 905–911.
44. Bhattacharjeea, S., Sil, S. and Chakrabarti, A. (2013), "*Evaluation of Power Efficient FIR Filter for FPGA Based DSP Applications*", Procedia Technology, **10**, pp. 856–865.
45. Ambede A., Simtha, K.G. and Vinod, A.P. (2013), "*A Low-Complexity Uniform and Non-uniform Digital Filter Bank Based on an Improved Coefficient Decimation Method for Multi-standard Communication Channelizers*", Circuits, Systems and Signal Processing, **32** (2), pp. 2543–2557.
46. Qaisar, S. S., Fesquet, L. and Renaudin, M. (2014), "Adaptive Rate Filtering a Computationally Efficient Signal Processing Approach", Signal Processing, 94, pp. 620–630.
47. Ambede, A., Shreejith, S., Vinod, A. P. and Fahmy, S. A. (2015), "Design and Realization of Variable Digital Filters for Software Defined Radio Channelizers using Improved Coefficient Decimation Method", IEEE Transaction Circuits System II, 63(1), pp. 59–63.

Analysis of Safety Applications in VANET for LTE Based Network

Arjun Arora, Nitin Rakesh and Krishn K. Mishra

Abstract Safety applications in VANET are fundamentally dependent on communicating safety messages. Such messages are delay-critical and transmission parameters need to be kept in check for effective communication among vehicles. As long term evaluation (LTE) is infrastructure based direct communication is not possible among vehicles, such messages need to pass through the infrastructure. In case of safety applications extra care needs to be taken as high speed vehicles are more reactive to delay than slow speed vehicles. This is because high speed vehicles change their positions more frequently. Higher priority needs to be given to such vehicles in order to lower the delay. Characterization of delay requirements is needed which in turn can be used to compute path loss, fading, transmission range, rate and access probability in order to satisfy the delay requirements. In this paper a scheduler is proposed for LTE network suitable for safety applications in VANET. This scheduler takes into account the vehicular speed (VS) and Vehicular Location (VL) to assign priority and resources. High speed vehicles receive a higher priority for allocation of network resources. Simulation of the proposed scheduler (PS) shows that it performs better in comparison to largest weighted delay first (LWDF) algorithm which is considered to be the best choice for development of safety applications in VANET.

Keywords VANET · NS3 · Simulation · LTE network · Vehicular communication

A. Arora (✉)
Department of CSE, UPES, Dehradun, India
e-mail: a.arora@ddn.upes.ac.in; arjunarora06dit@gmail.com

N. Rakesh
Department of CSE, ASET Amity University Campus, Noida, India
e-mail: nrakesh@amity.edu

K.K. Mishra
Department of CSE, MNNIT, Allahahbad, India
e-mail: kkm@mnnit.ac.in

G.M. Perez et al. (eds.), *Networking Communication and Data Knowledge Engineering*, Lecture Notes on Data Engineering and Communications Technologies 3, https://doi.org/10.1007/978-981-10-4585-1_12

1 Introduction

Over the past few years a lot of research has been done in VANETs, the main objective always being safety applications and their efficiency. This is so to provide the next level of transportation systems with highest levels of safety, security and efficiency. VANETs make use of Dedicated Short Range Communications (DSRC) in order to provide communication among participating vehicles. IEEE 802.11p is the recently approved standard for communication in VANETs and has been used for a large number of safety applications however it suffers from issues of unbounded delay and limited range. On the other hand IEEE 802.11p has not yet been extensively used in real world applications [1, 2]. As safety is the main concern an analytical approach is required for adjusting parameters related to communication.

Recent studies have revealed that LTE can act as a substitute for IEEE 802.11p in VANET based applications. This paper focuses on safety applications based on periodic message transmission. Such messages have information about VL and VS which help in awareness of other vehicles and environmental objects. This information is then used by the proposed scheduler (PS) to predict network resource allocation and gives higher priority to high speed vehicles which are more delay reactive. Critical regions on the road map such as intersections are given higher priority if high speed vehicles are present in that region. As LTE is infrastructure based all communicating vehicles [3] need to pass messages through the road side units (RSUs) and no direct communication is possible. This may lead to congestion at the RSU as network resources are limited. Congestion leads to high latency which cannot be accepted in case of safety applications.

Therefore there is a need for a scheduler at the RSU. Many schedulers are available for LTE most popular being LWDF which shows the most promising results for safety applications however LWDF does not take into account Quality of Service (QoS), channel condition and gives equal priority to all vehicles irrespective of high speed or low speed. The modified LWDF considers QoS requirements and provides priority but QoS parameters are taken to be constant which is not suitable for real world safety applications as QoS parameters [4, 5] may vary with time in the real world. By carefully studying the vehicle equation for motion, delay requirements for a given safety application can be computed using poisson distribution and stochastic geometry.

LTE networks consist of a core network and access network with a radio access component called as the Evolved Universal Terrestrial Radio Access Network (E-UTRAN), consists of User Equipment (UE), base stations (eNode). The network core termed as Evolved Packet Core (EPC). Protocol stack for LTE consists mainly of four layers- packet data convergence protocol (PDCP), medium Access Control (MAC), Radio Link Control (RLC) and Physical (PHY). The proposed scheduler is deployed in the MAC layer of LTE and is responsible for resource allocation to vehicles in the network [6–8].

The PHY layer of LTE has radio access support with system bandwidth ranging from 1.5 to 20 MHz. LTE makes use of Orthogonal Frequency Division Multiplexing (OFDM) for downlink and for uplink it makes use of Single Carrier Frequency Division Multiple Access (SC-FDMA) [9, 10]. The frequency channel is divided into sub-carriers of 180 kHz each. The time frame is divided into intervals of 1 ms. Smallest resource allocated in LTE to a vehicle is called Physical Resource Block (PRB) which consists of total 12 number of carriers arranged consecutively.

2 Analysis, Design and Related Works

The proposed scheduler (PS) is implemented in the MAC layer. The main objective of the PS is allocating of PRBs. For each PRB allocated, computation is carried out based on the information from control channel (CCH) and optimization parameters. The working of the MAC layer within the network is based upon slotted ALOHA [11, 12]. In which at every given slot of time, a given node transmits with a probability of 'ρ'. Assuming the distance to be 'd' from the neighbouring vehicle say 'k + 1' for a given vehicle 'k' before initiation of communication between the said two vehicles.

If we analyse the motion of the vehicles, then considering delay in communication to be 'β_c' satisfying the following equation to prevent collision between the two:

$$\beta_c \leq \frac{d}{n_i + 1} + \frac{n_i^2 - n_i^2 + 1}{2ln_i + 1} - \beta_{RT} \tag{1}$$

Such that n_i, $n_i + 1$ being the respective speed of vehicles 'k' and 'k + 1', 'l' is the stopping rate of vehicle and β_{RT} being the reaction time. In such a case the vehicle safety application needs to deliver at least a single packet within 'β_c' seconds to the 'k + 1'th vehicle. Now say 'Q' is the rate of transmission and 'v' is the length of the packet. Then total number of transmission allowed would be:

$$E = [\frac{\beta_c Q}{v}] \tag{2}$$

The probable success rate after 'E' transmissions would be T_w^E and is called as bounded delay success probability [13, 14]. The criteria for avoiding collision is simply to target the probability of success i.e. 1 − ε or higher for a given vehicle 'k + 1' within given time slots 'E'. Then for any given vehicle time slots 'E'. Then for any given vehicle the collision probability would be 'ε'. Now let's assume that probability is independent then we can say that:

$$T_w^E = 1 - (1 - g(1-g)E_w)^E \tag{3}$$

Such that 'E_w' is simply the probability of success for a given vehicle transmitting information for a receiver at distance 'h'. Then for a given set of poisson distribution with Rayleigh Fading:

$$E_w = e^{-p\zeta} \tag{4}$$

where

$$\zeta = 2Iu\gamma^{\frac{1}{\beta}} \frac{\Pi}{\alpha} csc \frac{\Pi}{\alpha} \tag{5}$$

Such that 'I' is the density of vehicles and 'α' is the exponent of loss in transmission. Now as ζ and E are both functions of γ and T_w^E [15, 16]. We can say that 'p' depends on β. Now in case $T_w^E = 1 - \varepsilon$ is not possible for a 'k + 1' vehicle, adjustment parameters need to be taken care of such that the probability of collision between 'k' and 'k + 1' vehicles is lowest. On differentiating Eq. 3 we get optimum probability for access:

$$\rho* = \frac{1}{\zeta} + \frac{1}{2}(1 - \sqrt{1 + \frac{4}{\zeta^2}}) \underset{\approx}{>} \frac{1}{2+\zeta} \tag{6}$$

Now $\rho^*(o) = \frac{1}{2}$ and $\rho^* = \Theta(\frac{1}{\zeta})$ with $\zeta \to \infty$ we substitute Eq. 6 in Eq. 3 and get:

$$T_w^{E*} \underset{\approx}{>} 1 - (1 - \frac{1+\zeta}{(2+\zeta)^2} e - \frac{\zeta}{2+\zeta})^{\mathrm{E}} \tag{7}$$

For selection of User Equipment (UE) the value of PRB is computed and that UE is selected which has highest computed value of PRB. Consider the computation of 'x' PRB for 'y' user as $C_{x,y}$ and the PS selects the nth UE with maximum computed value of PRB:

$$\mathrm{n} = \max\{C_{x,y}\} \tag{8}$$

Based on parameters used for computation of scheduling the schedulers can be aware or unaware of the channel. In case of unaware schedulers such as LWDF which do not consider channel quality, conditions and Quality of Service (QoS). Allocate PRB to users on the basis of:

$$C_{x,y} = \frac{1}{W_i(f-1)} \tag{9}$$

where $\overline{W_i(f-1)}$ is:

$$\overline{W_i(f-1)} = \alpha \overline{W_i(f-1)} + (1-\alpha) q_f(f) \tag{10}$$

In case of aware scheduler such as M-LWDF which consider condition of channel, quality and Quality of Service (QoS). They allocate PRB to users on the basis of:

$$C_{x,y} = \frac{h_{i,k}(f)}{\overline{W_i(f-1)}} \tag{11}$$

where $h_{i,k}(f)$ is the maximum attained throughput for a given 'x'th UE with 'y'th PRB and $\overline{W_i(f-1)}$ is previous average throughput of 'x'th UE.

As LTE is infrastructure based, direct communication among vehicles is not possible. The vehicles need to communicate using network infrastructure [17, 18]. Each vehicle which needs to communicate needs to do so with its neighbouring vehicles by first communicating with the base station and then the message is forwarded to the vehicles neighbours. Figure 1 shown in the previous section shows how transfer of safety message takes place in LTE. A server in the LTE communicates with each vehicle through EPC. The server receives all safety messages and decides how re-transmission of these messages would take place with vehicles interested in communication [19, 20]. The server should be able to compute the

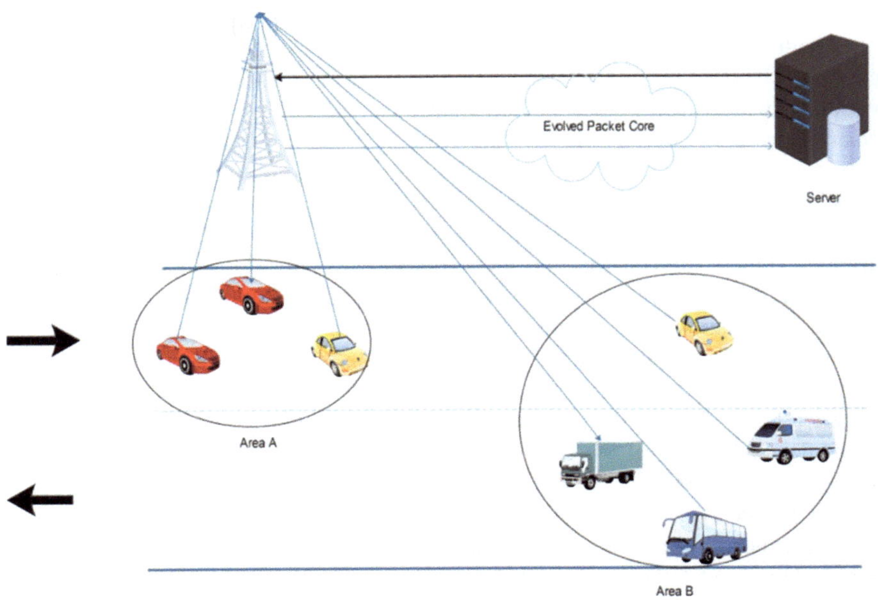

Fig. 1 Safety message communication among participating vehicular nodes

location of neighbouring vehicles as per the messages received earlier. In such a case unicasting is not possible as resources available are limited. Multicast approach is also inefficient as additional computation is required. Thus the server needs to simply combine all the information safety messages it receives and then broadcast the information to all vehicles present in that block. Vehicles on receiving this safety information only collect relevant information and discard the rest.

3 Proposed Scheduler for VANET Based Safety Application

In this section a LTE scheduler is proposed that can be used for VANET based safety applications. LTE being an infrastructure based network, all safety messages need to pass through the road side unit (RSU). A typical safety message is small in size and is transmitted in a periodic fashion. As the vehicle density increases in a given road segment such safety messages can create congestion especially in case of a LTE network as all are routed to the RSU or base unit [21, 22]. This congestion results in delay at the receiving of such messages to communicating vehicles. Safety applications in case of VANETs are sensitive to delay and vary from vehicle to vehicle, road segment to road segment [23, 24]. This is so because high speed vehicles travel a greater distance in a small amount of time in comparison to slow speed vehicles. This makes certain road segments more sensitive in comparison to others such as intersections this is so because the vehicles direct line of sight is not clear. Thus such regions require greater accuracy.

Schedulers which are unaware of the channel are not suitable for safety applications to be used in VANETs this so because resource allocation is done by them without considering channel quality, fading etc. Schedulers that are aware of the channel are also not suitable for safety applications in VANET as they provide maximum throughput but do not consider delay in messages which is not accepted in delay-critical safety applications. Therefore there is requirement of a scheduler which could be implemented in safety applications for VANETs that takes into account vehicle information and then allocates resources among communicating vehicles [25, 26]. The proposed scheduler uses a control channel which carries only critical information of a communicating vehicle to another vehicle. The communicating vehicle can also send channel quality; buffer status and other information using this control channel however capacity of this channel is very low. Separate parameters are defined for capturing and sending vehicle speed and location. The scheduler makes use of the core LTE server for gathering information of vehicle speed and location. In case of a situation where the proposed scheduler (PS) is unable to receive information from the server or the information received is out-dated then in such a case the PS makes use of a pre-defined value J_D for vehicle speed. Then speed can be estimated for a given vehicle 'k' as:

$$\widehat{J_k} = \begin{cases} \left(\frac{\tau}{r}\right)J_D + \left(1 - \frac{\tau}{r}\right)J_{lr}^k & \tau < r \\ J_D & \tau > r \end{cases} \tag{12}$$

where τ is simply defined as the time of last received information about vehicle 'k', 'r' is time at which received information becomes out-dated and J_{lr}^k is the last received information on speed of a vehicle 'k'.

Two metrics have been used for allocation of resources, one on the basis of vehicle speed and the other on the basis of vehicle location and speed. In case of scheduling on the basis of vehicle speed as we know that vehicles having higher speed change position more quickly in comparison to vehicles having lower speed [27, 28]. This makes high speed vehicles more delay critical thus the proposed scheduler (PS) takes into consideration the speed of a vehicle. However this is not sufficient for allocation of network resources because this way only higher speed vehicles would only get resources and slower vehicles may not get considered by the scheduler for resource allocation.

In order to balance the situation time duration between scheduled time and current time is taken into consideration, which is then computed by the scheduler using:

$$r_k = r_n - r_{ls}^k \tag{13}$$

where r_k is simply the time at which the 'k'th vehicle does not have updated transmission information. Then scheduler does calculation by:

$$C_{x,y} = \theta_k r_k t_k = \theta_k (r_w - r_{ls}^k) t_k \tag{14}$$

where θ_k equals to 0 if the 'k'th vehicle is scheduled else it equals to 1. '$r_k t_k$' simply being the estimation of 'k'th vehicles maximum distance it can travel in time 't_k'.

Now considering the road segments which are more crucial such as intersections in comparison to other road segments. Vehicles in these road segments require more accuracy, thus the proposed scheduler gives greater priority to vehicles in these road segments [29]. '$H_{k,m}$' is assumed to be distance of 'k'th vehicle normalised to 'm'th region of the road segment with higher importance. Then '$H_{k,m}$' can be computed as:

$$H_{k,m} = \begin{cases} \frac{|\chi_m - \chi_k|}{H} & |\chi_m - \chi_k| < D \\ 1 & |\chi_m - \chi_k| > D \end{cases} \tag{15}$$

where χ_m is centre of 'm'th region of higher importance and χ_k is location of 'k'th vehicle with H being the radius of region of importance. Normalised distance from point of interest to region of importance is computed as:

$$H_k = \min_m H_{k,m} \tag{16}$$

Equation 14 can be re-written as:

$$C_{x,y} = \beta(H_k)\theta_k r_k t_k = \beta(H_k)\theta_k(r_w - r^k_{ls})t_k \quad (17)$$

where ‘β’ is simply a decrement function having higher value for a vehicle which is nearer to the higher importance road segment given as ‘H_k’.

4 Simulation Explanation and Results

This section gives explanation of the results and simulation carried out with the PS for VANET based safety applications. The simulation has been done using NS3. The parameters taken in the simulation are shown in Table 1. The vehicle speeds in the simulation vary from 10 to 120 km/h. Every participating vehicle in the communication sends out safety information in the form of messages. These messages contain information on location and speed of a vehicle. This information is relayed to the base unit/server which in turn broadcast these messages in the region of requirement. Frequency for transmission of these safety messages used is 10 Hz.

The performance of the PS is compared to LWDF mainly on the basis of two [30] metrics. One computes average time of arrival between two consecutive messages. More the gap between two messages means more the error as location of vehicle changes often especially in case of high speed vehicles. The other computes average error in position of a vehicle between current position and previous recorded position. Figure 2 shows a graph with average position errors versus total number of vehicles. Initially as there is negligible congestion the error in position of the two schedulers is almost same. With increasing number of vehicles congestion starts to begin. In such a case the role of scheduler becomes important for allocation of network resources which are limited. High speed vehicles have worse channel [31, 32] conditions in comparison to low speed vehicles. As LWDF does not give priority to high speed vehicles, there occurs more delay ultimately resulting in greater error in position. However in case of the PS priority is given to higher speed vehicles which results in lesser position error of vehicles participating in communication.

Figure 3 shows the average time that two messages take when sent consecutively for a road segment assuming to have 50 vehicles. Now as seen in the graph

Table 1 Simulation parameters used

Simulation parameters used	
Algorithms used as scheduler	LWDF, PS
Number of vehicles	50, 150, 300
Message size	40 bytes
Transmission frequency	10 Hz
Cell radius	800 m
Simulation time	150 s

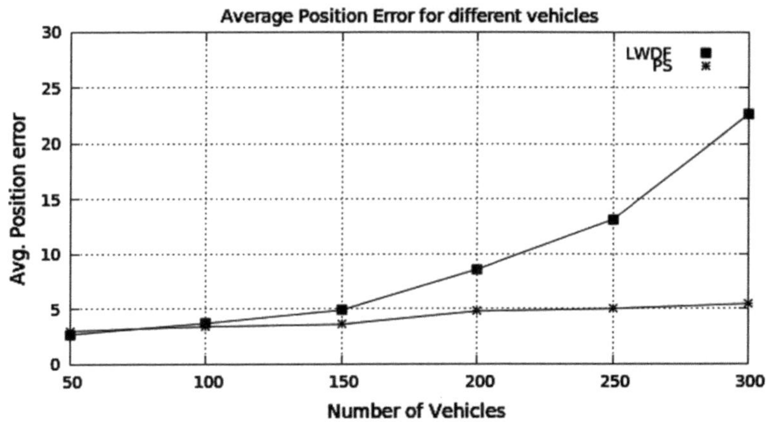

Fig. 2 Average position error versus number of vehicles

initially when there is no congestion messages are received in a timely manner and performance of both LWDF and PS is similar. However when the vehicles increase in number and reach up-to 150, congestion starts to occur and performance difference of the schedulers can be seen clearly. Figures 4 and 5 similarly show the average error in position and average time that two messages take for a set of communicating vehicles at different speeds when total vehicle number in a given road segment is 150. The PS can be seen showing better performance as the delay between messages is nearly constant for both high speed and slow speed vehicles. On the contrary the LWDF metrics show delay for high speed vehicles to be higher in comparison to slow speed vehicles resulting in greater error in position which is

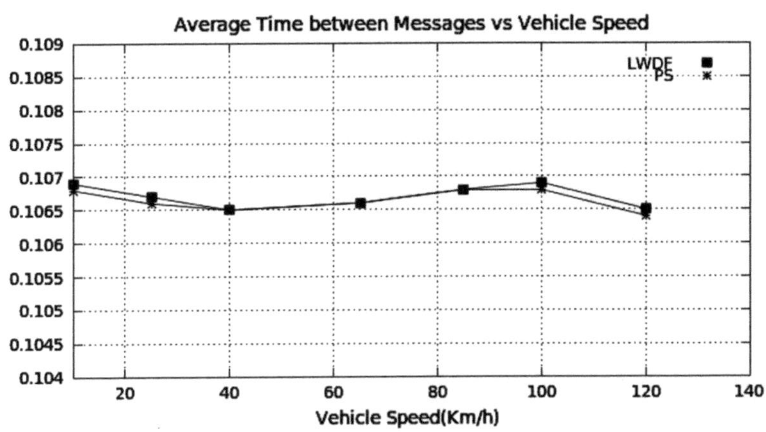

Fig. 3 Average time between two messages versus speed of vehicles for 50 vehicles

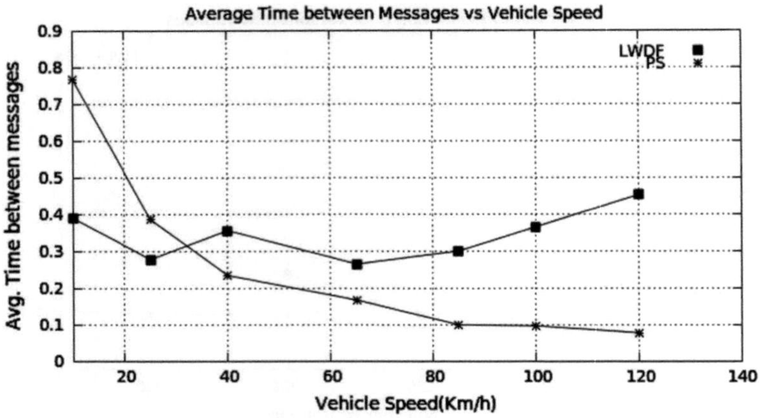

Fig. 4 Average time between two messages versus speed of vehicles for 150 vehicles

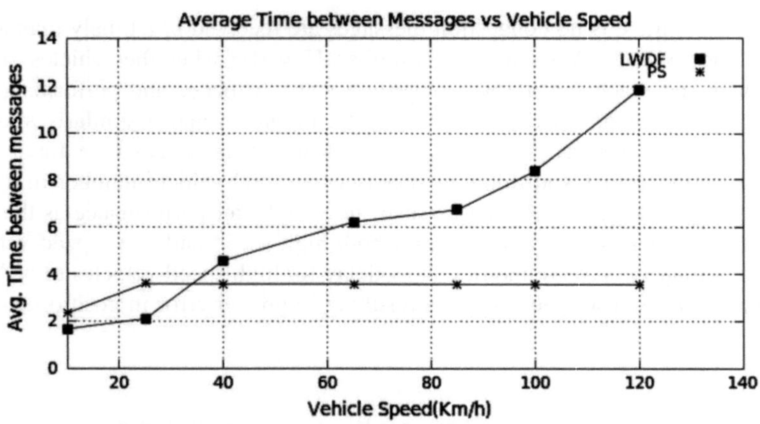

Fig. 5 Average position error versus number of vehicles for 150 vehicles

not acceptable in case of safety applications for VANETs. Figures 6 and 7 show performance of the two schedulers with 300 vehicles with network being fully congested the results can be seen more prominently.

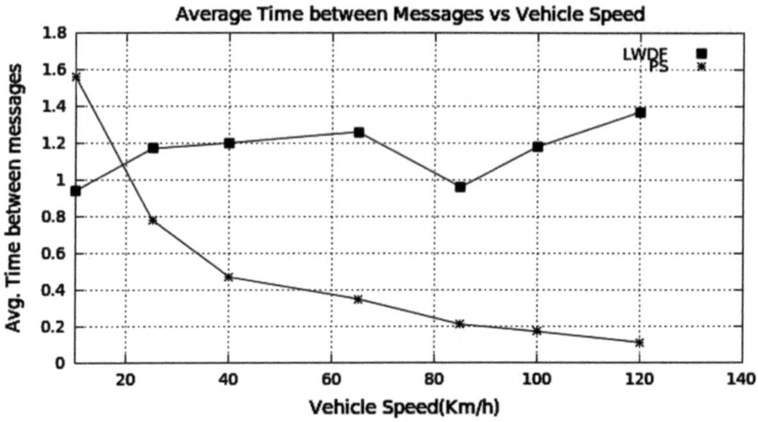

Fig. 6 Average time between two messages versus speed of vehicles for 300 vehicles

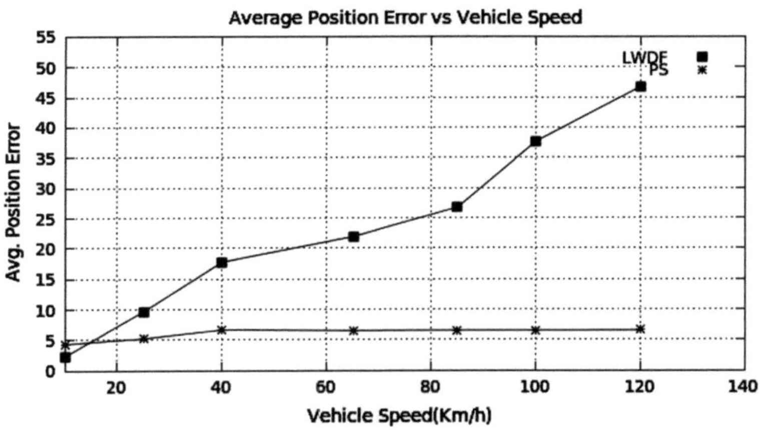

Fig. 7 Average position error versus number of vehicles for 300 vehicles

5 Conclusion

VANETs make use of safety messages for transmitting crucial information and in case of a LTE network there may be congestion in the network which may result in high levels of delay which is not acceptable. In such a scenario a scheduler becomes important in allocation of resources to the vehicular nodes especially in case of sensitive safety applications which are delay critical. This paper proposes a scheduler which is aware of the location and speed of the vehicles in a given road segment. This scheduler is implemented in LTE network which is practically more suited for VANET type networks.

As in case of high speed vehicles which are delay critical both speed and location are taken into account in order to assign resources in the least amount of time. Analysis of the simulation shows that the PS performs better in comparison to LWDF which is considered as the top scheduler for development of safety applications in case of VANETs. This is so because the PS uses metrics which assign priority to high speed vehicles preventing delays and reducing position errors.

References

1. S. Wang, B. Yang, F. Gu and X. Shi, "A novel reliable broadcast protocol for VANET's safety applications," *2016 6th International Conference on Electronics Information and Emergency Communication (ICEIEC)*, Beijing, China, 2016, pp. 282–286.
2. S. E. Carpenter, "Obstacle Shadowing Influences in VANET Safety," *2014 IEEE 22nd International Conference on Network Protocols*, Raleigh, NC, 2014, pp. 480–482.
3. S. Soeno and S. Sugiura, "Speed-dependent autonomous beamwidth variation for VANET safety applications," *Vehicular Networking Conference (VNC), 2015 IEEE*, Kyoto, 2015, pp. 127–130.
4. Di Felice, M., Ghandour, A. J., Artail, H., and Bononi, L. 2012. "On the Impact of Multi-channel Technology on Safety-Message Delivery in IEEE 802.11 p/1609.4 Vehicular Networks". In Computer Communications and Networks (ICCCN), 2012 21st International Conference on (pp. 1–8). IEEE.
5. A. Bessani, M. Correia, B. Quaresma, F. Andr, and P. Sousa, "Dep-Sky: dependable and secure storage in a cloud-of-clouds," Proc. Sixth conference on computer systems (EuroSys'11), ACM, 2011, pp. 31–46.
6. S. Sehrish; G. Mackey; P. Shang; J. Wang; J. Bent, "Supporting HPC Analytics Applications with Access Patterns Using Data Restructuring and Data-Centric Scheduling Techniques in MapReduce," Parallel and Distributed Systems, IEEE Transactions on, vol. 24, no. 1, pp. 158, 169, Jan. 2013.
7. K. George; V. Venugopal, "Design and performance measurement of a high-performance computing cluster," Instrumentation and Measurement Technology Conference (I2MTC), 2012 IEEE International, vol., no., pp. 2531, 2536, 13–16 May 2012.
8. N. Khaitiyakun, T. Sanguankotchakorn, and A Tunpan. "Data dissemination on manet using content delivery network (cdn) technique". In *Information Networking (ICOIN), 2014 International Conference on*, pages 502–506, Feb 2014.
9. Jinwoo Nam, Seong-Mun Kim and Sung-Gi Min, "Extended wireless mesh network for VANET with geographical routing protocol," *11th International Conference on Wireless Communications, Networking and Mobile Computing (WiCOM 2015)*, Shanghai, 2015, pp. 1–6.

10. L. Ramakrishnan, R. S. Canon, K. Muriki, I. Sakrejda, and N. J. Wright. "Evaluating Interconnect and virtualization performance for high performance computing", ACM Performance Evaluation Review, 40(2), 2012.
11. Y. Bai, D. Xie, S. Wang and M. Zhong, "Multi-path transmission protocol in VANET," *2015 International Conference on Connected Vehicles and Expo (ICCVE)*, Shenzhen, 2015, pp. 204–209.
12. I. B. Jemaa, O. Shagdar, F. J. Martinez, P. Garrido and F. Nashashibi, "Extended mobility management and routing protocols for internet-to-VANET multicasting," *2015 12th Annual IEEE Consumer Communications and Networking Conference (CCNC)*, Las Vegas, NV, 2015, pp. 904–909.
13. K.C. Lee, U. Lee, and M. Gerla, Oct. 2009. "Survey of routing protocols in vehicular ad hoc networks". *Advances in Vehicular Ad-Hoc Networks: Developments and Challenges, IGI Global.*
14. I. O. Chelha and S. Rakrak, "Best nodes approach for alert message dissemination in VANET (BNAMDV)," *RFID And Adaptive Wireless Sensor Networks (RAWSN), 2015 Third International Workshop on*, Agadir, 2015, pp. 82–85.
15. D. Jiang, & L. Delgrossi, (2008). "IEEE 802.11p: towards an international standard for wireless access in vehicular environments". In Proceedings of 67th IEEE vehicular technology conference on vehicular technology (pp. 2036–2040), May 2008.
16. F. Granelli, G. Boato, and D. Kliazovich, 2006. "MORA: A movement-based routing algorithm for vehicle adhoc networks". *1st IEEE Workshop on Automotive Networking and Applications (AutoNet 2006)* 2006. San Francisco, USA.
17. J. Zhao, and G. Cao, 2008. "VADD: Vehicle-Assisted Data Delivery in Vehicular". *Vehicular Technology, IEEE Transactions on*. 57.3. 1910–1922.
18. G. Yan, D.B. Rawat and B.B. Bista, "Towards Secure Vehicular Clouds," Proc. Complex, Intelligent and Software Intensive Systems (CISIS), 2012 Sixth International Conference on, 2012, pp. 370–375.
19. S. Pearson and A. Benameur, Privacy, "Security and Trust Issues Arising from Cloud Computing" Proc. Cloud Computing Technology and Science (CloudCom), 2012 IEEE Second International Conference on, 2012, pp. 693–702.
20. T. Atéchian, L. Brunie, J. Roth, and J. Gutiérrez. "DGcastoR: direction-based geo-cast routing protocol for query dissemination in VANET". *IADIS International Telecommunications, Networks and Systems*. 2008.
21. A. Festag, (2009). "Global standardization of network and transport protocols for ITS with 5 GHz radio technologies". In Proceedings of the ETSI TC ITS workshop, Sophia Antipolis, France, February 2009.
22. R.K. Shrestha, S. Moh, I. Chung, and D. Choi 2010. "Vertex-based multi-hop vehicle-to infrastructure routing for vehicular ad hoc networks". In *IEEE proceedings of 43rd Hawaii International Conference on System Sciences (HICSS).*
23. L. Kristiana, C. Schmitt and B. Stiller, "Survey of angle-based forwarding methods in VANET communications," *2016 Wireless Days (WD)*, Toulouse, 2016, pp. 1–3.
24. S. Grafling, P. Mahonen, & J. Riihijarvi. "Performance evaluation of IEEE 1609 WAVE and IEEE 802.11 p for vehicular communications". In Ubiquitous and Future Networks (ICUFN), 2010 Second International Conference on (pp. 344–348). IEEE.
25. Praveen K. Gupta and Nitin Rakesh, "Different Job Scheduling Methodologies for Web Application and Web Server in a Cloud Computing Environment", 2010 3rd International Conference on Emerging Trends in Engineering and Technology, IEEE, Pages 569–572, Goa, India, Nov. 19–21.
26. Nitin Rakesh and Vipin Tyagi, "Linear-code multicast on parallel architectures," Elsevier Advances in Engineering Software, vol. 42, pp. 1074–1088, 2011.
27. Nitin Rakesh and Nitin, "Analysis of Multi-Sort Algorithm on Multi-Mesh of Trees (MMT) Architecture", Springer Journal of Supercomputing, vol 57, no 3, pp. 276–313, 2011.
28. Nitin Rakesh and Vipin Tyagi, "Linear Network Coding on Multi-Mesh of Trees using All to All Broadcast," International Journal of Computer Science Issues, vol 8, no 3, pp. 462–471, 2011.

29. Nitin Rakesh and Vipin Tyagi, "Parallel architecture coding: link failure–recovery mechanism (PAC: LF – RM)," Springer International Journal of System Assurance Engineering and Management, vol 4, no 4, pp. 386–396, 2013.
30. Sandeep Pratap Singh, Shiv Shankar P Shukla, Nitin Rakesh, Vipin Tyagi, "Problem reduction in online payment system using hybrid model", International Journal of Managing Information Technology (IJMIT) Vol. 3, No. 3, pp. 62–71, August 2011.
31. Praveen K Gupta, Nitin Rakesh, "Different job scheduling methodologies for web application and web server in a cloud computing environment", 2010 3rd International Conference on Emerging Trends in Engineering and Technology (ICETET), pp. 569–572, 2010, India.
32. Arjun Arora, Nitin Rakesh, K.K. Mishra, "Scrutiny of VANET Protocols on the basis of Communication and Implementation of WAVE 802.11p/1609.4 with NS3 using SUMO", International Conference on Computer, Communication and Computational Sciences (ICCCCS), 2016, India. [Accepted].

Broadcast Storm Problem—A Hidden Consequence of Content Distribution in Content Delivery Networks

Darothi Sarkar, Nitin Rakesh and Krishn K. Mishra

Abstract Content caching at the Internet edge is becoming very popular because of flash crowd problem. Content Delivery Networks (CDN) was evolved for content caching with a complete solution for network reliability, scalability and performance enhancement. A number of researches have been focused on CDN issues like replica server placement, content caching, request routing and accounting management. But still some more issues are yet to be solved. This paper focuses on the concept of Broadcast Storm Problem (BSP) in CDN due to content distribution and request routing. Several approaches are available for content distribution and content caching. When there is any update in any surrogate, the same has to be communicated to all other servers over the network to avoid data inconsistency. Simple flooding or gossiping is generally used for the same, but these approaches are accompanied with BSP. Numerous BSP algorithms have been evolved, but the main concern is the wireless sensor networks. In this paper a comparison among several BSP algorithms has been shown which reveals that counter-based approach is much simpler and can be applied to any network. In addition to the comparative analysis, the counter-based scheme gets modified for CDN.

Keywords Content delivery networks · Content replication
Broadcast storm problem · Broadcast protocols · Flooding protocol
Probabilistic broadcasting · Counter-based broadcasting

D. Sarkar (✉) · N. Rakesh
Department of Computer Science and Engineering, Amity University, Noida, Uttar Pradesh, India
e-mail: darothissarkar@rediffmail.com

N. Rakesh
e-mail: nitin.rakesh@gmail.com

K.K. Mishra
Department of Computer Science and Engineering, Motilal Nehru National Institute of Technology, Allahabad, India
e-mail: kkm@mnnit.ac.in

G.M. Perez et al. (eds.), *Networking Communication and Data Knowledge Engineering*, Lecture Notes on Data Engineering and Communications Technologies 3, https://doi.org/10.1007/978-981-10-4585-1_13

1 Introduction

From the last decade there has been an increased popularity of World Wide Web which leads to the bottleneck towards the server due to the sudden rise in client requests [1]. The server gets overloaded and blocked because of the bandwidth limitations. One possible solution is to cache the content from the origin server to several strategic locations which helps the users to access their data from the nearby repository with minimum latency. This concept is implemented in the name of CDN which is one of the thrust areas of current research. In CDN the origin server stores all the data and they get replicated in several edge servers deployed by the content provider at different geologically dispersed Point of Presence (POP) [2].

Replication of data helps in yielding significant performance gain in terms of user latency and ensures almost 100% data availability. In current scenario almost all the social networking sites and video generating sites distribute their massive data through CDN. For example YouTube generates 30% of the total internet traffic [3]. So it became a big challenge to manage the huge number of user request with a considerable latency. As a result, YouTube caches all its frequently accessed i.e. hot content in edge servers. The recent advancement is Distributed Online Social Networks where there is no central repository and the data are getting stored in different surrogates using a P2P infrastructure [4]. The challenging areas of CDN are surrogate placement, content distribution, cache management, request routing and accounting management. The primary concern is surrogate placement i.e. how and where to deploy replica servers over the network in order to place the data near to the users. A number of replica server placement algorithms have been proposed with the intension of optimizing the performance in terms of latency and bandwidth [5]. Next area to be focused on should be content distribution and caching the content in replica servers. Efficient content distribution is one of the key factors for reducing server load and user latency [6]. CDN is categorized into pull-CDN and push-CDN based on this content distribution. In push-CDN, data is being pushed to the surrogates before they get accessed. In contrast, pull-CDN pulls data from origin server when they receive any request for the first time or in case of cache miss. The detailed study on content distribution and caching is discussed in literature survey.

In this paper we will introduce a concept called Broadcast Storm Problem in CDN and its probable solutions. In Cooperative-pull based CDN, when there is a cache miss in any surrogate, it will send request to its nearby replica servers for the data. But if the nearest one is not having the data, it will again send request to its nearest one and the process continues which is very much time consuming. So the better approach is to broadcast the request in case of cache miss. But the broadcasted and rebroadcasted signals may overlap with each other which may cause redundancy, collision, contention which are generally referred as BSP [7]. BSP may also occur when there is any change in any surrogate and it has to update all other edge servers including the origin server. Numbers of algorithms for avoiding BSP have been researched. The detail study on BSP is discussed in literature review section.

The rest of the paper is divided into four sections. The literature review on content placement and BSP is reported in Sect. 2. Section 3 formulates a problem and it comprises of a comparative study on existing BSP algorithms proposed by several authors and proposes a modified BSP for CDN. Section 4 draws a conclusion based on the comparative analysis done in Sect. 3.

2 Literature Survey

Literature is reviewed on two important aspects namely content distribution and replication and BSP.

2.1 Content Distribution and Selection

Content distribution and replication is a challenging issue in CDN. Once the surrogates are placed over the global network, the next question is how data will be distributed to the surrogates. Based on content distribution the CDN is either Pull-CDN or Push-CDN. In Pull-CDN, first time the content is retrieved from the origin to the surrogate servers. The surrogates cache the data for future use. In case of cache miss, it may send request to the origin server or to the nearby edge servers. The former is known as Uncooperative Pull-Based and the latter one is called as Cooperative Pull-Based [6]. Cooperative Push-Based CDN pre-fetches data to the surrogates before they are accessed. Server-initiated Push-Based provides comparatively better performance than client-initiated Pull-Based replication in terms of latency and availability.

The subsequent concern is which data is to be replicated in which surrogate? Content replication is being done in two ways, either in full-site replication or in partial-site replication scheme [6, 8]. In full-site replication, the entire data set from origin server gets replicated in all surrogates which ensures almost 100% data availability from replica servers. But the main drawback is the storage capacity. All edge servers should be equipped with huge memory to accommodate all the data from origin. One more problem is with the frequent updates in data. In contrast, the partial-site replication only replicates the embedded objects in replica servers and the base HTML page is accessed from content provider. Several approaches are available for partial-site replication. The main plus point is that the embedded objects are getting updated infrequently. The comparison between full and partial site is shown in Table 1.

Table 1 Full-site and partial-site replication

Replication approach	Advantages	Disadvantages
Full-site replication	i. Simple implementation ii. Almost 100% data availability	i. Large storage capacity ii. Frequent update in data
Partial-site replication a. Empirical-Based b. Object-Based c. Cluster-Based i. Per website-Based ii. Per URL-Based d. Popularity-Based	i. Infrequent update in embedded data ii. Less storage capacity	i. User latency may be more than Full-site

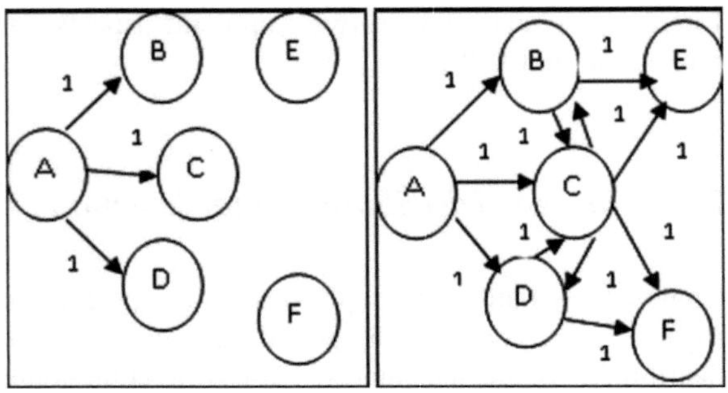

Fig. 1 Broadcast storm problem due to flooding

In CDN content is replicated to all the surrogates using any of the approaches listed in Table I. If there is any update in the origin, it has to inform each and every surrogate in the network. Conversely, if any data gets updated in any surrogate, the information has to be passed to all the surrogates including the origin. The easiest way to disseminate the information is broadcasting. Most important part is the selection of broadcast algorithms as data gets lost because of collision and contention occurred in case of broadcasting which is called as BSP. The next part of literature review is carried out on different algorithms to suppress the effect of BSP.

2.2 Broadcast Storm Problem

Broadcasting is a very common process in networking. A blind flooding is a very simple approach for the same but it results in high collision, contention and redundancy. Figure 1 shows the way in which flooding causes Broadcast Storm Problem. In this figure node A is broadcasting a message to its 1-hop neighbors B, C and D. Now these B, C and D in turn broadcast the same message to their 1-hop neighbors which results collision between broadcasted and rebroadcasted messages. A number of algorithms have been proposed for broadcasting in different areas of networking. In [7] authors mentioned 5 different approaches namely Probabilistic,

Counter-based, Distance-based, Location-based and Cluster-based scheme. After receiving a broadcast message, Probabilistic scheme broadcasts that message with probability p for the first time. In Counter-based scheme a threshold value C is considered. If a host hears the same message equal to or more than C times, it hinders rebroadcasting the message. Distance-based scheme considers a distance threshold d_{min}. If the distance between source and destination of a message is less than the threshold, the destination host will not rebroadcast the message. In Location-based scheme the metric is coverage threshold. If a host receives a message it will first calculate the coverage area which will receive the rebroadcasted message. If the calculated value is less than the threshold, it will not rebroadcast. The Cluster-based scheme is basically based on graph modeling. In this scheme only the gateway members can rebroadcast the message.

The recently evolved broadcast algorithms are based on the above mentioned basic five schemes. Reachability, Saved ReBroadcast (SRB), average latency, Collision rate, Packet delivery fraction, Route-Length Ratio are the performance measure of the network when BSP is applied. The detailed comparative study of the existing BSP algorithms is given in Sect. 3.

2.3 Solutions to BSP

Broadcasting can be achieved in several ways based on different parameters. The existing broadcast algorithms follow one of the following schemes-Simple Flooding, Probabilistic Broadcasting, Distance-based Broadcasting and Neighbor Knowledge-based broadcasting [9].

Simple Flooding: According to the authors in [10] the origin node will broadcast a packet to all other neighbor nodes and the receivers of the packet will again rebroadcast the same packet to their neighbors. Each node should broadcast the same packet only once. Probability of BSP is very high in flooding as each and every node is taking part in broadcasting without any constraint.

Probabilistic Broadcasting: Probabilistic broadcasting is an advanced form of flooding. In this scheme each node rebroadcasts a received packet with a predefined probability p [7]. Gossip-Based routing is a good example of Probabilistic broadcasting. In this scheme the probability using which the nodes will gossip depends on the number of neighbors of the nodes. If the number of neighbors is higher, the probability gets reduced, else it gets increased.

Gossip1(p): Each node except the source will broadcast a packet exactly once with probability p to its neighbor and if the similar packet is received for the second time, the packet gets rejected with $1-p$ probability [11]. The sender will broadcast the packet with probability 1.

Gossip1 (p, k): In this scheme, the first k hop neighbors will gossip with probability 1 and rest will continue their gossip with predefined p probability.

Gossip2 (p_1, k, p_2, n): This is a two threshold method. Here first k hop nodes will gossip with $p = 1$. For remaining nodes, if a node is having greater than n nodes then will gossip with p_1, otherwise gossip with p_2 where $p_2 > p_1$.

Gossip3 (p, k, m): It is similar to Gossip (p, k) with a modification that the nodes which should not broadcast a received message, will ultimately broadcast if they do not receive the same packet from at least m nodes.

Counter-based Broadcasting: Counter-based broadcasting is another example of probabilistic method. This scheme considers expected additional coverage as one of the performance evaluator parameters [7]. A threshold is defined that should be the maximum number of times a node can receive the same packet. Initially a counter is set to 0 and it gets incremented every time a same message is heard. If the counter is greater than the threshold within a prescribed time period, the node will not rebroadcast the packet.

Density-based Probabilistic Counter Scheme: This is a hybrid approach where the advantages of probabilistic and counter-based schemes have been merged. According to this scheme, each and every node can rebroadcast message with a predetermined forwarding probability p. This probability can be determined by the local density information i.e. 1-hop neighbor information. Each node is assigned a counter which is set to 0. It gets incremented every time the node receives the same message and the node removes the duplicate messages. This technique is aimed at reducing the collision and contention due to broadcasting [12].

Distance-Aware Counter-based Broadcast: Authors in [13] introduced distance into counter-based broadcast scheme. Based on a distance threshold two distinct Random Assessment Delays (RADs) are assigned to border and interior nodes. Smaller RAD is assigned to the nodes closer to the borders compared to the interior nodes. This scheme is efficient in terms of reachability and rebroadcasting irrespective of network density.

Dynamic Probabilistic Counter-based Broadcast: According to the approach, each node set a counter on receives of a packet for the first time. Till the RAD expired, the counter gets incremented as it receives the similar message R_n. If the value of the counter exceeds the threshold, a low forwarding probability is generated compared to the situation where the counter value is less than threshold. But whether a packet will be rebroadcasted or not, depends on a random number generated within a range of [0, 1]. If probability is greater than the R_n, the packet will be resent, otherwise gets dropped [14].

Distance-Based Broadcasting: In Distance-based Broadcasting, each and every node calculate its own forwarding probability based on the node-to-node distance [15].

Weighted p-persistence Broadcasting: Node x will rebroadcast a message received from node y with a probability $p_{xy} = d_{xy}/R$ if the message is received for the first time. d_{xy} denotes the distance between node x and y and R denotes the average transmission range.

Slotted-1 Persistence Broadcasting: When a node receives a packet for the first time, it will rebroadcast it with probability 1, if the node does not receives any duplicate packet within a time slot T_s, else the packet will be discarded. T_s is calculated as $S_{ij} \times \tau$, where S_{ij} is the predefined time slots and τ is the estimated 1-hop delay.

Slotted-P Persistence Broadcasting: When a node receives a packet for the first time, it will rebroadcast it with probability p, if the node does not receive any duplicate packet within a time slot T_s, else the packet will be discarded.

Neighbor Knowledge-Based Broadcasting-Flooding with Self Pruning: This approach considers that each node should add its neighbor list at the header of a packet before broadcasting [16]. The receiver checks and compares its neighbor information with the one included in the packet. If there is any extra node at the receiver's neighbor list, then only the packet will be rebroadcasted, otherwise dropped.

Scalable Broadcast Algorithm: Each node should have knowledge of their neighbors up to 2 hop radius [17]. Upon receiving a packet, a node will compare the sender's neighbors with its own. If any extra node will be reached after rebroadcasting, then only the node will wait for a RAD. If it does not hear any duplicate packet within the RAD, then the packet will be rebroadcasted, otherwise dropped.

Dominant Pruning: This scheme considers that each node consists of its 2 hop neighbor information. When a node broadcasts a packet, it adds the subset of neighbor nodes' addresses in the header of the packet. Only selected nodes will then rebroadcast the packet selecting their own subset of neighbors [16].

Double-Covered Broadcast: In this scheme, only some selected 1 hop nodes of the sender will retransmit the received packet. Retransmission should ensure that all 2 hop nodes of the sender will be covered and all 1 hop nodes (forwarding or nonforwarding) will be covered by at least two forwarding nodes [18].

3 Comparative Analysis

3.1 Problem Formulation

Consider a CDN consisting of one origin server and M replica servers. The network follows the Cooperative push-based scheme for content distribution. According to that the origin server will push the content to the surrogates before they are accessed. A mapping between the data and the surrogates will be maintained at content provider. Whenever there is any update at the origin, it informs all M replicas through broadcasting. Conversely, if there is any change in a surrogate; say there is an update in the content of K th surrogate by the client; the R_K will inform it to its neighbor by gossiping. The neighbor will again gossip about this change so that the origin server will get updated. The broadcasting and gossiping may cause data loss due to BSP. Now the algorithms for broadcasting and gossiping should be selected in such a way that it diminishes the effect of BSP.

3.2 Comparative Study

All broadcasting algorithms mentioned in 2.3 have been used to diminish the effect of BSP. The algorithms are featured with some advantages and disadvantages. A comparative study among the algorithms is shown in Table 2.

From Table 2 it is clear that simple flooding increases the possibility of BSP. Distance-based broadcasting is mainly applicable to mobile networks like Vehicular Ad hoc Networks (VANET) where neighbour information can be gained through satellite. Reachability goes higher through the use of Neighbour Knowledge-based broadcasting. Consequently, the complexity will also get high. The Probabilistic method, especially Gossip and counter-based broadcasting are efficient enough with acceptable complexity can be applied in any type of networks.

Table 2 Broadcasting algorithms with their pros and cons

Broadcasting algorithms	Advantages	Disadvantages
Simple flooding	Simple highest reachability	High collision, contention and redundancy
Probabilistic Broadcasting		
Gossip1 (p)	Simple	Gossip depends on the number of neighbours of the source node. If it is low, then there is a possibility that the gossip will die
Gossip1 (p, k)	Higher reachability than Gossip1 Reduces traffic upto 35% compared to flooding	Chance of premature message death. So p has to be large enough (0.65–0.75)
Gossip2 (p_1, k, p_2, n)	Less chance of message expiry Less redundancy as probability is controlled	No impact on regular networks
Gossip3 (p, k, m)	Prevents premature gossip death	Latency of retransmission may be an issue
Counter-based	Simple Inherent adaptability to local topology	Applicable in dense network In sparse network, it becomes a simple flooding and saved rebroadcast (SRB) decreases sharply
Density-based Probabilistic Counter scheme	Minimizes end-to-end delay and maximizes delivery ratio compared to counter-based	As every node is allowed to retransmit, it increases the channel contention
Distance Aware Counter-based	High reachability (almost 95%) as the primary parameter is Expected Additional overage (EAC) Not sensitive to the network topology	More complex as in addition to the two threshold values Counter and Distance, two different RAD values and two different probabilities for interior and border nodes have to be computed

(continued)

Table 2 (continued)

Broadcasting algorithms	Advantages	Disadvantages
Dynamic Probabilistic Counter-based	Dynamic probability computation method is useful for all types of networks Reduces channel contention even at high broadcast injection rate	Considers a situation and compute the forwarding probability where number of receiving of duplicate messages exceeds the threshold value
Distance-Based Broadcast		
Weighted *p*-Persistence	Light weight technique High reachability	Mostly applicable to mobile networks
Slotted-1 Persistence	High reachability as it assigns higher probability to the nodes away from the broadcaster	Mostly applicable to mobile
Slotted-*p* Persistence	High reachability Almost 90% reduction in packet loss	Mostly applicable to mobile
Neighbour Knowledge-Based Broadcast		
Flooding with Self Pruning	Simple	Not efficient for all types of networks
Scalable Broadcast	Uses neighbour knowledge efficiently	To enhance the gain in performance, a higher number of rebroadcasting nodes are required
Dominant Pruning	High reachability by using Greedy Set Cover algorithm	Difficult to use in mobile environment as it does not use local information to determine next rebroadcasting nodes
Double-Covered	Reliable retransmission. Sensitive to nodes' mobility as when mobility is high, transmission rate is very low because of wrong neighbour information	Number of retransmission can be higher as the sender will keep on broadcasting if it does not receive maximum number of acknowledgment from the forwarding nodes

3.3 BSP in CDN

Several solutions to BSP are shown in Table 2, but they are executed only in wireless sensor networks especially Mobile Ad hoc networks (MANET) and Vehicular Ad hoc Networks (VANET). The problem of BSP can also be solved in CDN by applying the existing solutions with some modifications. Current research has revealed that the clustering approach became popular and efficient for replica server placement in CDN. Consider, a CDN consisting of N number of nodes which are grouped into K clusters. All of K replica servers should maintain a record of other replica servers, their location and distance as required redirecting the client requests in case of cache miss. The distance to the replica servers is an important aspect in request routing. The broadcasting in CDN can be modified using counter-based scheme.

Say, there is an update in replica server R_i, it will inform all other replica servers as well as the origin. If it will follow simple flooding, may is the information will not reach to many of the surrogates due to BSP. So the better approach can be R_i will broadcast the message to the surrogates within its transmission range. The receiver will check the sender ID. If it is the originator of the message, they will send an acknowledgement. If the source does not receive at least 3 acknowledgements within a RAD, it will rebroadcast the message. The reason being, if a message gets replicated in three different servers, the probability of information loss becomes less.

The receiver of the message will wait for the specified RAD period. If the number of times it receives the same message exceeds the threshold value C, it refrains from broadcasting. Otherwise it broadcasts the message. Several research works reveals that performance becomes enhanced with $C = 3$.

4 Conclusion

Several solutions for Broadcast Storm Problem have been analyzed in this paper. Broadcast Storm Problem is a serious concern and it should get avoided in CDN to prevent the data inconsistency and data loss. But the existing solutions for BSP were mainly proposed for MANET and VANET. But there is a little concern about BSP in CDN though BSP is obvious in case of content update either in any surrogate or in origin. The comparative study reveals that the probabilistic approach for broadcasting is much more efficient in any type of networks. And counter-based scheme is supposed to be more appropriate to the CDN scenario as all the surrogate will have the information of their neighbors. So the counter-based scheme is slightly modified to solve the BSP in cluster-based CDN which is expected to enhance the performance of CDN in terms of less packet loss with high reachability.

References

1. H. Yin, X. Liu, G. Min, and C. Lin, "Content Delivery Networks: a Bridge between emerging applications and Future IP Networks," IEEE Network, vol. 24, no. 4, pp. 52–56, July–August 2010.
2. Darothi Sarkar, Nitin Rakesh, Krishna Kumar Mishra, "Problems in Replica Server Placement (RSP) over Content Delivery Networks (CDN)", In proceeding of 6th ACM International Conference on Computer and Communication Technology, pp. 175–179, 2015.
3. V. Gehlen, A. Finamore, M. Mellia, M. M. Munafò, "Uncovering the big players of the web", in proc. of 4th Int. Conf. TMA, pp. 15–28, 2012.
4. L. Han, M. Punceva, B. Nath, S. Muthukrishnan, L. Iftode, "Socialcdn: Caching techniques for distributed social networks", in Proceedings of the 12th IEEE International Conference on Peer-to-Peer Computing (P2P), pp. 191–202, 2012.
5. M. H. Al-Shayeji, S. Rajesh, M. Alsarraf, and R. Alsuwaid, "A Comparative Study on Replica Placement Algorithms for Content Delivery Networks," in proc. of IEEE Second International

Conference on Advances in Computing, Control and Telecommunication Technologies, pp. 140–142, 2010.
6. M. Pathan, R. Buyya, "A Taxonomy of CDNs", in Content Delivery Networks, R. Buyya, M. Pathan, A. Vakali, Eds. Berlin: Springer, 2008, pp. 33–77.
7. S.-Y. Ni, Y.-C. Tseng, Y.-S. Chen and J.-P. Sheu, "The broadcast storm problem in a mobile ad hoc network", In Proceedings of the IEEE/ACM MOBICOM, pp. 151–162, 1999.
8. E.P. Markatos and C.E. Chironaki, "A Top 10 Approach for Prefetching the Web," Proc. INET '98: Internet Global Summit, July 1998.
9. B. Williams and T. Camp, "Comparisons of Broadcasting Techniques for Mobile Ad hoc Networks", In Proceeding of third ACM International Symposium on Mobile Ad hoc Networking and Computing (MOBIHOC), pp. 194–205, 2002.
10. C. Ho, K. Obraczka, G. Tsudik and K. Viswanath, "Flooding for reliable multicast in multi-hop ad hoc networks", In Proceeding of the International Workshop on Discrete Algorithms and Methods for Mobile Computing and Communication (DIALM), pages 64–71, 1999.
11. Z. J. Haas, J. Y. Halpern and L. Li, "Gossip-based ad hoc routing", IEEE/ACM Transaction of Networking., vol. 14, no. 3, pp. 479–491, 2006.
12. A. S. Kumar, K. S. Ramaswami, "A New Density Based Probabilistic Counter Scheme for Solving Broadcast Storm in Mobile Ad Hoc Networks", Journal of Theoretical and Applied Information Technology, Vol. 29, No. 2, pp. 451– 459, 2014.
13. C. Chen, C.-K. Hsu and H.-K. Wang "A distance-aware counter-based broadcast scheme for wireless ad hoc networks," Military Communications Conference. IEEE, vol. 2, pp. 1052–1058, 2005.
14. A. Mohammed, M. Ould-Khaoua, L. Mackenzie and J. D. Abdulai, "Dynamic Probabilistic Counter-Based Broadcasting in Mobile Ad Hoc Networks", in Proceedings of Second International Conference on Adaptive Science Technology (ICAST ',09), pp. 120–127, 2009.
15. O. Tonguz, N. Wisitpongphan, J. Parikh, F. Bai, P. Mudalige and V. Sadekar, "On the Broadcast Storm Problem in Ad Hoc Wireless Networks", in Proceedings of Third International Conference on Broadband Comm., Networks and Systems (BROADNETS',06) pp. 1–11, Oct. 2006.
16. H. Lim and C. Kim., "Multicast tree construction and flooding in wireless ad hoc networks", in Proceedings of the ACM International Workshop on Modelling, Analysis and Simulation of Wireless and Mobile Systems (MSWIM), pp. 61–68, 2000.
17. W. Peng and X. Lu., "On the reduction of broadcast redundancy in mobile ad hoc networks", In Proceeding of first ACM International Symposium on Mobile Ad hoc Networking and Computing (MOBIHOC), pp. 129–130, 2000.
18. W. Lou and J. Wu, "Toward broadcast reliability in mobile ad hoc networks with double coverage", IEEE Transaction on Mobile Computing, vol. 6, no. 2, pp. 148–163, 2007.

Part II
Artificial Intelligence and Evolutionary Algorithms

Attribute Reduction Method Using the Combination of Entropy and Fuzzy Entropy

Rashmi, Udayan Ghose and Rajesh Mehta

Abstract The enormous size datasets are being used in various fields such as administration, engineering, management and so on. For information retreival from these datasets more time is being consumed. Fewer attribute datasets takes lesser time for computation, and are more understandable and intelligible. Attribute reduction is a tool for feature selection as it transforms data into knowledge. A new method using the combination of entropy and fuzzy entropy is proposed for removal of redundancy and irrelevant attributes which results in reducing the dataset size. The functioning of the proposed method is examined on standard datasets such as Sonar, Spambase and Tick-tack-Toe. Experimental results performed on various datasets show that proposed method gives significant improvement in attribute reduction. In this work, nearest neighbor classifier is used to examine the classification accuracy on original and reduced dataset.

1 Introduction

Attribute reduction is a common problem in large size datasets like pattern recognition, data mining and machine learning. In recent years, instances and dimensionality of features in datasets have grown dramatically. In many real world application data is stored in the terms of thousands of attributes [1]. It is well known that attributes irrelevant to recognition tasks may deteriorate the performance of learning algorithms. In other words, storing and processing all attributes (both relevant and

Rashmi (✉) · U. Ghose
USICT, Guru Gobind Singh Indraprastha University, Dwarka, New Delhi, India
e-mail: rashmibehal@gmail.com

U. Ghose
e-mail: udayan@ipu.ac.in

R. Mehta
Department of Computer Science and Engineering, Amity School of Engineering and Technology, New Delhi, India
e-mail: rajesh2010usit@gmail.com

G.M. Perez et al. (eds.), *Networking Communication and Data Knowledge Engineering*, Lecture Notes on Data Engineering and Communications Technologies 3, https://doi.org/10.1007/978-981-10-4585-1_14

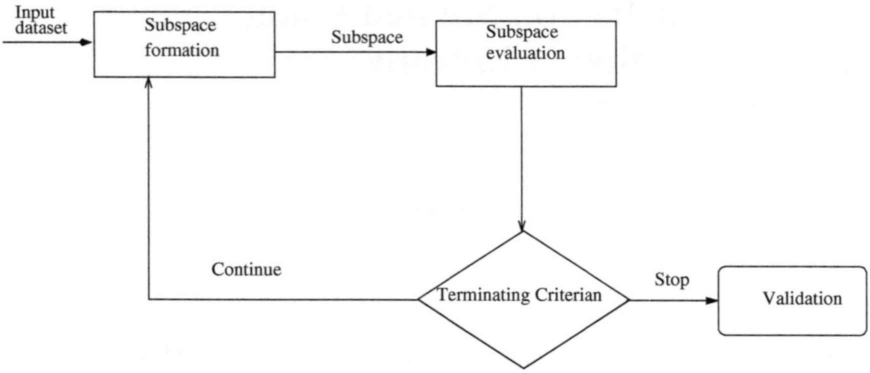

Fig. 1 Feature selection process

irrelevant) could be computationally very expensive and impractical. The classification accuracy is not affected by the irrelevant attributes, so they can be removed. Therefore, the removal of some attributes could be not only tolerable but also even desirable relative to the computational costs involved. The large heap of data is in the form medical reports, videos, images, and social sites data. Attributes are removed on the basis of feature relevance. Feature relevance helps to extricate the relevant features. Most semantics-preserving dimensionality reduction (or feature selection) approaches tend to be domain specific, however, relying on the use of well-known features of the particular application domains [2]. In dimension reduction feature selection is one of the important techniques and has become an important component in machine learning. Feature selection is a process that selects an optimal subset of features according to a certain criterion [3]. It is used to reduce the dataset size and reduced dataset is used for learning algorithm. Considering the number of relevant or non-redundant features increases the computation speed of a learning algorithm. For the large dataset learning is not so accurate as that of reduced dataset. To improve the performance in pitch of accuracy, speed, simplicity of rules feature selection has to be done. It enhances the quality of data by relevance feature. A relevance feature is the feature if it is removed; the performance of the decision system using remaining features will deteriorate in terms of performance. The significance measures such as dependency, distance, consistency and information gain are used to select attributes. On the basis of data set characteristics like data type, data size, and noise there are different types of feature selection method exists [4]. There are four basic steps in a typical feature selection method [5]:

1. a generation procedure to generate the next candidate subspace;
2. the subspace is evaluated by an evaluation function
3. to stop a generation procedure a terminating criteria is specified; and
4. the validity of the subspace is examined using a validation procedure (Fig. 1).

Wrappers, filters, and combination of both strategy are general stratagies to find the selected attribute subsets from the given dataset in feature selection. Filter strategy includes input feature set, subspace selection, and learning algorithm and wrapper strategy includes input feature set, subspace generation, and learning algorithm.

In filter approach features are selected independently of any learning algorithm that is is completely a separate preprocessor. After learning (induction) algorithm relevant attributes are obtained. Filter strategy is not bounded to any particular learning so it can be used in most of the domains. Wrapper approach is like classification as the evaluation is done on the basis of learning algorithm. A filter approach is better than wrapper approach because it has independent evaluation criteria for subset evaluation. A wrapper approach is less scalable then filter approach which makes filter approach more useful [6].

Son and Kim [7] presented a attribute reduction method based on Shanon entropy. Luukaa [8] introduced feature selection method based on fuzzy entropy. The combined approach of entropy and fuzzy entropy is used in this paper for attribute reduction.

The rest of the paper is organized as follows: fuzzy entropy and nearest neighbor classifier are introduced in Sect. 2. The proposed method for attribute reduction is explained in Sect. 3. The experimental results, discussion and analysis are described in Sect. 4. Finally, the conclusion is drawn in Sect. 5.

2 Priliminaries

2.1 Entropy Fuzzy Entropy

Entropy measures the degree of randomness in a dataset or how much information is stored in the dataset. Dataset randomness can be measured using Shanon entropy [9], fuzzy entropy [10], and rough set entropy [1]. In the existing Literature it is found that fuzzy entropy gives better result in terms of data reduction as compared to shannon entropy [7, 8]. Shanon entopy measures information of each attribute by the formula given as:

$$H(x) = -\sum_{i=1}^{N} P_i log P_i \tag{1}$$

In fuzzy set theory, degree of belongingness of each element is assigned by fuzzy membership function for classification whereas in classical set theory, each element either belong or does not belong to the particular set [11]. The concept of membership function is a generalisation of characteristic function of a classical set. The value of the generalised function lies within the unit interval [0, 1]. For all fuzzy set R defined in universal set U, its cardinality is represented by |R| and defined as

$$| R |= \sum_{a \in U} \mu_R(a) \tag{2}$$

where $\mu_R(U)$ represents the membership value of each element of R. The intersection and union operation are used to compute fuzzy entropy. The formulas are defined by the two fuzzy sets R and S is:

$$R \cup S = max\{\mu_R(a), \mu_S(a)\} \tag{3}$$

$$R \cap S = min\{\mu_R(a), \mu_S(a)\} \tag{4}$$

The degree of randomness of a fuzzy set R is defined by the fuzzy entropy as:

$$FE(R) = \frac{R \cap R'}{R \cup R'} \tag{5}$$

2.2 *Nearest Neighbor*

Nearest Neighbor (NN) is a lazy learning classifier as it does not do anything while training the dataset [12]. In NN more computation time is required as it computes the distance between each pair of training data and unassigned case Y. The NN of Y in training data are found and labelled to Y. An euclidean distance is used to define the nearest neighbors. For the two class problems positive and negative class is represented by p and n respectively, and where the samples are points in a two dimensional space. The k-nearest neighbor algorithm is used to find the approximate solutions [13]. NN is used to classify the sample points which are not assigned of the nearest of a set of previously classified points. The training of the NN algorithm is completed by taking 50% of the input sample which contains the classification information [12].

In this work, the classification accuracy on original and reduced dataset (after proposed method) is computed using nearest neighbor. The number of nearest neighbors k is one, distance is measured by euclidean distance and the sample classification is done on the basis of nearest one.

3 Proposed Method

In the proposed method, we focus on the attribute reduction using the combination of Shanon entropy and fuzzy entropy of each individual attribute. This is due to the ability of fuzzy entropy to discriminate data distribution under noise corrupted and redundant condition. Attribute extraction using the combination of entropy and fuzzy entropy are more robust. The proposed method is used to select the relevant attributes which are essential in decision making (Fig. 2).

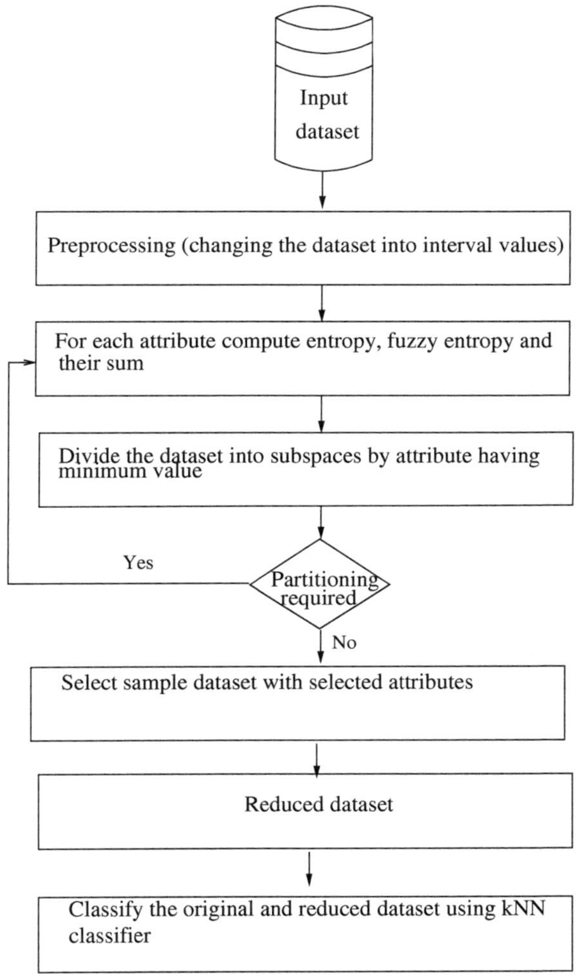

Fig. 2 Proposed method flowchart

Algorithm
Input: A decision table(DT)
Output: A reduced table(RT)
reduct_ entropy_ dataset(DT input)

1. Convert decision table into interval values using fuzzy triangular membership function.
2. Compute the entropy of each individual attribute of the input dataset.
3. Divide the dataset on the basis of minimum entropy value and that attribute is flagged as reduct attribute.

4. do while attributes in input dataset $\neq \emptyset$

or

all attributes have the same entropy

   (a) Compute combined entropy for each attribute in each class
   (b) Go to step 2.

5. For each attribute in reduced dataset

   (a) Compute the final sample by using the euclidean distance
   (b) Select the sample case having the minimum distance and store it in output

6. Return reduced_dataset

4 Experimental Results

In this section, the proposed method is evaluated by performing a number of experiments on different datasets. All the experiments are performed using Intel core (TM) i7-4500M CPU 1.80 GHz × 4 ubuntu machine with 8 GB RAM in MATLAB 8.0.1 platform. Here, input samples are randomly split into two groups with half of the input samples are used for training set and rest half of the samples are used for testing. Firstly, the input datasets are changed into interval values by fuzzy triangular membership function. Secondly, sum of entropy and fuzzy entropy is computed to find the relevant attributes which are essential in decision making. Thirdly, sample selection is done on the basis of closeness. Training results are used to classify the test samples using nearest neighbor classifier. The accuracy of the classifier is defined as:

$$accuracy = \left(\frac{acc\_classified}{total\ number\ of\ sample} \right) \times 100 \tag{6}$$

where acc_classified is number of accurately classified samples. In this work, three high dimensional standard datasets [2] from different fields like Computer Science (Spambase), Physical Science (Sonar mines and rocks sm/r) and Gaming (Tick-tack-Toe) of size 4601×57, 208×60 and 958×9 respectively are selected from UCI repository [2]. The performance of the method explained in this work is examined separately using (i) entropy (ii) fuzzy entropy (iii) combination of entropy and fuzzy entropy, ten times and then the mean classification accuracy value is computed. Classification accuracy, reduced dataset size, and computation time of these selected datasets are listed in Tables 1, 2, 3, 4 and 5. Significant number of attributes are reduced of the tested dataset by the method proposed in this paper. The results of the experiments on tested datasets are explained as:

Sonar data set: In the reduced dataset, 68.76% average classification accuracy is achieved with Shanon entropy, 85.80% classification accuracy is achieved with the fuzzy entropy, and 91.5% average accuracy is achieved using the proposed method. While the average classification accuracy on Sonar data set using these three method are 75.63%, 75.63%, and 75.63% respectively.

Spambase data set: The average classification accuracies on reduced dataset by using entropy, fuzzy entropy, and combination of entropy and fuzzy entropy are 99.87%, 97.57%, and 87.55% respectively. The proposed method reduces the attribute on selected dataset significantly and also scraps the execution time. Still, Shanon entropy method gives the best accuracy that is 99.87%. It is inferred from results listed in Tables 1, 2, 3, 4 and 5 that using fuzzy entropy and the proposed method gives significant amount of reduction in attributes and time (in secs) is achieved. The average classification accuracies on original dataset by using entropy,

Table 1 Original datasets description

Dataset	Sample size	Attributes size	Classes
Sonar	208	60	2
Spambase	4601	57	2
Tick-tack-Toe	958	9	2

Table 2 Reduced datasets description

Dataset	Entropy	Fuzzy entropy	Proposed method
Sonar	22×17	16×14	15×13
Spambase	47×41	63×54	35×28
Tick-tack-Toe	13×9	18×9	15×9

Table 3 Accuracy before reduction

Dataset	Entropy	Fuzzy entropy	Proposed method
Sonar	0.7563	0.7563	0.7563
Spambase	0.7569	0.7585	0.7585
Tick-tack-Toe	0.7995	0.7995	0.7995

Table 4 Accuracy after reduction

Dataset	Entropy	Fuzzy entropy	Proposed method
Sonar	0.6814	0.8580	0.9150
Spambase	0.9987	0.9757	0.8755
Tick-tack-Toe	0.8471	0.8750	0.8889

Table 5 Execution time in seconds

Dataset	Entropy	Fuzzy entropy	Proposed method
Sonar	19.35	7.10	3.65
Spambase	921.20	330.27	215.79
Tick-tack-Toe	18.24	20.80	13.06

fuzzy entropy, and combination of entropy and fuzzy entropy are 75.69%, 75.85%, and 75.85% respectively.

Tick-tack-Toe data set: The improved average classification accuracies on reduced dataset by using entropy, fuzzy entropy, and combination of entropy and fuzzy entropy are 84.71%, 87.50%, and 88.89% respectively while on original dataset is 79.95% for each dataset. The reduction rate is also improved by using proposed method as listed in Tables 1 and 2.

From the experimental results performed on Sonar, Spambase, Tic-tack-Toe dataset it is inferred that with the combination of entropy and fuzzy entropy a significant amount of classification accuracy is achieved.

5 Conclusion

A new attribute reduction method using combination of entropy and fuzzy entropy is presented in this paper. Significant amount of attribute reduction with high average classification accuracy as measured by nearest neighbor classifier is achieved using the proposed method. Fuzzy triangular membership function is used to convert input dataset into a range of values. The importance of attribute is known by the minimum of combined entropy values, selected attribute is taken as reduct. The redundant instances are also removed to reduce the datasize on the basis of closeness. Finally, the classification accuracy of reduced datasets accuracy is computed by nearest neighbor classifier. From the experimental results it is observed that the proposed method gives large amount of attribute reduction with better classification accuracy. However, the proposed method is suitable for two class classification problem. In future, we intend to develop a feature reduction algorithm for multi class datasets using rough set theory.

References

1. Qian, Yuhua, et al. An efficient accelerator for attribute reduction from incomplete data in rough set framework. Pattern Recognition 44.8, 1658–1670 (2011).
2. Merz, Christopher J., and Patrick M. Murphy. UCI repository of machine learning databases. Internet: http://www.ics.uci.edu/~mlearn/MLRepository.html, University of California, Irvine, Department of Information and Computer Sciences (1998).
3. Seo, Jinwook, and Ben Shneiderman. A rank-by-feature framework for unsupervised multidimensional data exploration using low dimensional projections. Information Visualization, 2004. INFOVIS 2004. IEEE Symposium on. IEEE, 2004. doi:10.1109/INFVIS.2004.3.
4. Chandrashekar, Girish, and Ferat Sahin. A survey on feature selection methods. Computers and Electrical Engineering 40.1, 16–28 (2014).
5. Jagadev, A. K., S. Devi, and R. Mall. Soft Computing for Feature Selection. Knowledge Mining using Intelligent Agents, 6–217 (2011).
6. Gheyas, Iat A., and Leslie S. Smith. Feature subset selection in large dimensionality domains. Pattern recognition 43.1, 5–13 (2010).

7. Son, Seung-Hyun, and Jae-Yearn Kim. Data reduction for instance-based learning using entropy-based partitioning. International Conference on Computational Science and Its Applications. Springer Berlin Heidelberg, 2006. doi:10.1007/11751595_63.
8. Luukka, Pasi. Feature selection using fuzzy entropy measures with similarity classifier. Expert Systems with Applications 38.4, 4600–4607 (2011).
9. Shannon, Claude E., and Warren Weaver. The mathematical theory of communication. University of Illinois press, 2015. Available at http://www.amazon.com.
10. Al-Sharhan, Salah, et al. Fuzzy entropy: a brief survey. Fuzzy Systems, 2001. The 10th IEEE International Conference on. Vol. 3. IEEE, 2001. doi:10.1109/FUZZ.2001.1008855.
11. Mehta, Rajesh, Navin Rajpal, and Virendra P. Vishwakarma. Robust image watermarking scheme in lifting wavelet domain using GA-LSVR hybridization. International Journal of Machine Learning and Cybernetics, 1–17 (2015). doi:10.1007/s13042-015-0329-6.
12. Cover, Thomas, and Peter Hart. Nearest neighbor pattern classification. IEEE transactions on information theory 13.1, 21–27 (1967).
13. Michalski, Ryszard S., Jaime G. Carbonell, and Tom M. Mitchell, eds. Machine learning: An artificial intelligence approach. Los Altos, CA: Morgan Kaufmann (2013), Available at http://www.springer.com/in/book/9783662124079.

Evaluation of Factors Responsible for Road Accidents in India by Fuzzy AHP

Shivam Nanda and Shalini Singh

Abstract India is a developing country fighting against its population growth. There has been an exponential growth in the number of automobiles in India to meet the needs of growing population. Road safety is one of the major concern in our country. Although the Indian government has been trying to tackle this issue for several years, yet there is a substantial increase in the number of road crashes which has become the major cause of death. The Methodology adopted for identifying the most crucial factor of road accidents is based upon Fuzzy-AHP technique. A multi-criteria decision making (MCDM) model is constructed which takes seven different criteria as inputs from different literature reviews and practical investigations to assign different priorities/weights to the seven criteria. The weights of criteria are usually characterized by fuzzy numbers. In this paper data for 35 States was collected from National Crime Records Bureau, Ministry of Road Transport and Highway, Global status report on road safety 2013. Finally, the decision is made by the computational process and effectiveness of Fuzzy AHP.

Keywords MCDM · AHP · Fuzzy AHP · Accident rates in India

1 Introduction

The frequency of road accidents in India is very high and is increasing at a very rapid rate, due to which India ranks second in road accidents. Every hour 16 die on Indian roads while one major accident takes place every minute and most of them are due to drink and drive cases. In 2013 over 1,37,000 individuals were dead in road accidents. There were about 1.7 lakh crashes due to over-speeding which lead

S. Nanda
Dehradun Institute of Technology, Dehradun, India
e-mail: ishivamnanda@gmail.com

S. Singh (✉)
Graphic Era Hill University, Dehradun, India
e-mail: shalini080793@gmail.com

G.M. Perez et al. (eds.), *Networking Communication and Data Knowledge Engineering*, Lecture Notes on Data Engineering and Communications Technologies 3, https://doi.org/10.1007/978-981-10-4585-1_15

to 49,000 deaths. As indicated by street activity wellbeing specialists, the real number of setbacks might be higher than what is recorded, the same number of auto collisions goes unreported. States like Rajasthan, Uttar Pradesh, Maharashtra, Tamil Nadu and Karnataka contributed more than 40% of the casualty. Among 53 uber urban areas, Delhi recorded the most number of fatalities at 2,199 while Chennai enlisted 1,046 similar accidents. Bhopal and Jaipur ranked third and fourth asserting 1,015 and 844 deaths respectively. By far most of most climate-related accidents happen on wet asphalt and amid precipitation: 73% on wet asphalt and 46% amid precipitation. A much littler rate of climate-related accidents happen amid winter conditions: 17% amid snow or hail, 13% happen on frosty asphalt and 14% of climate-related accidents occur on cold or slushy asphalt. Just 3% happen within the sight of mist. Hence, it becomes essential to analyze various factors responsible for road accidents but it is a very intricate task to make comparisons in relative road safety between various factors accountable for road accidents. An MCDM based approach i.e. Fuzzy-AHP is used which provides techniques for comparing and ranking different features. The study contains fifteen different states and cities of India, with seven different criteria. Few of the criteria used are drink and drive cases, road and weather condition, vehicle design, driving over the speed limit etc. Among these states and cities, the one which is most prone to road accidents will be selected at the end of the study. There are various sections in this report like literature review, methodology, case study including problem definitions, calculations and result. Finally, the conclusion is discussed in the end.

2 Literature Review

This study proposes fuzzy AHP technique to check the most influential factor responsible for road accidents. The AHP is a technique by Saaty where one can solve multiple criteria decision problems by assigning weights to different parameters. The study has been done to find out the cities which are highly prone to deaths due to road accidents and what were the reasons behind the accidents. The World Health Organization estimates that worldwide each year 1.3 million people are killed and between 20 and 50 million are injured in road crashes. Street mischances are basically brought on by uncalled for cooperation's between vehicles, amongst vehicles and other street clients and/or roadway highlights. The circumstance that prompts uncalled for connections could be the consequence of the mind-boggling transaction of various components, for example, asphalts qualities, geometric elements, movement attributes, street client. Hence, it becomes essential to determine the factor which is most responsible for road accidents in order to minimize road accidents. A number of research conducted among which few researchers have been discussed as follows.

Chen et al. [1] analysis build up a statistically based risk evaluation model, that focuses on accident hazard on street bends. The adopted approach utilizes records of past accidents from insurance agencies to decide most important factors responsible for road accidents.

Gonzalez et al. [2] explains the various drawbacks in rural areas where care may not come at proper times due to late discovery and high transport times reason being very long distances or weather conditions.

Garber and Ehrhart [3] usage of variables of parameters like average speed, SD of average speed, number of vehicles on each lane, width of a lane based on it predicting accident rates with the help of linear regression analysis.

Ahmed [4] identifying the errors in road geometry by remote sensing technique and determining areas most prone to road accidents in Pakistan.

Masaoe [5] data of road accidents is collected by traffic police of Tanzania regarding deaths and injuries. Factors responsible for such accidents are identified and different weights are assigned to them based on experts judgements. The computational results obtained after this evaluation reveal safety standards and help in the improvement of road safety. Also, these results can help in predicting losses by the insurance companies and based on which cost analysis can be done.

Abdullah and Zamri [6] ranking of different factors responsible for road accidents in Malaysia with the help of correlation analysis, MCDM i.e. Multi-Criteria Decision Making and Fuzzy Technique.

Cafiso et al. [7] determining safety standards of rural roads in terms of variables like exposure, road geometry with the help of Global Positioning System (GPS) and manual inspection in order to prevent future accidents.

Ma et al. [8] making performance indicators which determine road safety based on the application of fuzzy Delphi technique.

Pirotti et al. [9] road safety analysis at a particular segment of road with the help of past accidents data that helps in assisting multi-user to access the data around the world.

Du et al. [10] determining the highway road traffic safety in one of the coldest region of Heilongjiang Province—China with the help of five most influential factors affecting safety level: environmental factor during winters, low coefficient of friction on snowy road, automobile performance at low temperatures, poor vision, other driving factors.

Qirjako et al. [11] checked the occurrence of critical road accidents in Tirana, Albania, and defined their parameters responsible for accidents. This work consisted of data related to all road traffic accidents and with the help of binary regression analysis evaluation of different factors was done in order to predict future accidents.

Hejar et al. [12] analyse occurrence of road traffic accidents involving motorcar occupants by the help of cross-sectional sampling study. Establishing relation between safety factors and car occupants in Selangor, Malaysia.

Fridulv [13] explained drowsiness as a critical factor contributing to 3.9% of total road traffic accidents observed during late hours. Night time causes sleep to drivers which adversely affect the concentration thus leading to road accidents.

James and Scott [14] reported that rate of road accidents of teenagers are ten times that of adults per mile. A study was done to determine the past records of teenage driver accidents. The major factors leading to higher number of accidents caused by youth was because of their lack of attention, high-speed driving, risky behaviour of driving.

3 Methodology

Here, Fuzzy-AHP technique is used where the problem is considered as a Multi-Criteria Decision Making (MCDM) model. Seven different criteria's are served as inputs whose weights are calculated by using Fuzzy-AHP, based on which the final ranking is done. The advantages of using AHP are AHP is simpler-Unique and quantitative pieces of information reported, Unique as opposed to evaluate weights/scores straightforwardly, Unique systems to handle more difficult issues.

3.1 AHP and Fuzzy AHP Approach

In AHP the pairwise examinations for every level as for picked criteria and after that went to a size of 1–9. The thought about qualities are dependably in genuine

Table 1 Seven criteria's along with their definitions

No.	Criteria	Definitions
1	Condition of vehicle	Depending upon the condition of a vehicle there can be various defects in the vehicle like rupture of brake shoe, uncontrolled acceleration, failure of steering, insufficient headlights, overloading, damage of clutch plate etc.
2	Fault of cyclist	Cyclist faults can be overspeeding, lack of attention, cycling in the middle of the street, disobeying traffic rules etc.
3	Fault of pedestrian	It is due to lack of attention, disobeying traffic rules like entering a street or highway while intoxicated, crossing in the middle of the street, walking along highways, bridges, or causeways where pedestrian access is prohibited etc.
4	Fault of driver	Fault of driver plays a crucial role in accidents now a days like distracted driving, overspeeding, drunk driving, reckless driving, drowsy driving, disobeying traffic rules etc.
5	Defect in road condition	There can be numerous defects in the condition of road like misleading or damaged signs, blind curves and poor geometry of roads, broken guardrails, inappropriate road materials, lack of appropriate road markings etc.
6	Weather condition	Weather condition plays an important role it includes ice, heavy rain, slippery road surface, landslides etc.
7	All other causes	All other factors include avoiding safety gears like seat belts and helmets, animals on the road, absence of street and traffic lights on roads etc.

Table 2 Common data

States/UTs	C1	C2	C3	C4	C5	C6	C7
Andhra Pradesh	504	378	876	33222	1339	392	5813
Arunachal Pradesh	24	23	10	79	23	15	77
Assam	0	85	213	6184	0	0	53
Bihar	368	159	143	6225	283	457	2685
Chhattisgarh	682	158	93	9178	316	243	2841
Goa	6	65	116	3954	1	4	166
Gujarat	69	605	2648	23980	32	37	578
Haryana	146	118	306	7890	142	131	1332
Himachal Pradesh	15	4	15	2662	34	0	169
Jammu and Kashmir	69	6	335	4350	28	6	1915
Jharkhand	266	120	182	3374	239	252	1278
Karnataka	477	259	837	36545	241	263	5826
Kerala	0	7	21	36142	0	0	4
Madhya Pradesh	1112	425	847	40975	818	365	6668
Maharashtra	442	447	1991	56418	315	21	6682
Manipur	211	0	0	0	102	18	440
Meghalaya	117	0	22	23	32	37	252
Mizoram	0	0	0	110	0	0	0
Nagaland	6	0	0	5	0	0	31
Odisha	33	143	90	8781	19	10	209
Punjab	163	125	82	3804	92	162	1913
Rajasthan	28	12	21	21939	209	30	730
Sikkim	0	0	0	11	0	16	131
Tamil Nadu	107	735	1495	57507	1072	291	6550
Tripura	20	5	42	705	0	37	79
Uttarakhand	21	3	9	1013	43	0	383
Uttar Pradesh	1482	1803	1313	12759	756	824	11035
West Bengal	669	179	478	6191	660	919	3194
A and N Is	0	0	0	236	0	0	0
Chandigarh	0	0	0	419	0	0	0
Dadra and Nagar	0	0	0	85	0	0	0
Daman and Diu	0	0	0	40	0	0	10
Delhi	0	0	0	0	0	0	6937
Lakshadweep	0	0	0	3	0	0	0
Pondicherry	0	0	0	1125	0	0	56

numbers. Despite the fact that AHP has effortlessness and straightforwardness in its basic leadership and gives out a tolerable result however when the multifaceted nature of the issue expands its capacity to give more exactness reduces, and now and then accuracy solely turn into an imperative trademark. Hence traditional AHP is by all accounts lacking for this exploration fill in that capacity definite pair shrewd examination is verging on difficult to decide for this situation where the data and information are unverifiable. Consequently, for such case new MCDM procedures are more exact simply like Fuzzy AHP. In Fuzzy set hypothesis, the individuals work between the range of real numbers [0, 1]. Every part capacity and data portrays Fuzzy set. The Fuzzy set components are gone between a range which is normally [0, 1]. In the Fuzzy set, each individual is assembled yet no sharp limits are said.

4 Case Study

4.1 Problem Definition

In our study, first of all, we have decided the criteria are responsible for the road accidents. These criteria are chosen on the basis of various reviews and surveys. Criteria along with their definition are explained in Table 1. After that, Table 2 consists of common data which has been collected from National Data Sharing and Accessibility Policy (NDSAP) indicating the number of accidents and their causes. Then, we prepared the Fuzzy AHP data matrix with the help of surveys conducted

Table 3 Fuzzy data matrix

	C1	C2	C3	C4	C5	C6	C7
C1	1	3	0.333	0.142	3	5	0.2
C2	0.333	1	0.333	0.111	0.333	5	0.2
C3	3	3	1	0.142	3	5	0.333
C4	7	9	7	1	7	7	5
C5	0.333	3	0.333	0.14	1	3	0.333
C6	0.2	0.2	0.2	0.14	0.333	1	0.142
C7	5	5	3	0.20	3	7	1

Table 4 Results obtained with fuzzy AHP

Criteria	Weights	λ_{max}, CI, RI	CR
C1	0.091		
C2	0.049	$\lambda_{max} = 7.168$	
C3	0.118	CI = 0.028	CR = 0.02
C4	0.456	RI = 1.32	
C5	0.063		
C6	0.026		
C7	0.197		

Table 5 Final rank matrix

States/UTs	C1	C2	C3	C4	C5	C6	C7	∑(C1–C7)	Rank
Andhra Pradesh	0.030267	0.008805	0.039036	0.25419	0.061	0.010664	0.119052	0.523014	5th
Arunachal Pradesh	0.001441	0.000536	0.000446	0.000605	0.001048	0.000408	0.001577	0.00606	27th
Assam	0	0.00198	0.009492	0.047315	0	0	0.001085	0.059873	19th
Bihar	0.0221	0.003704	0.006372	0.047629	0.012892	0.012432	0.05499	0.160119	12th
Chhattisgarh	0.040957	0.003681	0.004144	0.070223	0.014396	0.00661	0.058185	0.198196	11th
Goa	0.00036	0.001514	0.005169	0.030253	4.56E-05	0.000109	0.0034	0.040851	20th
Gujarat	0.004144	0.014093	0.118	0.183477	0.001458	0.001007	0.011838	0.334016	7th
Haryana	0.008768	0.002749	0.013636	0.060368	0.006469	0.003564	0.02728	0.122833	14th
Himachal Pradesh	0.000901	9.32E-05	0.000668	0.020368	0.001549	0	0.003461	0.02704	21st
Jammu and Kashmir	0.004144	0.00014	0.014928	0.033283	0.001276	0.000163	0.03922	0.093153	17th
Jharkhand	0.015974	0.002795	0.00811	0.025815	0.010888	0.006855	0.026174	0.096612	15th
Karnataka	0.028646	0.006033	0.037298	0.279615	0.010979	0.007155	0.119318	0.489044	6th
Kerala	0	0.000163	0.000936	0.276531	0	0	8.18E-05	0.277712	8th
Madhya Pradesh	0.06678	0.0099	0.037744	0.31351	0.037265	0.009929	0.136563	0.611691	3rd
Maharashtra	0.026544	0.010413	0.088723	0.431668	0.01435	0.000571	0.136849	0.709118	2nd
Manipur	0.012671	0	0	0	0.004647	0.00049	0.009011	0.026819	22nd
Meghalaya	0.007026	0	0.00098	0.000176	0.001458	0.001007	0.005161	0.015808	24th
Mizoram	0	0	0	0.000842	0	0	0	0.000842	32nd
Nagaland	0.00036	0	0	3.82E-05	0	0	0.000635	0.001033	31st
Odisha	0.001982	0.003331	0.004011	0.067185	0.000866	0.000272	0.00428	0.081927	18th
Punjab	0.009789	0.002912	0.003654	0.029105	0.004191	0.004407	0.039179	0.093237	16th
Rajasthan	0.001681	0.00028	0.000936	0.16786	0.009521	0.000816	0.014951	0.196045	10th
Sikkim	0	0	0	8.4E-05	0	0.000435	0.002683	0.003202	29th
Tamil Nadu	0.006426	0.017121	0.06662	0.44	0.048836	0.007916	0.134146	0.721066	1st

(continued)

Table 5 (continued)

States/UTs	C1	C2	C3	C4	C5	C6	C7	∑(C1–C7)	Rank
Tripura	0.001201	0.000116	0.001872	0.005394	0	0.001007	0.001618	0.011208	25th
Uttarakhand	0.001261	6.99E-05	0.000401	0.007751	0.001959	0	0.007844	0.019286	23rd
Uttar Pradesh	0.089	0.042	0.05851	0.097622	0.034441	0.022416	0.226	0.569988	4th
West Bengal	0.040176	0.00417	0.021301	0.047369	0.030067	0.025	0.065414	0.233496	9th
Andaman and Nicobar	0	0	0	0.001806	0	0	0	0.001806	30th
Chandigarh	0	0	0	0.003206	0	0	0	0.003206	28th
Dadra and Nagar	0	0	0	0.00065	0	0	0	0.00065	33rd
Daman and Diu	0	0	0	0.000306	0	0	0.000205	0.000511	34th
Delhi	0	0	0	0	0	0	0.142072	0.142072	13th
Lakshadweep	0	0	0	2.3E-05	0	0	0	2.3E-05	35th
Puducherry	0	0	0	0.008608	0	0	0.001147	0.009755	26th

among the different group of people and accident victims. After that weights of each criterion were calculated along with the maximum eigen value (λ_{max}), CI, RI and CR with the help of Fuzzy AHP technique shown in Table 4. In the next step normalized data is calculated by dividing the each column value with its maximum value and then carry on by multiplying normalized data with their criteria weights. Finally, the summation of each row is done. The row that gives maximum value is indicated as rank 1st and so on indicated in Table 5. All this data is calculated with the help of Fuzzy AHP technique (Table 3).

5 Conclusion

This study analyses the application of the MCDM technique which is Fuzzy-AHP to find out the most influential factor responsible for accidents in India. Application of this MCDM technique enables a comparative analysis of alternative ranking and weights of criteria's based on the reviews and survey from different groups of people and accident victims. The selection of Fuzzy-AHP was done due to its capability to give the most précised result.

Firstly, we collected the data of the rate of accidents of different states and cities. The values of fuzzy tables are input by the survey and decisions. Then, the fuzzy data matrix is calculated with the help of Fuzzy AHP methodology. The normalized data table as well as the weights of each criteria are calculated by fuzzy AHP technique. Following the procedure as mentioned in problem definition the final ranking is done.

Finally, our research concluded the final rank of states/cities which are having higher accident rates. In this study, Tamil Nadu ranks 1st which has the highest accident rate due to fault of driver. Maharasthra and Madhya Pradesh ranked 2nd and 3rd respectively. The lower weight criterion which is weather condition is least responsible for any accidents in India. So from our research, we conclude that the fault of the driver is solely responsible for almost 45% of accidents in India. The fault of driver may include the reason such as poor driving skills, drink and drive, undertrained drivers and overspeeding at narrow roads or at the place where the speed limit is given. This criterion can be reduced by setting up the proper checking booths to ensure that whether the driver has drunk or not especially at night, by setting up the proper street lights, ensuring the optimum speed at various accident prone areas by constables etc.

References

1. Chen, S., Rakotonorainy, A., Sheehan, M., Krishnaswamy, S. and Loke, S. 2006. Assessing Crash Risks on Road Curves. Australian Road Safety Research, Policing and Education Conference, Gold Coast, Queenland, 25th–27th October.

2. Gonzalez RP, Cummings G, Mulekar M, Rodning CB. Increased mortality in rural vehicular trauma: identifying contributing factors through data linkage. J Trauma. 2006; 61: 404–409.
3. Garber, N.J. and Ehrhart, A.A. 2007. Effect of Speed, Flow and Geometric Characteristics on Crash Frequency for Two-Lane Highways. Transportation Research Record, Vol. 1717, pp. 76–83.
4. Ahmed, A. 2007 Road Safety in Pakistan, National Road Safety Secretariat, Government of Pakistan, Islamabad.
5. Masaoe, E.N. 2007. Study on Road Accidents in Mainland Tanzania [Online]. Available: http://www.sumatra.or.tz. [2013, September 25].
6. Abdullah, L. and Zamri, N. 2010. Ranking of the Factors Associated with Road Accidents using Correlation Analysis and Fuzzy TOPSIS. Australian Journal of Basic and Applied Sciences, Vol. 4(2): 314–320.
7. Cafiso, S., Di Graziano, A., Di Silvestro, G. and Persaud, B. 2010. Development of comprehensive accident models for two-lane rural highways using exposure, geometry, consistency and context variables. Accident Analysis and Prevention, Vol. 42 (4), pp. 1072–1079.
8. Ma, Z., Shao, C., Ma, S. and Ye, Z. 2011. Constructing road safety performance indicators using fuzzy Delphi method and grey Delphi method. Expert systems with applications, Vol. 38, pp. 1509–1514.
9. Pirotti, F., Guarnieri, A. and Vettore, A. 2011. Collaborative Web- GIS Design: A case study for Road Risk Analysis and Monitoring. Transactions in GIS, Vol. 15(2), pp. 213–226.
10. Du, X.J, Wang, Z.Y, Guo, H.H. and Li, J.P. 2013. Traffic Safety Evaluation Model for Highway in Cold Region Based on Fuzzy Theory. Research Journal of Applied Sciences, Engineering and Technology, Vol. 6(6), pp. 1048–1052.
11. Qirjako Burazeri, B., G., H. Hysa and E. Roshi, 2008. Factors Associated with Fatal Accidents in Tirana Albania: Cross-sectional Study. Croatian Medical Journal, 49(6): 734–740.
12. Hejar, A.R., M.Z., Afiah, S. Kulantayan and T.H. Law, 2005. Car Occupants Accidents and Injuries among Adolescents in a State in Malaysia. Proceedings of the Eastern Asia Society for Transportation Studies, 5: 1867–1874.
13. Fridulv S. 1998. Road Accident Caused by Drivers Falling Asleep. Accident Analysis and Prevention. 31: 639–649.
14. James M.A. and Scott M. A. 2003. Young Novice Drivers: Careless or Clueless. Accident Analysis and Prevention. (35): 921–925.
15. Avikal, S., Mishra, P.K. and Jain, R. (2013) 'An AHP and PROMETHEE methods-based environment friendly heuristic for disassembly line balancing problems', Interdisciplinary Environmental Review, Vol. 14, No. 1, pp. 69–85.
16. Avikal S., Mishra P.K., & Jain R.: A Fuzzy AHP and PROMETHEE method-based heuristic for disassembly line balancing problems. International Journal of Production Research, 52(5), 1306–1317 (2013).
17. Avikal S., Jain R., & Mishra P.K.: A Kano model, AHP and TOPSIS method-based technique for disassembly line balancing problems under fuzzy environment. Applied Soft Computing 25 (2014) 519–529.
18. Guite K (2012) UT wise total number of road accidents in India classified according to cause of accidents. In: Open Government Data (OGD) Platform India. https://data.gov.in/catalog/state-ut-wise-total-number-road-accidents-india-classified-according-cause-accidents

Development of Knowledge Capsules for Custom-Tailored Dissemination of Knowledge of Seismic Data Interpretation

Neelu Jyothi Ahuja, Ninni Singh and Amit Kumar

Abstract Knowledge management system is a repository of factual information. Seismic data interpretation is a field of exploration geophysics, which deals with interpretation of seismic images, to infer subsurface geology and provide information regarding hydrocarbon accumulation. This knowledge of interpretation is rare, expensive and largely individualistic. Lack of formal interpretation rules, causes seismic experts to use their own expertise gained over years of experience, leading to uncertainty. In current work a knowledge management framework is proposed, which initiates with the knowledge engineer gathering tacit knowledge from seismic experts, followed by a knowledge manager, synchronizing, sequencing, formalizing and organizing it in explicit form to develop a knowledge capsule, to facilitate its sharing through tutoring. Knowledge capsules have been refined to effectively suit different levels and knowledge grasping preferences of novice seismologists.

Keywords Tacit knowledge · Explicit knowledge · Knowledge management Seismologist

1 Introduction

Knowledge is considered as a summary or conclusion of true belief. Generally people may have some misconception about knowledge, considering knowledge merely as information or data. Data is unprocessed facts while information is the

N.J. Ahuja · N. Singh (✉) · A. Kumar
Department of Center for Information Technology, University of Petroleum and Energy Studies, Dehradun 248007, Uttaranchal, India
e-mail: ninnisingh1991@gmail.com

N.J. Ahuja
e-mail: neelu@ddn.upes.ac.in

A. Kumar
e-mail: amitanit007@gmail.com

G.M. Perez et al. (eds.), *Networking Communication and Data Knowledge Engineering*, Lecture Notes on Data Engineering and Communications Technologies 3, https://doi.org/10.1007/978-981-10-4585-1_16

processed data and knowledge is the understanding of information which is obtained through experience [1].

Knowledge is considered as one of the powerful and valuable assets of life [2]. It is effective acquisition of conclusion from facts and information. Its sharing always enhances it. Individuals who may be holding knowledge on a given subject matter may be termed as experts of that particular subject domain. At times, the experts may not be comfortable, sharing it, for the fear of losing their importance [3, 4].

The focus of the work is the subject matter domain of "Seismic data interpretation". Seismologists interpret seismic images to understand the structure of earth bed. This interpretation forming basis of petroleum exploration is largely dependent on the interpreter's expertise. This expertise comes with experience. There may be a case of uncertainty when same set of seismic snaps get interpreted differently by different interpreters. This is because there are no thumb rules, each expert uses his or her own expertise that they have gained over years of experience. Thus this interpretation knowledge is present in 'tacit' form. Knowledge accumulation with interpreters is not an acceptable scenario, primarily because of two reasons, one this essential body of knowledge is unable to percolate to the younger and novice seismologists, two the knowledge is statistically and tacitly available with experienced seismologists, without existing in an explicit repository form for use by everyone. An ideal situation is effective knowledge share facilitating the young seismologists to use the rules of interpretation generated by experienced seismologists. But most challenging task is the identification of quality knowledge and its conversion from tacit form to explicit form [5].

2 Related Work

Literature review indicates efforts in knowledge management, primarily from organizational knowledge repository development perspective. Further there has been very limited work or negligible work available in terms of understanding the nature of knowledge, computation of degree of tacit-ness (t_d), and utilizing it for future knowledge dissemination. There has been no effort towards developing knowledge capsules and delivering them, to knowledge seeking individuals in a custom-tailored manner, best suitable to their learning preferences and prior-knowledge levels. This, as focus of the present work is its novelty.

However, some work done in the past concerning tacit knowledge and related knowledge frameworks, has been presented in the section below.

Jones and Leonard [1] implement a knowledge management (KM) framework of transformation from tacit knowledge to organizational knowledge, integrating several scenarios [6, 7]. Their proposed model is a summation of both organizational and initiative features. Organizational includes two characteristics i.e. collaborative and innovative. The initiative includes four characteristics i.e. Top management support, Formal knowledge management staff, incentives as per quality and discussion about knowledge management with staff or employee. Innovation culture under organization feature invites new innovations and gives full support for the successful implementation and employees get suitable incentives for their new creativity. Collaborative culture under organization feature plays a vital role in the knowledge management implementation. Sharing knowledge helps extracting the tacit knowledge from the several employees to form organizational knowledge. Top management in initiative feature support is important for any type of new creation. Employee must understand the views of top management related to new creation or innovation i.e. are they convinced with the new creation or not. As knowledge management is itself a big task it requires staff for moving knowledge from tacit knowledge to organizational knowledge. Giving incentives as per quality is really important because it encourages other employees also to contribute their knowledge. Discussion about the knowledge management is important because it helps the employees to understand their contribution in the organization.

Al-Qdah and Salim [8] proposed a framework for the management of tacit knowledge based on media richness theory. They state that tacit knowledge can be made available or can be transfered from expert to layman by classifying knowledge into many levels based on the degree of tacit-ness. Expert based on his/her experiences can identify the degree of tacitness (low, high, medium). After identification of this degree expert selects the most suitable ICT (Information and communication technology) technique which helps him to deliver the knowledge in an effective manner.

Nonaka and Krogh [9, 10] proposed an article which elaborated the knowledge creation theory. Their contribution helps the scholars to understand the critical issues of management of organizational knowledge. They quoted six fundamental questions presented below:

(1) Ask the justification of belief.
(2) Ask the basis of tacit or explicit knowledge.
(3) Importance of tacit knowledge in an organization.
(4) What will be the outcomes and how it is beneficial socially?
(5) How tacit knowledge is related to the social practice?
(6) What will be the outcome after knowledge conversion?

Tennis [11] proposed a classification of Knowledge organization. Knowledge organizational research framework is classified into epistemology, theory and methodology. They briefly present the deviation in cognate of knowledge organization.

Desel [12] suggest a methodical generation of model. They elaborate on the correlation between the process involved in knowledge management and the related run (automatic initiated run).

This section below presents the knowledge management framework and design of knowledge capsules.

3 Proposed Framework

The scope of this paper is limited to the subject matter of "Seismic Data Interpretation". The objective of seismic interpretation is to obtain the structural map of earth bed by using reflection technique. In order to discover under earth, oil and gas deposits, a variety of exploration methods have been developed over the years. Seismic reflection methods yield most precise and accurate results [13], and play a prominent role in search for suitable geological structures, where commercially viable amounts of oil and gas deposits may have been lodged. Interpretation of seismic images enables seismic experts to infer subsurface geology and generate information regarding possible accumulation at depths of hundreds or thousands of feet below the ground surface. Geophysicists use their training and experience (i.e. their prior knowledge) to apply a concept (or rarely, to generate a new one) to construct an interpretation. Lack of formal knowledge-base of interpretation rules, causes dependency on human experts, additionally, hindering much-needed training/imparting of knowledge to forthcoming generation. Same seismic snaps interpreted differently by different seismologists leads to uncertainty. This is because there are no thumb rules and each seismologist uses his or her own expertise that they gained over years of experience. Thus this information is present in tacit form. The transformation from tacit to explicit form to facilitate knowledge dissemination is much needed for effective percolation to younger generation and utilization for years to come.

To overcome this kind of uncertainty a framework is proposed and flow diagram for the same is shown in Fig. 1. Seismic data interpretation involves a set of steps outlining the procedure in which a given seismic snap (image) depicting subsurface geology is interpreted. Having been in process of interpreting seismic images over several years, the seismologists are very well-versed and uncover many less-known or unknown details. This knowledge is very valuable and deserves to be solicited and preserved for future use. A knowledge management framework for handling

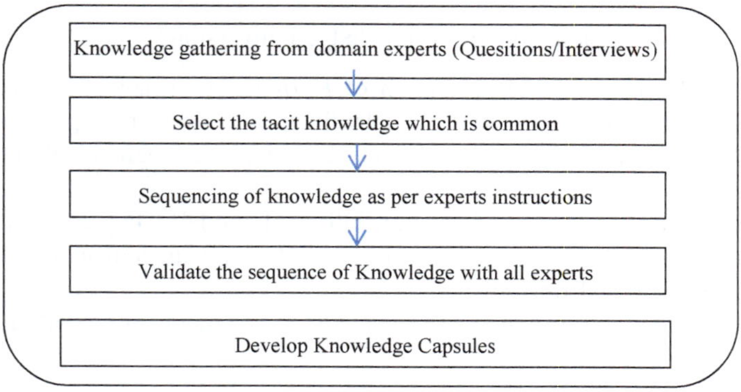

Fig. 1 Flow diagram of proposed framework

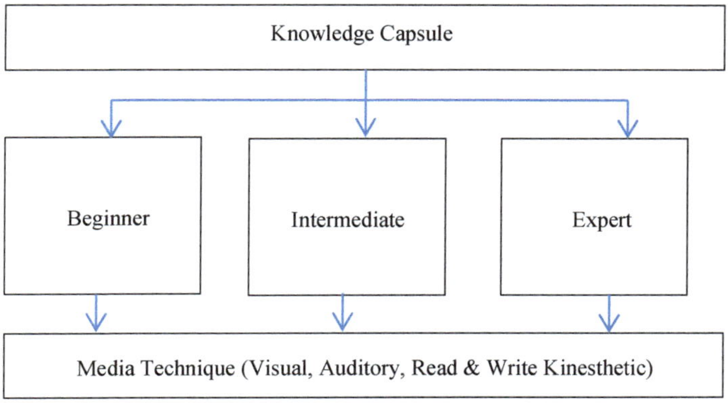

Fig. 2 Knowledge capsules

this knowledge is proposed. A methodology for the flow of knowledge through this framework is presented in the subsection below.

Knowledge engineer solicits knowledge from the identified subject matter experts, (seismologists) by tools such as questionnaires, interviews and discussions. After detailed discussion and consensus, knowledge engineer selects the commonly agreed knowledge.

After extraction of knowledge, the role of a knowledge engineer, transforms into that of a knowledge manager. Knowledge manager sequences and classifies the gathered knowledge as per seismologist's guidelines and the degree of tacit-ness (t_d). t_d can be defined as the extent to which a given piece of knowledge is tacit and undocumented. Knowledge manager then validates the sequenced knowledge through ongoing consultation with seismologist and develops knowledge Capsules or individual units of knowledge (Fig. 2).

Hence, the knowledge capsule is made available to novice seismologists in different learning styles. Knowledge manager maintains the same knowledge concept in different styles so that it could be offered to different individuals in a manner that best suits the individual. A novice seismologist having preliminary prior knowledge and preference for 'visual' knowledge may be offered the knowledge capsule of a particular topic of 'seismic data interpretation' at a 'beginner' level with visual style. Thus the grasp of an individual looking for knowledge can be better satiated, by delivering tailored subject matter.

4 Experimental Details

This section represents the implementation of knowledge capsules for 'Seismic Data Interpretation'. As discussed above, for gathering 'Tacit knowledge' from experts, several meetings with domain experts (from anonymous government

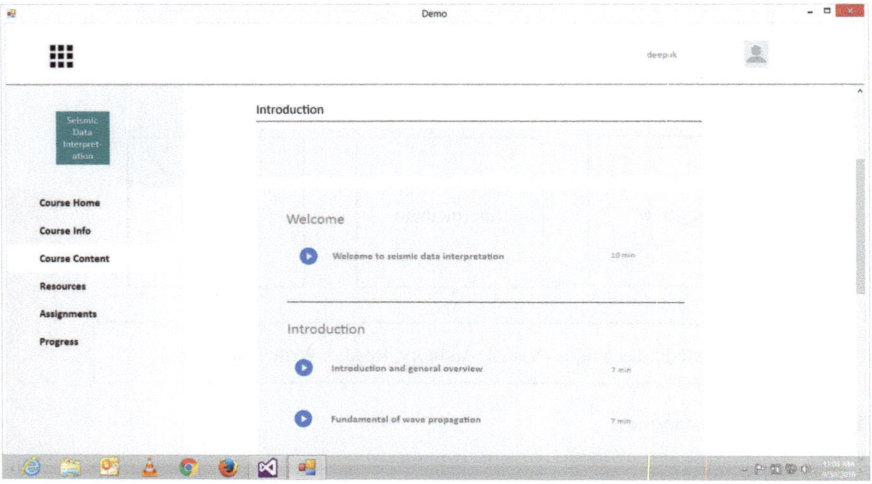

Fig. 3 Course plan

organization) were conducted. The gathered knowledge was worked upon to develop a ‘course delivery plan’ (course plan) outlining the topics and subtopics. Over a series of meetings, this plan was refined and fine-tuned. Figure 3 presents the final course plan as being offered through the dashboard of the tutoring software.

As discussed, the current scope is three learner profiles (‘Beginner’, ‘Intermediate’, ‘Expert’) and four learning style (‘Visual’, ‘Auditory’, ‘Read and Write’,

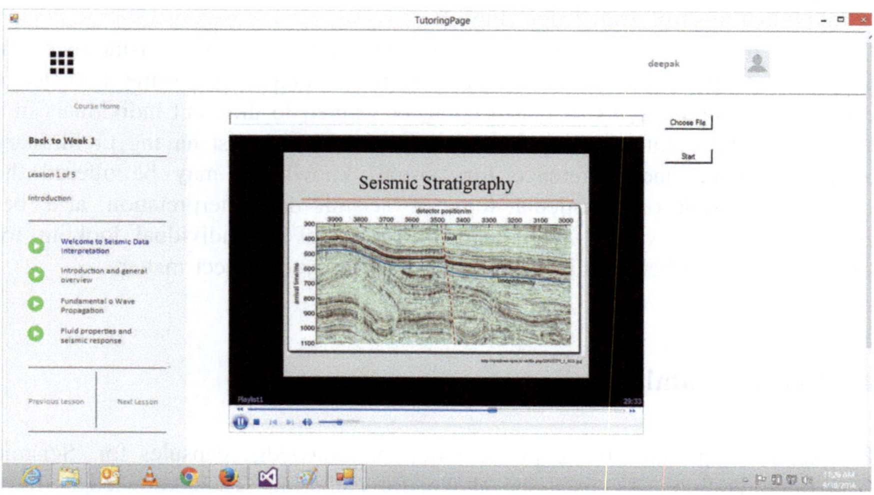

Fig. 4 Working of knowledge capsules

'Kinesthetic'), adding up to twelve tutoring strategies. The learner dashboard forms an interactive interface to deliver the knowledge capsules that is the learning material, (subject matter of seismic data interpretation) in twelve distinct tutoring strategies, offering the learner a customized learning experience. Figure 4 presents the working model of the system.

5 Conclusion

The knowledge capsules are effective and customized solution facilitating smooth percolation of rare seismic interpretation knowledge to the next generation of seismologists. The collection of knowledge capsules can be made available to learners, as a knowledge repository, which is an asset. It can be also made part of the tutoring modules to improve its access to Knowledge seekers at large.

Acknowledgements This work is being carried out at University of Petroleum and Energy Studies with the reference number SR/CSI/140/2013. The authors thankfully acknowledge the funding support received from Cognitive Science Research Initiative, Department of Science and Technology for the project. The Authors thank the management of University of Petroleum and Energy Studies, for Supporting the work and granting permission to publish it.

References

1. Kiku Jones and Lori N. K. Leonard, "From Tacit Knowledge to Organizational Knowledge for successful KM", Knowledge Management and Organizational Learning, Vol. 4 of the series Annals of Information Systems pp 27–39, 2009.
2. X. Ye and Zhang Huirong, "Literature Review on Enterprise Tacit Knowledge Management", International conference on, pp. 1844–1847, 2010.
3. Jana Kratka, "Tacit knowledge in stories of Expert Teachers", Procedia Social and Behavioral Sciences, Vol 171, pp. 837–846, 2015.
4. Ragna Seidler-de Alwis and Evi Hartmann, "The use of tacit knowledge within innovative companies: knowledge management in innovative enterprise", Journal of Knowledge Management, vol. 12, pp. 133–147, 2008.
5. N. J. Ahuja, P. Diwan, "An Expert System for Seismic Data Interpretation using Visual and Analytical tools", International Journal of Scientific & Engineering Research, 2012, Vol. 3(4), pp. 1–13.
6. Jon Arild Johannessen, Johan Olaisen, and Bjorn Olsen, "Mismanagement of tacit knowledge: the importance of tacit knowledge, the danger of information technology, and what to do about it.", International journal of information management, vol. 21, pp. 3–20, 2001.
7. Teresa A. Wasonga and John F. Murphy, "Learning from tacit knowledge: the impact of the internship", International Journal of Educational Management, Vol. 20(2), pp. 153–163, 2006.
8. Mohammad Sh. Al-Qdah, Juhana Salim, "A Conceptual Framework for managing tacit Knowledge through ICT Perspective", 4th international conference on electrical engineering and informatics pp 1188–1194, 2013.

9. Ikujiro Nonaka and Georg Von Krogh, "Tacit knowledge and knowledge conversion: controversy and Advancement in Organizational Knowledge Creation Theory", Perspective Organization Science, Vol 20(3), pp. 635–652, 2009.
10. Ikujiro Nonaka, "A dynamic theory of organizational knowledge creation" Organization science, Vol 5, pp. 14–37, 1995.
11. Joseph T. Tennis, "Epistemology, theory and methodology in knowledge organization: toward a classification, metatheory and research Framework", in Knowledge organization Vol. 35(2/3) pp. 102–112, 2008.
12. Jorg Desel, "from human knowledge to process Models", Information Systems and e-Business Technologies, Vol. 5, pp 84–95, 2008.
13. Mamdouh R. Gadallah and Ray L. Fisher "Applied seismology: a comprehensive guide to seismic theory and application", PennWell Corporation. Pp. 473, 2005.

Design of Low Cost Blood Glucose Sensing System Using Diffused Reflectance Near-Infrared Light

Jyoti Yadav, Asha Rani, Vijander Singh and Bhaskar Mohan Murari

Abstract The present work proposes a low cost and portable Non-Invasive Blood Glucose Measurement (NIGM) system based on Near-infrared (NIR) light. In vitro system using single LED (940 nm) with an array of photodetectors is fabricated. Regression analysis is carried out to study the relationship between detector output voltage and actual glucose concentration. Low RMSEC (reflectance: 12.87 mg/dl, transmittance: 15 mg/dl) of in vitro measurement motivated to design a sensor patch for non-invasive in vivo glucose measurement. The accuracy of our indigenous device was tested by comparing non-invasively estimated blood glucose with invasively measured blood glucose. To estimate the glucose concentration from the detector voltage signal an ADAptive Linear NEuron (ADALINE) based Neural Network structure is used. The calibration model is prepared using data of 10 non-diabetic subjects. The observed RMSEP was 14.92 mg/dl with correlation coefficient (0.87) in the case of In vivo experiment. The prediction of glucose concentration is in the clinically acceptable region of the standard Clark Error Grid (CEG). The proposed design of NIR light based glucose measurement can be used to develop an NIGM system.

Keywords Non-invasive · Blood glucose measurement · Near-infrared spectroscopy · ADAptive linear NEuron (ADALINE) · Artificial neural network (ANN)

The original version of this chapter was revised: Reference 17 has been updated. The erratum to this chapter is available at https://doi.org/10.1007/978-981-10-4585-1_26

J. Yadav (✉) · A. Rani · V. Singh
Instrumentation and Control Engineering Division, NSIT, Dwarka, New Delhi, India
e-mail: bmjyoti@gmail.com

B.M. Murari
Department of Sensors and Biomedical, VIT University, Vellore, Tamil Nadu, India

G.M. Perez et al. (eds.), *Networking Communication and Data Knowledge Engineering*, Lecture Notes on Data Engineering and Communications Technologies 3, https://doi.org/10.1007/978-981-10-4585-1_17

1 Introduction

Diabetes is a chronic disease in which body fails to control the blood glucose concentration in the normal range. According to a survey made by the International Diabetes Federation (IDF) in 2014, 387 million people had diabetes; by 2035 this will rise to 600 million [1].

The diabetes complications can be prevented by regular blood glucose monitoring. At present, the available glucose measurement systems are invasive in nature. Invasive devices are painful and expensive in the long term. Therefore, development of a clinically accurate, Non-invasive Blood Glucose Measurement (NIGM) is needed for reducing the complications of diabetes [2]. The researchers have been exploring non-invasive technologies for many years. Still, it is in primal stages of development. There are various NIGM technologies available in the literature. Due to low absorption and higher penetration, Near-infrared Spectroscopy (NIRS) is a conceivable technique for NIBG measurement. Many studies demonstrate the possibility of NIR based NIBG measurement. In the year 1992 Robinson et al. [3] published the first report on NIBG measurement using NIRS. Heise et al. [4] tried many aspects of NIBG measurement through NIRS. Results showed the feasibility of NIBG measurement using diffuse reflectance. Liu et al. [5] presented NIBG monitoring through the palm. Satisfactory results were observed for a single person over a day. Danzer et al. [6] reported that over a short period of time, individual calibration can give satisfactory results. However, for long-term monitoring results are poor. Mauro et al. [7] devised an optical fiber probe to observe the forearm spectra and Standard Error of Prediction (SEP) observed were 32.2 mg/dl. The author reported that for diffuse reflectance measurements the effective optimal path length should be 1.3–2 mm. Moreover Burmeister et al. [8] observed that percentage of body fat affects the signal-to-noise ratio of the measurement and for reliable glucose sensing, measurement site with less body fat should be considered. The problems and prospects of NIR based NIBG measurement [2] are reviewed in detail recently. The review reveals various prospects of NIBG measurement using NIR technique.

The present work proposes a non-invasive methodology to detect blood glucose using NIR light source (940 nm), where glucose has a fairly good absorption as compared to other chromophores of blood. In addition, we have used ADALINE neural network to estimate blood glucose. The output of detector after suitable processing is taken as inputs and invasive measurements from Glucometer are considered as the target output of the neural network. Artificial Neural Network (ANN) is well suited for the characterization of complex non-linear systems like biological systems [9]. The main advantages of ANN are its learning capability for development of new solutions to problems which are not well defined. Therefore, it is used to model the relationship between blood glucose level and detector voltage.

The subsequent sections are organized as follows: Research methodology including the principle of glucose measurement is given in Sect. 2. Prototype design and implementation of in vitro and in vivo measurement system are also discussed in Sect. 2. The ANN based calibration model used to estimate the blood glucose concentration from optical detector response is also discussed in this section. The experimental results and discussion of in vitro and in vivo measurement systems are given in Sects. 3 and 4 respectively. Finally, Sect. 5 concludes the research work.

2 Research Methodology

The principle involved in the experimentation is based on the attenuation of NIR light by glucose concentration. Attenuation is not only caused by absorption but also due to scattering of light which can be described by the light transport theory [10, 11] given by Eq. (1).

$$I = I_0 \mathrm{e}^{-\mu_{\mathrm{eff}} \mathrm{L}} \quad (1)$$

where. I_0 is the intensity of incident light, I is the intensity of reflected light, optical path-length is L, and μ_{eff} is given in Eq. (2) [11].

$$\mu_{eff} = \left[3\mu_a\left(\mu_a + \mu_s^{'}\right)\right]^{\frac{1}{2}} \quad (2)$$

Light transport in tissue is expressed in terms of the absorption coefficient μ_a and reduced scattering coefficient $\mu_s^{'}$. The μ_a is related to the tissue chromophores concentration and is given by Eq. (3).

$$\mu_a = 2.303\,\varepsilon\,\mathrm{C}\,\mathrm{cm}^{-1} \quad (3)$$

where ε is molar extinction coefficient and c is molar concentration.

The absorption coefficient μ_a changes with variation in glucose concentration. The reduced scattering coefficient is given in Eq. (4).

$$\mu_s^{'} = \rho\sigma[1 - g] \quad (4)$$

where ρ is the density of scattering centers, σ is scattering cross section and g is anisotropy factor.

The intensity of light scattered from tissue changes with variation in blood glucose concentration. When blood glucose concentration increases, the glucose concentration of ECF increases while that of tissue cells remains constant and the refractive index of ECF(η_{ECF}) approaches that of tissue cells thus leading to the decrease in the intensity of scattering. In addition, the decrease in the intensity of scattering results in shorter optical pathlengths for transmission measurements while it results in longer path lengths for reflection measurement.

Different bands in the NIR region are combination overtone band (2000–2500 nm), first overtone band (1400–2050 nm) and second overtone absorption bands (750–1550 nm). Overtone and fundamental wavelengths of different bands are given in Table 1. Amerov et al. [12] observed that combination and first overtone regions give absorption information, whereas shorter wavelengths primarily gives scattering/reflectance information.

2.1 Light wavelength selection

To improve the accuracy of the glucose measurement, selected NIR LED should have high mean absorption. Glucose absorption spectra between 900 and 980 nm were studied as shown in Fig. 1. Table 1 shows that the glucose molecule exhibits peak absorption at 939 nm which is very close to 940 nm. Moreover, this wavelength range is chosen to utilize the advantage of "therapeutic window" where tissue absorption is low and light penetration is high. Although NIR radiation in this range has lower glucose absorption, but due to less attenuation of the optical signal by other constituents required depth of penetration can be obtained. Therefore, NIR LED of 940 nm wavelength is utilized in the present work.

The objective of the present work is also to use comparativelylow cost and easily available parts/components so as to offer a low-costsolution to invasive blood glucose testing. Such constraints resulted in the utilization of LED in place of higher power LASER usedin spectrometer. However, LED has a smaller size,

Table 1 Overtone and fundamental wavelengths of different bonds [13]

Overtone	Bond	Wavelength	Bond	Wavelength
Fundamental	vC-H	3377 nm	vO-H	2817 nm
First overtone	2vC-H	1688 nm	2vO-H	1408 nm
Second overtone	3vC-H	1126 nm	3vO-H	939 nm
Combination	C-H + O-H		1536 nm	

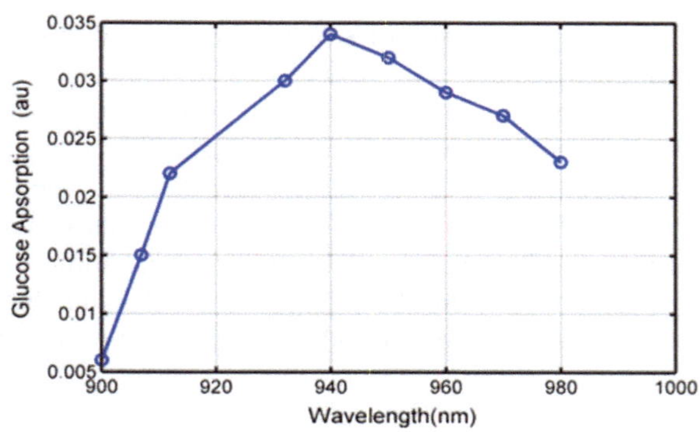

Fig. 1 Glucose absorption spectra between 900 and 980 nm [13]

lower cost, less power requirement and also causes no damage to tissue. Thus LED-basedNIGM system would be an economical option to LASER based systems and considered in the present work. The optical LED with optical power (45 MW) and spectral bandwidth of (50 nm) is used. As the LEDs are very sensitive optical component, therefore the performance and wavelength may change significantly due to fabrication and current variations, etc. Hence the spectrum is tested by varying the forward current of LED. At 100 mA current, the relative radiant power and detector output are found to be maximum and also verified with a data sheet of LED provided by the manufacturer. In order to supply the constant current to LED, regardless of the variations in voltage supply, BC548 (transistor) and VN2222 (MOSFET) based constant current circuit is also designed.

Photodetectors (OPT101) of Texas instruments are used to detect the attenuated light. The photodiode has high sensitivity in the range 300–1100 nm. The cost of LED as well as the detector in this wavelength (second overtone) range is relatively very less as compared to first and the combination overtone region. The cost is approximately 20 times less in the case of LED and 5 times less in the case of the detector for the second overtone region considered in this work. Further, the Arduino microcontroller (ATmega328) is used to control the switching time of LED's and photodiodes. It allows the user to modify and improve the design conveniently. The selection of these components in the prototype leads to an economic design of the sensor system. LM 324 IC is considered for filter design. It consists of four high-gain frequency compensated operational amplifiers, which can operate on a single supply. To assess the feasibility of proposed blood glucose measurement technique, the in vitro prototype is first designed.

2.2 *In Vitro Prototype Design*

The in vitro test setup is devised to examine the attenuation of NIR light by varying glucose concentration. The prototype is devised with a LED and an array of photodiodes is shown in Fig. 2. The test tube (for holding the glucose solutions) has a typical refractive index (1.42). The sample preparation is as follows:

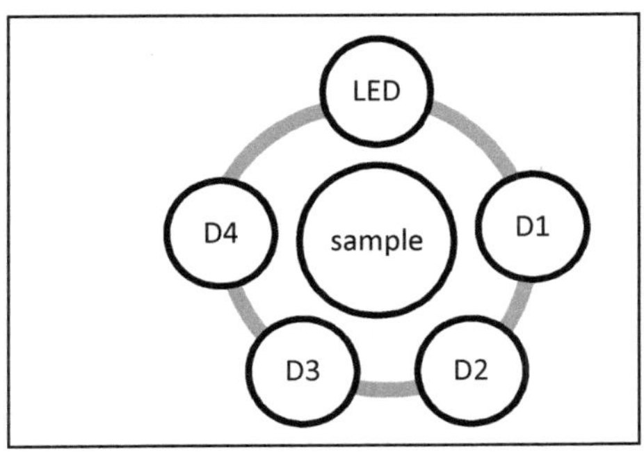

Fig. 2 *Top* view of in vitro test setup

Ringer solution which is a mixture of chemical salts in ringer solution is prepared using a standard protocol. This special salt solution is considered as synthetic blood plasma and is isotonic to physiological pH 7.4. It consists of 0.72 g NaCl, 0.017 g $CaCl_2$ and 0.037 g KCl in 100 ml of distilled water. The different concentrations of glucose are obtained by adding 40 mg/dl glucose concentration every time.

The test tube, LED and detector array are fixed on a base to minimize the movement of the setup. We have observed that a small variation in the position of the system leads to erroneous results. The average of the outputs obtained from photodiodes D2 and D3 give transmission information as they are placed at opposite ends of the centrally placed LED, whereas an average of the outputs from D1 and D4 photodiodes gives scattering information which measures the light scattered due to the presence of glucose (Fig. 2). Regression algorithm is used to build the best fit model between glucose concentration and the detector output for in vitro experimentation.

Regression Algorithm

Regression algorithm is utilized to establish the strength of the relationship between the series of independent variables (detector output x) and dependent variables (actual glucose concentration y). Linear regression is used to find the outcome y. The general form of regression is:

$$y = a + bx + u$$

where:

y is the variable to be predicted
x is the independent variable
a is the intercept
b = the slope
u = the regression residual

This technique is further extended to develop a sensor patch for in vivo NIBG measurements.

2.3 *In Vivo Prototype Design*

The block diagram of the in vivo testing system is shown in Fig. 3. It consists of sensor patch, signal conditioning circuit and a calibration model. The sensor patch further consists of a LED and four detectors detector along with driving circuit. LEDs are current-driven devices which require the constant-current source to keep the constant brightness. Transistor—MOSFET constant current driver circuit is used to supply constant current across LED regardless of the variation in voltage supply. Experimental setup for in vivo measurement is shown in Fig. 4. In order to

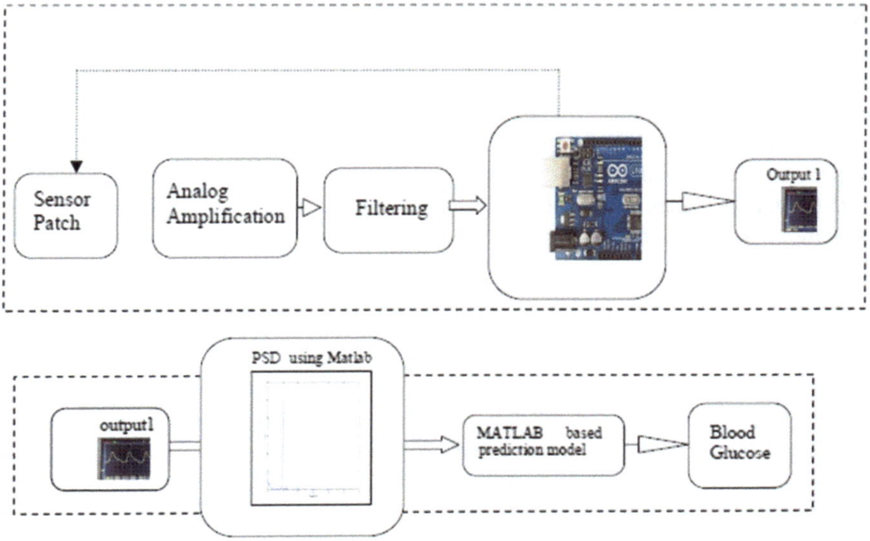

Fig. 3 Block diagram of in vivo blood glucose measurement system

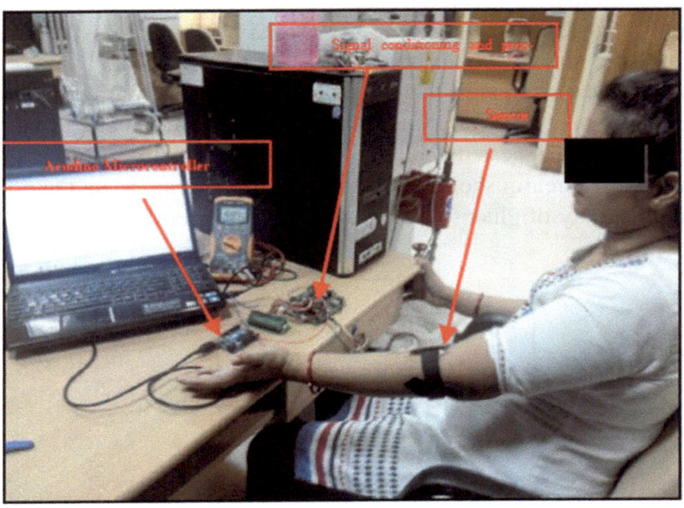

Fig. 4 Experimental setup for in vivo measurement of forearm

measure the signal effectively, a multi-sensor approach is considered. Thus the sensor patch is designed using four photodiodes to measure the attenuation of light due to glucose that incorporates a high degree of integration. It is proposed that combining these sensor signals will allow a more reliable NIBG monitoring.

The LED is placed at the center and four photodiodes are at 4 mm distance. The sensor patch is used to observe diffused reflectance spectra over forearm. Firstly the detector voltage signal is amplified and filtered and then acquired signal is saved for further processing. Later Matlab signal processing toolbox is used for signal analysis. The peak power of power spectral density (PSD) is used for calibration purpose to predict actual blood glucose.

During experimentation, the subjects are asked to sit in relaxed position and movement is prohibited during the optical measurement. It was observed that the slight variation in patch position causes variation in readings. Therefore, sampling position of the patch is marked on the forearm and is thus makes it strictly fixed for subsequent measurements. The arrangement of LED and photodiodes with the expected path of light is shown in Fig. 5.

The sensor patch is used to monitor diffused reflectance spectra over forearm. It is placed near "*brachial artery*" for measurement of glucose, covered with black covering in order to minimize the effect of stray light. The velcro strap is used to fasten the sensor patch around the forearm. Once the patch is placed on the skin the diffused spectra initially show fluctuations which stabilize after 30 s. This time is required to stabilize the detector output, whereas the time taken by calibration model to predict glucose is negligibly small.

Human skin has 3 layers: epidermis, dermis and subcutaneous. The reflected signal obtained from dermis contains abundant glucose information as compared to the epidermis and subcutaneous [12]. As the LED illuminates the forearm, the photons from the source propagate diffusely in the tissue and the trajectory of detected photons shows a banana-shaped region with a maximum depth of d/2 from the skin surface, where d is the distance between light source and optical detector [8].

Maruo et al. [8] suggested that the average optimal path length for glucose measurement over forearm should be 1.3–2 mm. In order to select the optimal path length, the sensitivity of glucose concentration is analysed for a different source to

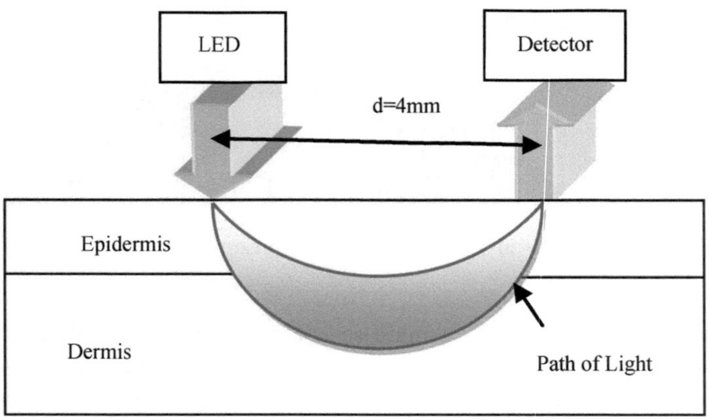

Fig. 5 Diagram of cross-sectional area of skin layers and light path

detector distances. In the present work LED to detector distance of 4, 3 and 2 mm are tested to find the optimal distance. The maximum change is observed with 4 mm distance. Therefore, the distance of 4 mm is considered between LED and photodiode to achieve optimal path length, whereas at 2 mm distance negligible variations are observed.

After necessary amplification, a low pass filter having 10 Hz cut off frequency is used to filter high-frequency components. High pass filter of 0.5 Hz cut off frequency is utilized to eliminate lower frequency signals due to various physiological noises such as respiration and thermoregulation etc. Further, a neural network-based calibration model is designed in MATLAB to obtain the glucose measurement from the detector output voltage.

Under the constant environmental conditions, the designed prototype is tested on ten non-diabetic subjects. Total 250 Samples are collected from ten volunteers (with consent) in the period of 2 weeks. As ANN can predict accurately within the range of the training data, thus the concentration of blood sugar sample should lie in that range. Informed consent is obtained from all participants and the study is approved by the Institutional Ethical Committee (IEC). Two measurement sites, i.e. forearm and wrist are considered for in vivo study. Datasets for each subject are acquired by taking one sample before a meal and three samples are taken after 60, 90 and 120 min of a meal at an interval of 30 min. The approximate nutrient content of a meal taken by the subjects during experimentation containsCalories: 250, Fat: 7.5 g, Protein: 5.1 g, Carbohydrates: 40.8 g (Table 2).

2.3.1 Calibration Models for Estimation of the Blood Glucose

NIR light from LED passes through the glucose sample or blood and is detected by photodiodes after attenuation. The detector converts the attenuated light into a voltage signal. This voltage signal is calibrated to obtain the glucose concentration in the blood sample. Glucose concentration is estimated from the photodiode output

Table 2 Characteristics of subjects

Subject	Gender	Age	BMI kg/m^2
1	Male	39	22
2	Female	38	19
3	Male	29	25
4	Female	29	22
5	Female	26	21
6	Female	26	23
7	Female	27	22
8	Male	23	24
9	Male	25	24
10	Male	25	25

Skin tone Normal Indian

with the help of ANN for in vivo measurement. There are several factors which affect the amount of blood glucose level, such as the amount of meal, physical work and the level of stress [13]. Moreover, the blood glucose level is not reflected properly in the measurements because of nonlinearity, complex dynamics and uncertainties in the human body which are person specific. ANN has properties that are ideal for approximating such type of functions [9]. Therefore, the predictive ability of ANN model is used in the present work to extract glucose concentration from the measured signal. There are various types of ANN model based on architecture, learning algorithm, and activation function. ADALINE is most widely used as it has very powerful learning law. It uses continuously predicted values (from the net input) to learn the model coefficients, which is more "powerful" since it tells us by "how much" we were right or wrong. Adaline uses the linear active function, in contrast to the unit function used in perceptron.

The ADALINE structure based ANN is used for estimation of blood glucose concentration from the outputs of photodiodes, which leads to ADALINE based in vivo system. The input data to the neural network is the voltage signal of four detectors and target output is blood glucose concentration which is obtained from Glucometer. The neural network weights are adjusted so that the output of the neural network is same as that of the target value. This trained ANN can now predict the glucose values for any detector voltage signal. The input-output patterns are prepared and are used for training and testing of the network.

Thennadil et al. [14] reported that for non-diabetic subjects, there is no significant difference between capillary blood glucose and glucose concentration at the dermis. Therefore time lag is not considered in the present work as experimentation is performed only on non-diabetic subjects. Two data sets of 240 input-output patterns each are prepared from forearm and wrist respectively. The data set is divided into training, validation and testing data sets. The data is divided into three parts: 80% of data sets (192 samples) are used for training, 10% (24 samples) used for validation and remaining 10% (24 samples) are used for testing of the neural network. Subject no. 7 and 8 data is used for validation and subject no. 9 and 10 data is used for testing purposes.

The neural networks are trained and tested using the acquired input-output data for estimation of blood glucose non-invasively. Training and testing of ANN are implemented using MATLAB 2011 (b) with real patient experimental data set. The computer configuration is: intelcoreTM i3-370 M, 2 GB RAM, 2.40 GHz processor.

Adaptive Linear Neuron Based Calibration Model

ADALINE Architecture

At the second stage of the research work, ADAptive Linear NEuron (ADALINE) network is used for estimation of blood glucose. ADALINE was developed by Bernard Widrow and Ted Hoff in 1960 [15]. Weights of ADALINE network are updated according to the weighted sum of inputs instead of transfer function [15]. Least Mean Square (LMS) algorithm is used for training of weights W and bias b of

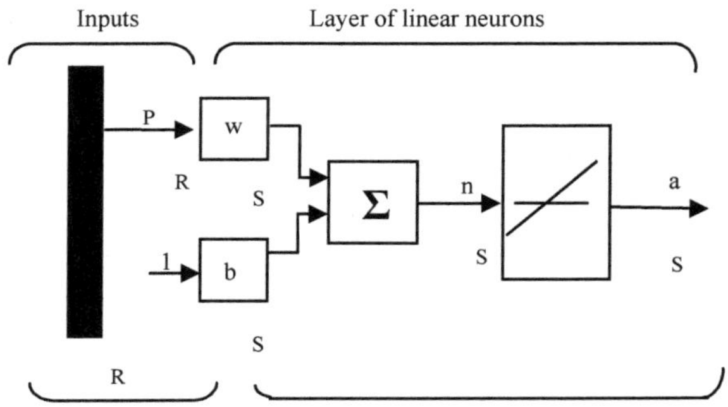

Fig. 6 ADALINE network

the network. The basic structure of ADALINE network is shown in Fig. 6. It is a single layer neural network, which has multiple nodes. Each node accepts multiple inputs and provides one output. The number of elements in input vector is R and the number of neurons in the layer is S.

The output of the network is given by

$$a = \text{pureline}\left(w_p + b\right) = \left(w_p + b\right) \tag{5}$$

where a is output, p is the input vector, w is the weight, b is bias and purelinis linear transfer function.

ADALINE Training:

The most commonly used training algorithm for ADALINE is least mean square error (LMS). To minimize the mean square error (MSE), the LMS algorithm adjusts the w and b of the neural network [15].

The error is the difference between the target output and the predicted network output. The objective is to minimize the mean of the sum of the square of these errors.

$$MSE = \frac{1}{n}\sum_{k=1}^{n} e(k)^2 = \frac{1}{n}\sum_{k=1}^{n} (t(k) - p(k))^2 \tag{6}$$

where p is input to the neural network, t is corresponding target output and e is the difference between target and actual value.

The LMS algorithm is performed by using an approximate steepest descent method. The partial derivative of the squared error with respect to the weights at the *k*th iteration is shown in Eq. (7).

This process is repeated, which successively cause a reduction in mean square error and consequently a local optimal value is achieved by the weight vector. At each iteration, gradient is calculated as $\nabla f(\mathrm{k}) = \nabla \mathrm{e}^2(\mathrm{k})$.

The first R elements of ∇e(k) are partial derivative w.r.t weights whereas (R + 1)st element is partial derivative w.r.t bias.

$$[\nabla e^2(k)]_j = \frac{\partial e^2(k)}{\partial w_{1,j}} = 2\text{e(k)}\frac{\partial e(k)}{\partial w_{1,j}} \tag{7}$$

For j = 1, 2,… R and

$$[\nabla \text{e}^2(\text{k})]_{R+1} = \frac{\partial e^2(k)}{\partial b} = 2\text{e(k)}\frac{\partial e(k)}{\partial b} \tag{8}$$

Now partial derivative of ∇e(k) w.r.t weights is calculated as follows:

$$\frac{\partial e(k)}{\partial w_{1,j}} = \frac{\partial [t(k) - p(k)]}{\partial w_{1,j}} = \frac{\partial}{\partial w_{1,j}}[t(k) - ({}_1w^T \times p(k) + b)] \tag{9}$$

$$\frac{\partial e(k)}{\partial w_{1,j}} = \frac{\partial}{\partial w_{1,j}}\left[t(k) - \left(\sum_{i=1}^{R} w_{1,i}p_j(k) + b\right)\right] \tag{10}$$

At the *k*th iteration p_j is the *j*th element of the input vector.

$$\frac{\partial e(k)}{\partial w_{1,j}} = -p_j(\text{k}) \tag{11}$$

Similarly, the final element of gradient w.r.t bias can be simplified as

$$\frac{\partial e(k)}{\partial b} = -1 \tag{12}$$

The gradient of error at kth iteration is

$$\nabla f(\text{k}) = \nabla \text{e}^2(\text{k}) = -2\text{e(k)p(k)} \tag{13}$$

The steepest gradient descent algorithm with α learning rate is

$$w(k+1) = \text{w(k)} - \alpha \nabla f(\text{x})|_{x=x_k} \tag{14}$$

After putting the value of ∇f(k) from Eq. (13) in Eq. (14) we get

$$\text{w(k + 1)} = \text{w(k)} + 2\alpha e(\text{k})\text{p(k)} \tag{15}$$

$$\text{b(k + 1)} = \text{b(k)} + 2\alpha e(\text{k}) \tag{16}$$

Here e is an error, w is weight, bis bias and α is learning rate. For larger values of α, learning takes place early, but the larger value of αmay lead to instability and error value might increase.

The α must be less in order to get stable learning. The value of α should be less than the reciprocal of the max eigenvalue of correlation matrix p^Tp of the input vector. The last two Eqs. (15) and (16) constitute the LMS algorithm which is also known as Widrow-Hoff learning algorithm.

The discussed neural networks are used for estimation of blood glucose concentration from the outputs of photodiodes. As the ADALINE network is a single layer network, it has four inputs and one output (4–1).

Validation of Experimental Results:

In this work, the results are validated by two methods: one with conventional validation (training, test, and validation) and another with k-fold cross-validation to verify the accuracy of ANN models [16]. The results obtained are also compared to the previously reported results. Generally in prediction problem's data set is divided into training, validation and testing subsets. This is the case when plenty of data is available. In the present work measured data set is small therefore only limited numbers of test samples are available for testing. Also, single training, validation, and test experiment may cause overfitting. These limitations are overcome by re-sampling methods through cross-validation. Therefore, the number of test samples is increased through cross-validation method. The present work *k*-fold cross-validation method is used to generalize the performance of estimator. The advantage of this method over repeated random sub-sampling is that eventually all the data sets are used for both training and validation. The ten-fold cross-validation method is utilized to train the network. The data set is randomly divided into 10 non-overlapping subsets. For each of 10 experiments, nine folds are used for training and the remaining one fold for testing purpose thus fitting a total of 10 models. For each fold, only 10% of data is used for testing, whereas 90% data is used for training.

Measures for Performance Evaluation:

The model performance of the in vivo system is evaluated in terms of widely used indices in the diabetes community: Root Mean Standard Error of Prediction (RMSEP) and Correlation Coefficient (r^2). A low RMSEP associated with a high r^2, is an indication of an efficacious estimation of blood glucose.

3 Experimental Results

The research work presents two types of analysis which aim at the development of NIBG measurement system: In vitro and In vivo experimentation.

3.1 In Vitro Experimentation

The designed in vitro prototype is used to measure the glucose concentration of the aqueous solution and ringer solution samples. The obtained glucose concentration is in terms of voltage signals obtained from the optical sensor. Regression is used to build the relationship between optical sensor output and actual glucose concentration.

The scattering and transmittance of NIR light through the samples is analyzed. The glucose concentration versus detector response for scattering as well as transmittance is shown in Fig. 7 ringer solution samples. It is observed from the results that for scattering as well as transmittance, a linear relationship is obtained between glucose concentration and optical sensor output for all the cases. The relationship is given by $y = c + b \times x$.

Where x is optical sensor output and y is actual glucose concentration. The b and c parameters of the above equation are for scattering and transmittance of NIR light in all the samples and are tabulated in Table 3. The Root Mean Square of Calibration (RMSEC) between the actual and predicted glucose concentration is also given in Table 3 for both the cases.

It is observed from the results (Fig. 7) that refractive index of the solution increases with the increase in glucose concentration of ringer solution. This leads to increase in NIR light scattering and hence outputs of D1 and D4 photodiodes increases. On the other hand, the increase in glucose concentration leads to decrease in transmittance. Therefore the outputs of D2 and D3 photodiodes decreases.

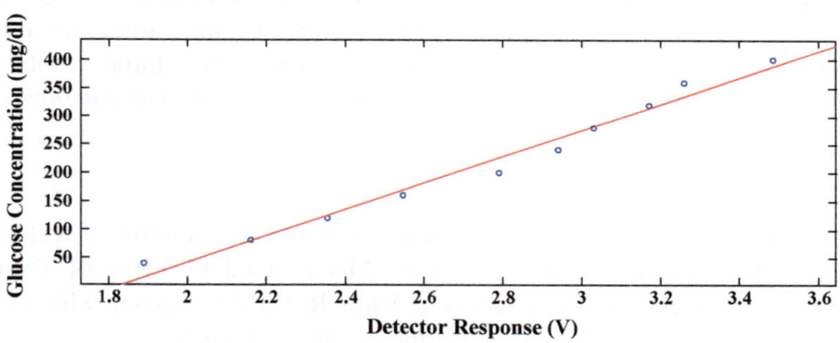

(a) Scattering

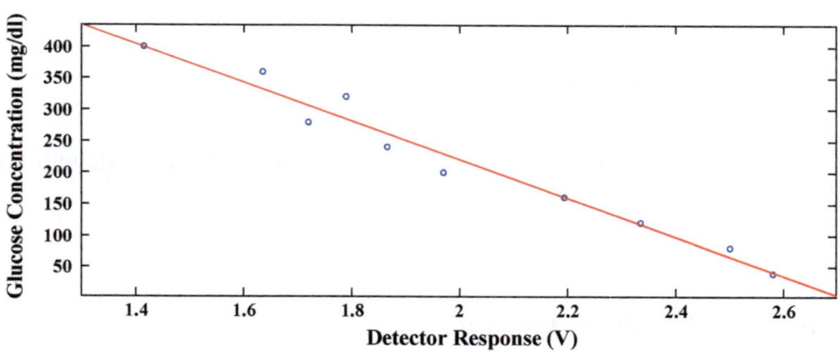

(b) Transmittance

Fig. 7 Response of detector for ringer solution

Table 3 In-vitro test results for two samples

Sample	Scattering			Transmittance		
	b	c	RMSEC	b	c	RMSEC
Ringer solution	233.36	−424.657	12.8715	−306.9128	834.0054	15.0099

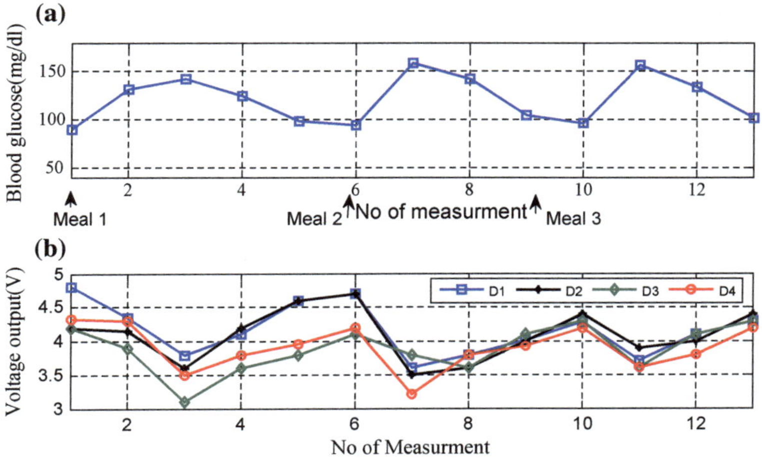

Fig. 8 Temporal variations in the blood glucose concentration (forearm): blood glucose measured with Glucometer (**a**) four detectors response (**b**)

The round shape of test tubes used in the prototype may be the source of scattering. The use of cuvette (rectangular and made up of low scattering quartz) may provide more accurate results. The use of cuvette may provide more accurate and repeatable results. The in vitro analysis reveals a strong correlation between optical sensor output and actual glucose concentrations. It is also observed that significant changes are noticed on account of variation in glucose concentrations.

The satisfactory performance of in vitro prototype provides the motivation to extend the devised methodology for in vivo measurements and is explained in the next section.

3.2 *In Vivo Experimentation*

In non-diabetic subjects, a calibration model is made using ADALINE neural network structure. Temporal variations in the blood glucose concentration (measured blood glucose and detector response) are shown in Fig. 8.

Figures 9 and 10 depict the results using the wrist and forearm as measurement sites respectively. Figure 11 depicts the Clark Error Grid Analysis (EGA) of in vivo analysis using ADALINE. The quantitative analysis of results for forearm and wrist are given in Table 4 which shows that forearm is better measurement site as compared to the wrist. EGA of in vivo analysis (Table 4) depicts that 83.33% points lie in region A, whereas 16.67% points lie in region B. It is also observed that the measurements made at forearm are more accurate as compared to the wrist. This is due to the reason that wrist arteries are narrower than the forearm and therefore lesser information is obtained regarding blood glucose. The results obtained from experimentation are compared with the results reported in the literature as shown in Table 5.

Fig. 9 In-vivo glucose measurement on forearm

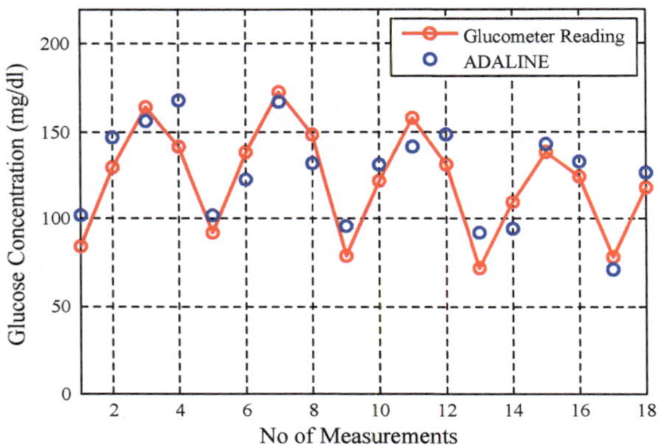

Fig. 10 In-vivo glucose measurement on wrist

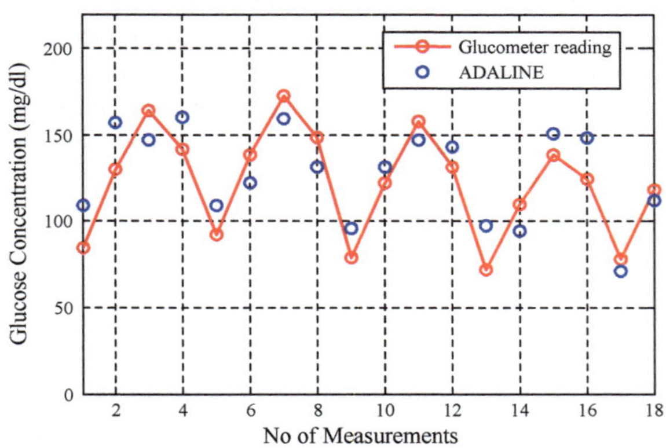

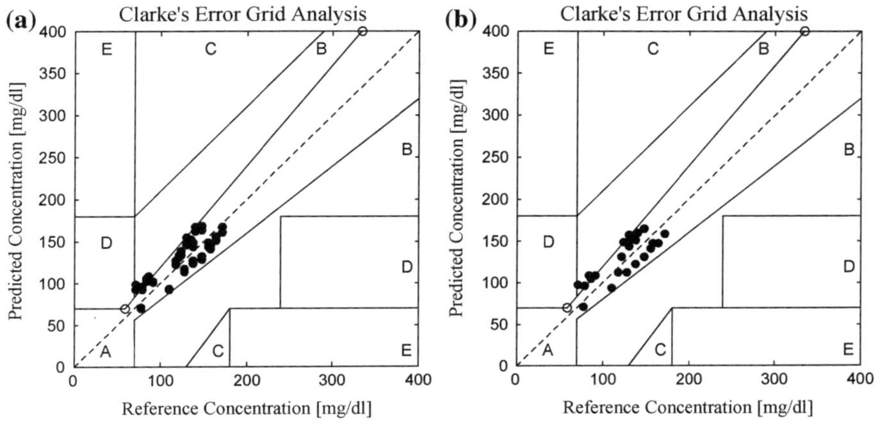

Fig. 11 EGA for ADALINE based in vivo system for **a** forearm and **b** wrist

Table 4 Performance indices in case of In vivo study

Measurement site	Method	Performance indices	ADALINE (mg/dl)
Forearm	Conventional testing	RMSEP (mg/dl)	14.92
		Correlation coefficient (R^2)	0.87
		Error grid analysis:	83.33% in A region, 16.67% in B region, None in C, D and E
	10-k fold validation	RMSEP (mg/dl)	16.87
Wrist	Conventional testing	RMSEP (mg/dl)	17.52
		Correlation coefficient (R^2)	0.81
		Error grid analysis:	79.17% in A region, 20.83% in B region, None in C, D and E
	10-k fold validation	RMSEP (mg/dl)	18.98

4 Discussion

The experimental results are compared with the results available in the literature. It is revealed from the analysis that a significant improvement is achieved in the results as compared to those available in the literature. In the case of continuous blood glucose monitoring, development of a hand-held portable device is required as the frequency of glucose measurements for diabetic patients. The designed prototype is LED based, therefore it may prove to be an economical, portable and convenient alternative to present invasive techniques. In addition to this, repetitive measurements can be made at low cost, as no reagent is required for testing.

Table 5 Comparison with other studies

Reference	Method	Summary	Specificity	Site
Heise et al. [4]	Diffuse reflectance	RMSEP = 36.4 mg/dl	1500–1600 nm	Oral mucosa
Chowdhury et al. [9]	Transmission	RMSEP = 23.76 mg/dl MAPE = 11.09% $R^2 = 0.76$	Modulated ultrasound and infraredlight	Finger
Guevara et al. [17]	Diffuse reflectance NIRS and impedance spectroscopy	RMSEP = 21.96 mg/dl	700–1000 nm (NIRS) and 1–200 MHz (impedance spectroscopy)	Forearm
Shu-Jen et al. [18]	Reflectance Diffuse reflectance	SEP < 27.0273 mg/dl Temp modulated from 22 and 38 °C	590, 660, 890 and 935 nm	Forearm
This study (Conventional Testing results)	Diffuse reflectance	RMSEP = 14.9248 mg/dl,	940 nm	Forearm
10-k fold validation		RMSEP = 16.875 mg/dl		

The implications of these preliminary studies show the prospects of a miniaturized low-cost NIR based instrument. In the present work the prototype is tested for short term measurements. The results are encouraging and in the near future, the prototype may be tested for long-term monitoring also.

During the experimentation, the movement of subjects is completely restricted to minimize motion artifacts. Therefore a cradle can be developed to position the forearm over the sample interface in a reproducible location with a reproducible amount of pressure. The primary requirement of this unit is to hold the forearm inaccurate geometrical position. The use of adaptive filters will also minimize the motion artifacts. Data of only 10 subjects are collected and tested because this is preliminary experimentation. Further diabetic subject's dataset can further validate our algorithms for long-term monitoring.

5 Conclusion

The research work focuses on the design of low-cost, LED-based portable NIBG measurement system using diffused reflectance spectra. In vitro experimentation depicts strong correlation with the glucose concentration in the sample. Further, a sensor patch is designed for in vivo measurement. ANN calibration models simulated in MATLAB are used to predict the blood glucose from detector voltage. It is observed for in vivo study that forearm is a better measurement site as compared to wrist for the present model. All the measurements lie in clinically acceptable A and

B zones of Clarke error grid analysis. The RMSEP and R^2 are significantly improved in comparison to the techniques reported in the literature. Hence it is concluded that the designed system may be conveniently used for continuous monitoring of blood glucose level.

References

1. 6th Edition of the Diabetes Atlas In: International Diabetes Federation. Foreword IDF Diabetes Atlas. 2014. http://www.idf.org/diabetesatlas/update-2014.
2. J. Yadav, V. Singh, A. Rani, B.M. Mohan, Prospects and limitations of non-invasive blood glucose monitoring using Near-infrared Spectroscopy, *Biomedical Signal Processing and Control*, 18: 214–227, 2015. doi:10.1016/j.bspc.2015.01.005.
3. M.R. Robinson, R.P. Eaton, D.M. Haaland, G.W. Keep, E.V. Thomas, B.R. Stalled, P.L. Robinson, Non-invasive glucose monitoring in diabetic patients: a preliminary evaluation, *Clin. Chem.* 38:1618–22, 1992.
4. H.M Heise, R Marbach, Th. Koschinsky and F.A Gries, Non-invasive blood glucose sensors based on near-infrared spectroscopy, Artificial Organs 18 issue 6 (1994) 439–447. DOI:10.1111/j.1525-1594.1994.tb02230.x.
5. R. Liu; W. Chen; K. Xu, "Next Step of noninvasive glucose monitor by NIR technique from the well controlled measurement condition and results," in Biophotonics, 2004. APBP 2004. The Second Asian and Pacific Rim Symposium pp. 22–23, 14–17 Dec. 2004, DOI:10.1109/APBP.2004.1412258.
6. K. Danzer, Ch. Fischbacher, K. U. Jagemann, K.J. Reichelt, Near infrared diffuse reflection spectroscopy for non-invasive blood glucose monitoring, *LEOS newsletter* 12(2) 9–11, 1998.
7. K. Maruo, M. Tsurugi, J. Chin, T. Ota, H. Arimoto, Y. Yamada, M. Tamura, M. Ishii, and Y. Ozaki, Noninvasive Blood Glucose Assay Using a Newly Developed Near-Infrared System, *IEEE Journal Of Selected Topics In Quantum Electronics*, 9(2), 2003, pp. 322–330. doi:10.1109/JSTQE.2003.811283.
8. J.J. Burmeister, M.A. Arnold, G.W. Small, Evaluation of measurement sites for non invasive blood glucose sensing near-infrared transmission spectroscopy, *Clin. Chem.*, 45(9): 1621–1627, 1999. http://www.ncbi.nlm.nih.gov/pubmed/10471677.
9. S. Ramasahayam, K.S. Haindavi, S.R. Chowdhury, Noninvasive Estimation of Blood Glucose Concentration Using Near Infrared Optodes, Smart Sensors- Measurement and Instrumentation Volume 12, 2015, pp 67–82. DOI- 10.1007/978-3-319-12898-6_4.
10. N.A. Bazaev, Yu. P. Masloboev, S. V. Selishchev, Optical Methods for Noninvasive Blood Glucose Monitoring, *Biomedical Engineering*, 45(6), 229–233, 2012. DOI- 10.1007/s10527-012-9249-x.
11. O.S. Khalil, Non-invasive glucose measurement technologies: an update from 1999 to the dawn of the new millennium, *Diabetes TechnolTher.* 6(5): 660–697, 2004. doi:10.1089/dia.2004.6.660.
12. A.K. Amerov, J. Chen, G.W. Small, M.A.Arnold, Scattering and absorption effects in the determination of glucose in whole blood by near-infrared spectroscopy, *Anal Chem.*, 77(14): 4587–94, 2005. DOI:10.1021/ac0504161.
13. S. Ramasahayam, S.H. Koppuravuri, L. Arora and S.R. Chowdhury, Noninvasive Blood Glucose Sensing Using Near Infra-Red Spectroscopy and Artificial Neural Networks Based on Inverse Delayed Function Model of Neuron, *J Med Syst*, 39 (1):166, 2015. doi- 10.1007/s10916-014-0166-2.
14. S. N. Thennadil, J.L. Rennert, B. J. Wenzel, K. H. Hazen, T. L. Ruchti, and M. B. Block, Comparison of Glucose Concentration in Interstitial Fluid, and Capillary and Venous Blood

During Rapid Changes in Blood Glucose Levels, Diabetes Technology & Therapeutics. September 2001, 3(3): 357–365. doi:10.1089/15209150152607132.
15. A. Rani, V. Singh, J.R.P. Gupta, Development of soft sensor for neural network based control of distillation column, *ISA Transactions*, 52:438–449, 2013. doi:10.1016/j.isatra.2012.12.009.
16. H. Temurtas, N. Yumusak, F. Temurtas, A comparative study on diabetes disease diagnosis using neural networks, Expert Systems with Applications, Volume 36, Issue 4, May 2009, Pages 8610–8615. doi:10.1016/j.eswa.2008.10.032.
17. E. Guevara, F.J. González, Joint optical–electrical technique for non-invasive glucose monitoring, *Mex. J. Phys.* 56(5): 430–434, 2010. doi:http://www.scielo.org.mx/scielo.php?script=sci_arttext&pid=S0035-001X201000050001.
18. Y. Shu-Jen, F.H. Charles, S.K. Omar, Monitoring blood glucose changes incutaneous tissue by tissue by temperature-modulated localized reflectance measurements, *Clin. Chem.* 49(6): 924–934, 2003. doi:10.1373/49.6.924.

Hybrid Chunker for Gujarati Language

Parneet Kaur, Vishal Goyal, Kritida Shrenik Shah and Umrinderpal Singh

Abstract Gujarati is a first language of Indian state of Gujarat and furthermore for union territories of Daman and Diu and Dadra and Nagar Haveli. Chunking is the basic technique for any language processing work. It is a method of identifying and splitting the text into correlated small pieces (or chunks). The objective for presented research work is to build Hybrid Chunker for Gujarati. Initially baseline system using Hidden Markov Model is developed and then correction rules are applied on it. The annotated corpus of one lakh sentences developed by TDIL, MoCIT, GoI in association with Department of Gujarati, Gujarat University, Ahmedabad is used. The system is trained on 80,000 sentences of chunked Gujarati text and tested on 20,000 sentences. Afterwards, Post-processing is done with rules on output of baseline system. By applying these rules, the system achieved improvement in overall accuracy and it is 95.31%.

Keywords TDIL corpus for gujarati language · Hidden markov model Hybrid approach · POS-tagged sentence · Post-processing rules

1 Introduction

Chunking is the basic technique for any natural language processing work. It is a method of identifying and splitting the text or concepts into correlated small pieces (called chunks). Chunking techniques are used for full parsing. It is used as a pre-processing step in machine learning as well as natural language processing so as to automatically extract linguistic knowledge from annotated corpora. It is engaged with sentence segmentation, named entity extraction, information extraction, POS tagging and parsing. Chunker identifies Noun Phrase (NP), Conjunction Phrase

P. Kaur (✉) · V. Goyal · U. Singh
Department of Computer Science, Punjabi University, Patiala 147002, India
e-mail: parneetmann.kaur@gmail.com

K.S. Shah
Department of Gujarati, Gujarat University, Ahmedabad 380009, India

G.M. Perez et al. (eds.), *Networking Communication and Data Knowledge Engineering*, Lecture Notes on Data Engineering and Communications Technologies 3, https://doi.org/10.1007/978-981-10-4585-1_18

(CCP), Verb Group Phrase (VGF), Adverb Phrase (RBP), Adjectival Phrases (JJP), Chunk for Negatives (NEGP) and Chunk for Miscellaneous words (BLK) in corpus. The chunking process is branched into two parts. First, Identification of Chunk Borderlines—develop a chunker which takes POS-tagged sentences as input, and second, labeling of Chunks—mark chunk borderlines with their well-formed linguistic or semantic groups [1].

The purpose of presented work is to develop a Hybrid Chunker for Gujarati which is based on Hidden Markov Model and a rule based chunker. At present, no such hybrid chunker is available in Gujarati.

2 Earlier Relevant Work

In presented work, various techniques for chunking matching with Hybrid, Statistical and Rule-based approach have been studied for different languages such as English, Hindi etc. Also several techniques for HMM (Hidden Markov Model) and linguistic rules have been studied.

The initial effort for chunking has been accomplished by Church K. in 1988 for English through statistical approach which is based on machine learning [2]. Skut and Brants modified it through standard HMM tagging methods in 1988 [3]. In 1995, Ramshaw and Marcus used transformation based learning (TBL) with help of large annotated English corpus [4]. In 2000, Veenstra, and Bosch achieved accuracy of 91.05% precision and 92.03% recall based on memory based phrase chunking for English [5]. In 2000, GuoDong Zhou, Jian Su and TongGuan Tey continued using the same methods and achieved 91.9 and 92.2% of precision and recall respectively using a contextual lexicon [6]. Support vector machines (SVM) technique has been used for chunking in English by Kudo and Matsumoto, in 2001, and achieved an accuracy of 93.48% accuracy [7].

In 2005, Akshay Singh, Sushma Bendre and Rajeev Sangal used Hidden Markov Model (HMM) for chunking in Hindi language with accuracy of 92.63% precision and 91.7% for chunk labeling with a recall of 100% [1]. Ashish Tiwari, Arnab Sinha and Dr. Sudeshna Sarkar presented HMM based Chunking for Hindi with a new approach. They used n-tag scheme with an accuracy of POS-tag's highest precision 89.02% and word_pos tag approach acquires precision and recall of 85.58 and 98.48% consequently [8]. In 2007, Samiran Mandal, Asif Ekbal, and Sivaji Bandhoyopadhyay presents POS tagging using HMM and Rule based Chunking for Bengali, Hindi and Tamil. This chunker has demonstrated 80.63%, 71.65% and 53.15% accuracy respectively with the unannotated test sets for Bengali, Hindi and Telegu [9]. Wajid Ali and Sarmad Hussain developed a verb phrase Chunker for Urdu using a Hybrid approach. They developed a rule based VP chunker using linguistic rules for automatic chunking and used an Urdu corpus of 1,00,000 words that is tagged manually with VP chunk tags. This system achieved 98.44% accuracy [10].

In this work, we present Hybrid Chunking for Gujarati. A union of Gujarati words and POS (Part-of-Speech) tags are given to the system as an input. The union of these two grants the finest result. Distinct mechanisms are used and correlate in order to gain prime results for labeling the chunks. We have followed Hidden Markov Model and rule-based for error correction is used to develop the Chunker.

3 Approach Used

A hybrid approach (Firstly HMM and afterwards Post-processing rules) is applied for chunking process. Both techniques are described as follows:

1. Hidden Markov Model

In case of Simple Markov Models, the state is easily noticeable to the observer, yet it is not true in case of Hidden Markov Models. A hidden Markov model (HMM) enlists both observed events and hidden (unobserved) events. Words are the observed events that we see in the input and POS tags are unobserved (or hidden) events. The state is not directly observable, but the output (dependent on the state) is visible. The word 'hidden' deals with the state sequence, by which the model passes through, not with the parameters of the model. Even if these parameters are known exactly, still this model is referred to as a 'hidden' Markov model. The precise interpretation of Hidden Markov Model (HMM) is

$$\lambda = (\mathrm{P}, \mathrm{Q}, \pi) \tag{1}$$

where P is a set of Probabilities of Event Transition; Q is a set of Probabilities of Event Emission; π is a set of Probabilities of Initial Event.

A first-order hidden Markov model conveys two simple inferences. First, the feasibility of a singular event depends only on the earlier event.

Markov Inference $P(t_j|t_i \ldots t_{j-1}) = P(t_j|t_{j-1})$ (2)

Second, the feasibility of an object observation ob_j depends only on the event which produced the observation t_j and not on any other event or any other observations.

Output Inference $P(ob_j|t_1 \ldots t_j, \ldots, t_T, ob_1, \ldots ob_j, \ldots, ob_T) = P(ob_j|t_j)$ (3)

Hidden Markov Model has some computational problems. First problem is to determine likelihood of sequence of observations. It can be determined as, injective function and the Markov Inference depicted in Eq. (2), for an exact hidden (or unobserved) event sequence $T = t_0, t_1, t_2, \ldots, t_T$ and an observation sequence $OB = ob_1, ob_2, \ldots, ob_T$; then the joint probability density is:

$$P(ob_1, \ldots, ob_M, t_1, \ldots, t_M) = P(t_1)\left[\prod_{m=2}^{M} P(t_m|t_{m-1})\right] * \prod_{m=1}^{M} P(ob_m|t_m) \quad (4)$$

where $P(t_1)\left[\prod_{m=2}^{M} P(t_m|t_{m-1})\right]$ and $\prod_{m=1}^{M} P(ob_m|t_m)$ expresses Transition Probabilities and Emission Probabilities respectively.

Transition probabilities are those which are based on bigrams; and Emission Probabilities are the word likelihoods. HMM uses $n-grams$ hierarchy technique such as unigram, bigram, and trigram. The discounting can help to determine the issue of zero probability density $n-grams$. Suppose, first do an attempt to calculate trigram probability $P(a_m|a_{m-1}a_{m-2})$ however, there is not equivalent trigram $a_{m-2}a_{m-1}a_m$ counts available, then approximate calculation of its feasibility density can be computed with the help of bigram probability $P(a_m|a_{m-1})$. Then as well, if there is no counts present to measure $P(a_m|a_{m-1})$, and then use the unigram probability $P(a_m)$. There are two approaches to determine this $n-gram$ hierarchy i.e. backoff and interpolation. Here, we use interpolation. In interpolation, combine the probability approximation calculations of all the $n-grams$ evaluators, such that do a weighted interpolation of trigram, bigram, and unigram counts.

In simple interpolation, associate various arrangements of $n-grams$ with help of linearly interpolating all the models. Hence, evaluate the trigram probability $P(a_m|a_{m-1}a_{m-2})$ through merging together the unigram, bigram, and trigram probabilities; and every probability density is weighted by δ:

$$P(a_m|a_{m-1}a_{m-2}) = \delta_1 P(a_m|a_{m-1}a_{m-2}) + \delta_2 P(a_m|a_{m-1}) + \delta_3 P(a_m) \quad (5)$$

such that summation of all $\delta' s$ is 1 i.e.$\sum_j \delta_j = 1$.

In complex form of interpolation, also, every δ weight is calculated in complex manner, through context condition weights. The following Eq. (6) for interpolation with context-conditioned weights:

$$P(a_m|a_{m-2}a_{m-1}) = \delta_1(a_{m-2}^{m-1}) P(a_m|a_{m-2}a_{m-1}) + \delta_2(a_{m-2}^{m-1}) P(a_m|a_{m-1}) + \delta_3(a_{m-2}^{m-1}) P(a_m) \quad (6)$$

HMM also works as a generative model as it generates observation sequence through traversing from one event to another in accordance with to the transition probabilities and emitting an observable i.e. from a discrete set of observables such as a finite alphabet, from each state inspected in accordance with the emission probabilities of the state [11]. The Learning algorithm is as follows:

```
Input data
Prepare an outline emission_prob, trans_prob, data
for each sentence in document
        prev = " * "        #Make the Sentence Start
        data = [prev]++
        divide sentence into tags_words with " "
        for each wt in tags_words
                divide wt into tag, words with "_"
                trans_prob[prev + " " + tag]++
                data[tag]++
                emission_prob[tag + " " + word]++
                prev = tag
        trans_prob[prev + " * "]++
#Print the Transition Probabilities
for each key, value in trans_prob
        divide key into prev, word with " "
        print "T", key, value/data[prev]
#For emission probabilities, do identical process with "E"
```

Hidden Markov Model consists of unobserved (hidden) variables. The job for searching a variable sequence underlying in some another observation sequence is known as decoding. This is another problem for HMM. The formal definition of Decoding is defined as:

Given Hidden Markov Model (HMM) input $\lambda = (P, Q)$ and observation's sequence $OB = ob_1, ob_2, \ldots, ob_T$, then the most probable sequence of states $T = t_1, t_2, \ldots, t_T$.

The frequent and better decoding HMM algorithm is Viterbi algorithm. Viterbi algorithm computes state sequence with the maximum probability. After decoding the entire sequences, the sequence of states is backtracked. Following are the steps for Viterbi algorithm:

Input a sentence $s_1 \ldots s_n$ and parameters $q(t_n|t_{n-1}), e(x_i|t_i)$.
Define k to set of all tags $k_{-1} = k_0 = (start), \pi(0, start, start) = 1$.
For $k = 1 \ldots n$

For $a \in k_{k-1}, b \in k_k$

$$\pi(k, a, b) = \arg\max(\pi(k-1, a, b) * q(a|b) * e(s_k|b))$$

end For.
Return $\arg\max(\pi(n, b) * q(stop|b))$

end For.

2. Rule based Technique

The Rule-based technique has been used in post-processing for error correction. Different linguistic rules have been used to perform chunking on the output of baseline system.

4 Architecture of Proposed System

Firstly, training is provided to the system. The system must be trained before chunking process. The system is trained by using the corpus of one lakh sentences provided by TDIL. Input given to the system for Chunking is POS tagged sentence. The input is formed in a particular format and every sentence consists of POS tag parallel to word in the sentence. The HMM model gives relevant chunk tag adjacent to every POS tag. Afterwards, post processing is implemented by error correction rules to intensify the accuracy of baseline system. Rules are applied to all wrong tagged chunks to correct them. The Fig. 1 shows Architecture for the proposed system. It contains mainly two modules. One is Training Process and another is Chunking Process.

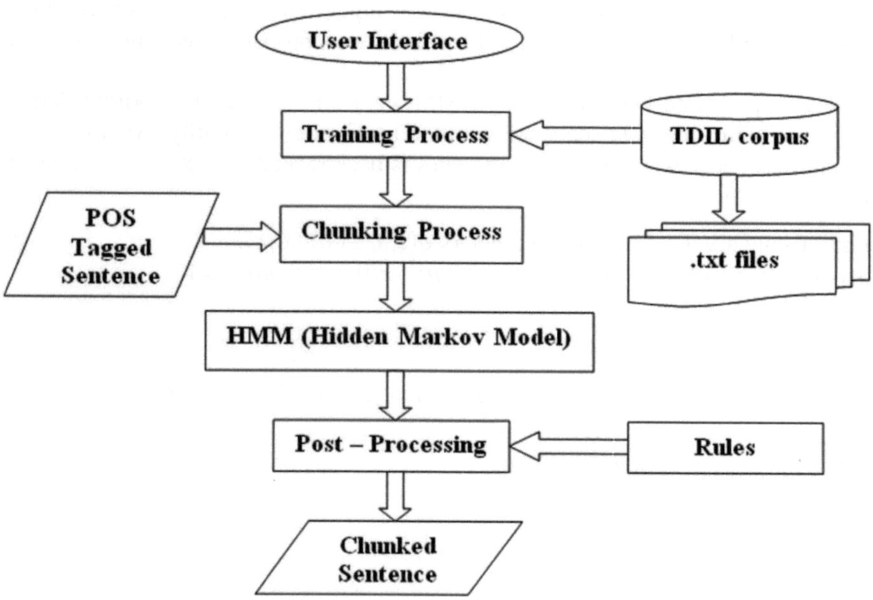

Fig. 1 Architecture of hybrid chunker

1. User Interface

It is a GUI interface provides information regarding how to interact with developed system. It includes a textbox area to which we provide input and gives us an output through different functions that are discuss in further section.

2. Training Process

It is necessary to provide training to the developed system and is done with the help of TDIL Gujarati Corpus. Corpus contains different sentences with which system gets trained. All these sentences are in chunked form. All sentences are saved in text files. This process has discussed in Training Algorithm.

3. Chunking Process

Firstly, chunking is done by HMM approach. It takes input in the form of POS (Part-of-Speech) tagged sentence such that POS tag is correspondent to the every word. Hidden Markov Model measures state sequence probabilities using Viterbi algorithm. HMM gives chunked sentence but with some errors. For these error corrections, some linguistic rules are used which is termed as error correction or post-processing with rules. After applying rules, accuracy improves along with the error corrections and also provides chunked sentences.

5 Experiments

Assume any POS tagged sentence containing sequence of words $W_m = w_i$ such that $w_i = (w_1, w_2, w_3, \ldots, w_m)$ and POS tags $T_m = t_i$ such that $t_i = (t_1, t_2, t_3, \ldots, t_m)$.

સમપૂરીઓના\N_NNP દર્શનથી\N_NN મળે\V_VM છે\V_VAUX મોક્ષ\N_NN .\RD_PUNC હિદુધર્મમાં\N_NNP તીર્થનું\N_NN ઘણું\QT_QTF મહત્વ\N_NN છે\V_VAUX .\RD_PUNC

Next task is to label the chunks. The components of chunks (i.e., POSs of tokens) use to allocate chunk label. Different chunk categories—NP, CCP, VGF, RBP, JJP, NEGP, BLK present in proposed system. To select chunk labels, machine learning based method is tried. Following is an example after chunk labeling.

[[સમપૂરીઓના\N_NNP દર્શનથી\N_NN]]_NP [[મળે\V_VM છે\V_VAUX]]_VGF [[મોક્ષ\N_NN]]_NP [[.\RD_PUNC]]_BLK [[હિંદુધર્મમાં\N_NNP તીર્થનું\N_NN]]_NP [[ઘણું\QT_QTF]]_JJP [[મહત્ત્વ\N_NN]]_NP [[છે\V_VAUX]]_VGF [[.\RD_PUNC]]_BLK

A rule based system used in post-processing for error correction. Here we discuss one, among different errors and corresponding rules for correction. The most frequent error occur in noun phrase as it gives problem while assigning N_NST tag to NP, when it comes before adjective, as it can come by itself. The whole tag pattern is scanned for correction purpose. The following rule makes correction.

If $POS(w_i)$ = JJ and $POS(w_{i-1})$ = N_NST then chunk tag for $w_{i-1} = NP$ and for w_i = JJP.

See below the bold tag as an example:

Before- [[આ\DM_DMD ધમ્મ\N_NN ચીનમાં\N_NNP ઈ.સ.\N_NN]]_NP **[[પૂર્વે\N_NST]]_JJP** [[બીજા\JJ]]_NP [[શતકમાં\QT_QTC]]_JJP [[પહોંચ્યા\V_VM હતા\V_VAUX]]_VGF [[.\RD_PUNC]]_BLK

After- [[આ\DM_DMD ધમ્મ\N_NN ચીનમાં\N_NNP ઈ.સ.\N_NN પૂર્વે**\N_NST]]_NP** [[બીજા\JJ શતકમાં\QT_QTC]]_JJP [[પહોંચ્યા\V_VM હતા\V_VAUX]]_VGF [[.\RD_PUNC]]_BLK

Some of rules are listed below

1. If $POS(w_i)$ = QT_QTC and $POS(w_{i-1})$ = N_NST then chunk tag for w_{i-1} = NP and for w_i = JJP.
2. If $POS(w_i) = RB$ and $POS(w_{i-1}) = N\_NN$ then chunk tag for $w_{i-1} = NP$ and for $w_i = RBP$.
3. If $POS(w_i)$ = QT_QTO and $POS(w_{i-1})$ = N_NN then chunk tag for w_{i-1} = NP and for w_i = JJP.

6 Results

There are two experiments which are performed. Firstly, base line system is accomplished using HMM technique with the help of testing data and achieve 91.23% accuracy. By applying some of rules on output of baseline system, it is observed that there is improvement in accuracy by approx. 4%. So, overall accuracy

Table 1 Result of experiments

Evaluation results					
HMM approach			Hybrid approach		
Precision	Recall	F-measure	Precision	Recall	F-measure
92.5%	90%	91.23%	96.66%	94%	95.31%

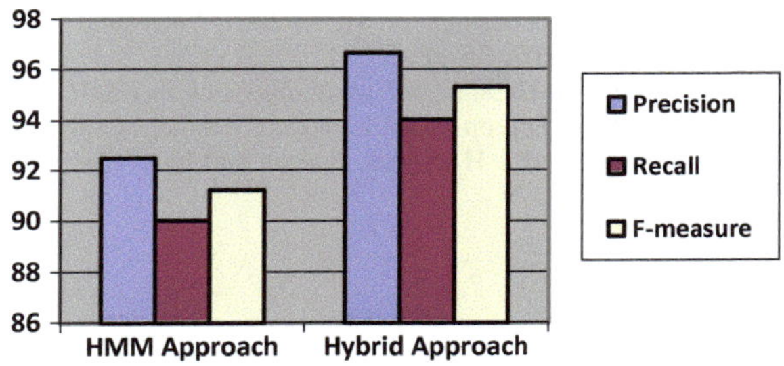

Fig. 2 Comparison of evaluation results

for both experiments is 95.31%. Proposed system is evaluated by using different notations such as Precision, Recall and F-measure (Table 1 shows results of both experiments and Fig. 2 shows their comparison).

7 Conclusions

Hybrid approach for chunking in Gujarati is presented which is based on HMM technique under statistical chunking and rule based error corrections afterwards. The provided rules are best to the available corpus developed by TDIL, MoCIT, GoI in association with Department of Gujarati, Gujarat University, Ahmedabad. These rules may vary when more corpus files will available along with differentiate in language as it can be used. Using hybrid approach, the best accuracy is 95.31%.

References

1. A. Singh, S. Bendre and R. Sangal, *HMM based Chunker for Hindi,* Proceedings of Second International Joint Conference on Natural Language Processing, 2005, 126–131.
2. K. Church, *A Stochastic Parts Program and Noun Phrase Parser for Unrestricted Text,* Proceedings of Second Conference on Applied Natural Language Processing, 1988, 136–143.
3. W. Skut and T. Brants, *Chunk Tagger,* Statistical Recognition of Noun Phrases - ESSLLI-1998, 1998.
4. L. A. Ramshaw and M. P. Marcus, *Text Chunking Using Transformation-Based Learning,* Proceedings of the 3rd Workshop on Very Large Corpora, 1995, 88–94.
5. J. Veenstra and A. Bosch, *Single-Classifier Memory-Based Phrase Chunking,* Proceedings of CoNLL-2000 and LLL-2000, 2000, 157–159.

6. G. Zhou, J. Su and T. Tey, *Hybrid Text Chunking,* Proceedings of CoNLL-2000 and LLL-2000, 2000, 163–165.
7. T. Kudo and Y. Matsumoto, *Chunking with Support Vector Machines*, Proceedings of NAACL 2001, 2001, 1013–1015.
8. A. Tiwari, A. Sinha and S. Sarkar, A New Approach for HMM Based Chunking for Hindi, *Term Project on Speech and Natural Language Processing,* 2005.
9. A. Ekbal, S. Mandal and S. Bandhoyopadhyay, *POS tagging using HMM and Rule based Chunking,* Proceedings of IJCAI-2007 Workshop on Shallow Parsing in South Asian Languages, 2007, 25–28.
10. W. Ali and S. Hussian, *A Hybrid approach to URDU verb phrase chunking,* Proceedings of the 8th Workshop on Asian Language Resources, 2010, 137–143.
11. D. Jurafsky and J. H. Martin, Speech and Language Processing, Prentice-Hall, 2014.

Sound Localization in 3-D Space Using Kalman Filter and Neural Network for Human like Robotics

Aakanksha Tyagi, Sanjeev Kumar and Munesh Trivedi

Abstract Sound Localization is the process of identifying direction (with distance) and location of the source from which the sound is detected. It is one of the important functions of human brain. In brain sound localization is done through the neurons present in it. The sound signals from the outside world are come inside the brain through the ear. In this paper, the process of Sound Localization activity performed by human brain that incorporates realistic neuron models is discussed and the accurate position of the sound sources by using the Kalman filter and neural network is examined. The results demonstrate that finding position in 3D is more accurate as compared in 2D as its average error gets reduced. This work can be used to detect the location of the sound sources in three dimensions and can be also implemented in robots and cochlear implants for treating hearing loss.

Keywords Sound localization · Neuron models · Kalman filter
Neural network

1 Introduction

One of an important function of the brain is sound localization and it would help to make smarter robots as well as other machines. The origin of sound source is found out on the basis of source and subject relative locations between them. Like in

A. Tyagi (✉) · S. Kumar
Computer Science & Engineering Department, Krishna Institute of Engineering and Technology, Ghaziabad, India
e-mail: aakankshard@gmail.com

S. Kumar
e-mail: sanjeevgreen@gmail.com

M. Trivedi
Computer Science & Engineering Department, ABES Engineering College, Ghaziabad, India
e-mail: munesh.trivedi@gmail.com

G.M. Perez et al. (eds.), *Networking Communication and Data Knowledge Engineering*, Lecture Notes on Data Engineering and Communications Technologies 3, https://doi.org/10.1007/978-981-10-4585-1_19

human beings the sound can be originated either from left or right hand side. The intensity of the waves produced by the sound originating object decides the same. Now if we can do the same using any tool, then the robots can react as per the sound.

It is very difficult to decide the origin of sound being produced. Like on a slow traffic road we came into situations when we were unable to decide the origin of the horn making vehicle. This situation motivates researchers to make a tool that can efficiently decide the same and react based on that as human brain does. So we have given an approach for Sound Localization in 3D space using Kalman filter and Neural Network inspired by the auditory cortex of human brain.

1.1 Neurotransmission

The signaling process of a neuron in the target cell is a process called neurotransmission. There are about 1 billion to 1 trillion cells called neurons in human brain that maintains the electric charge or nerve signals inside the brain. The structure of a neuron is shown in Fig. 1. Nerve signals are transmitted through nerve cells in brain that are present in neurons at up to 200 mph. Neurons are same as other cells in the body, but they are different electrochemically and due to this property neurons are able to send the messages and electrical signals between each other. A complex network of neurons forms the brain. These neurons function as the building blocks of the nervous system, transferring information to and from the brain and all over the body [1]. Neurons are the cells that are electrically excitable and these cells processed and send the information with the help of chemical and electrical signals. A neuron consists of three parts: First is cell body which is also termed as soma, in this cell nucleus lies that provides the instruction of what action is to be performed by the neuron. Second part is axon which is a slender projection of neuron that connects its cell nucleus with the other neuron's dendrite. Third part is dendrite which is a short projection of neuron that is responsible for sending neurotransmitters with the help of receptor sites that lies on it. The communication between neurons is done through the links between the axons and dendrites.

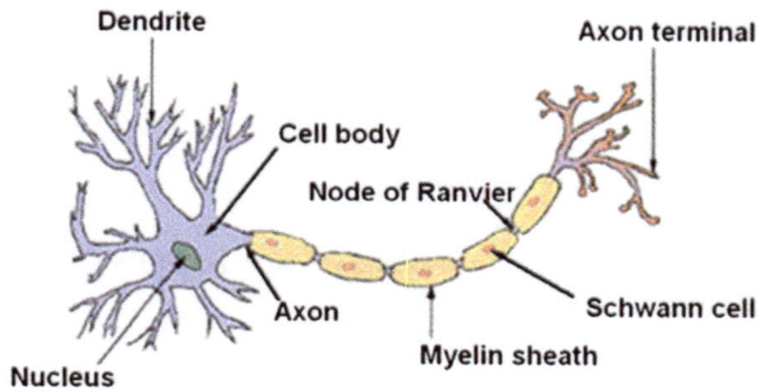

Fig. 1 Structure of neuron [14]

The metabolic activities of different ions such as Na, K, Cl and Ca present inside the neurons provide a voltage gradient on its membrane. The action potentials resulting from changes in the voltage have different charge and this can be seen and measured as a spike which is like a wave signal and also termed as brain wave or rhythm.

1.2 Sound Localization

The process of identifying direction (with distance) and location of the source from which the sound is detected is known as Sound localization. This helps to identify the intensity of the sound being detected. The organ by which the sounds are localized in human beings is known as ear. For identifying the direction and source of sound detected, the human brain extracts and processes different form of data. This includes the shape of the sound spectrum at the tympanic membrane, the different sound intensity for both left and right ear, the time of arrival to both left and right ear etc.

1.3 Types of Cues for Sound Localization

For Binaural sound localization, there use two sensors that detect the sounds coming from a sound source. In biological systems these sensors are ears and in technical systems, it is microphones. For sound localization, animals and humans used two major classes of cues: First one is Monaural Cues that can be extracted using one ear (or microphone). For the sound wave arriving at the eardrum, the

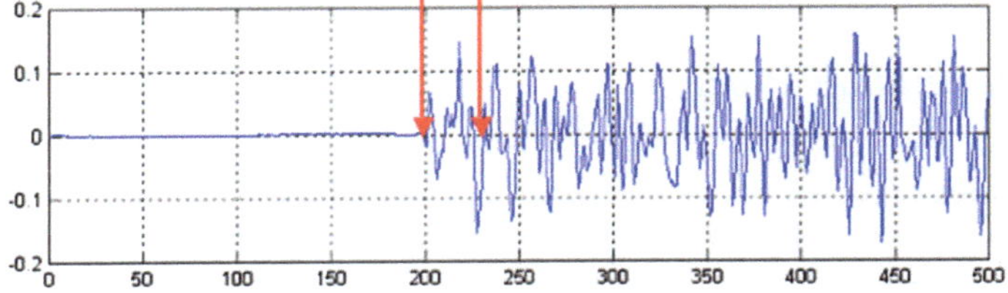

Fig. 2 Interaural time difference (ITD) [16]

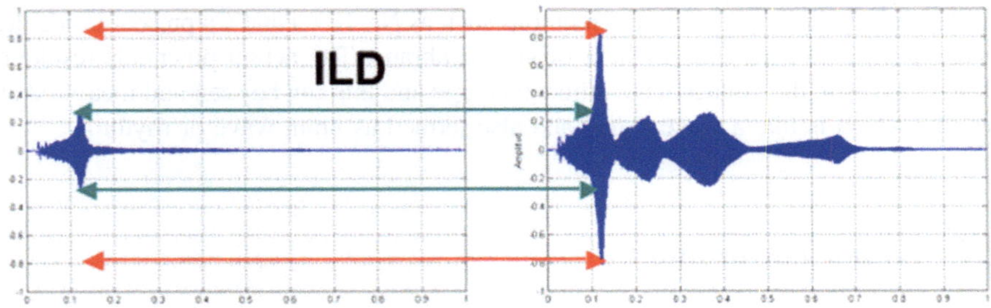

Fig. 3 Interaural level difference (ILD) [16]

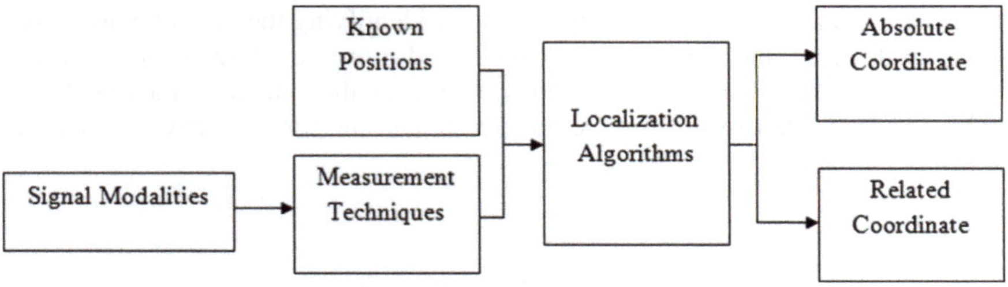

Fig. 4 Process of a localization algorithm

head and ear shape act as direction-dependent frequency filters which is known as head-related transfer function (HRTF). It differs between individuals (Fig. 2).

Other one is Binaural Cues that are extracted using two ears (microphones). They are of two types; first one is Interaural Time Differences (ITDs) which is due to the difference between the arrival time of sound from the source to each ear i.e. if a sound source is nearer to the left side ear, then the arrival time of sound reaching to the left ear is early in comparison to the right ear. In Fig. 3 x-axis represents the amplitude and y-axis represents the time. The sound source is 100 ms white noise from left.

Second one is Interaural Level Differences (ILDs) which is due to the level of sound loudness from the source to each ear. i.e. if a sound source is nearer to the left side ear, then the sound reaching to the left ear is loud in comparison to the right ear. This loudness differs between ears because of head's acoustic "shadow". Here sound source is a sweep from right.

1.4 Localization Process

In localization process basically there are two steps. First it senses some parameters which will be used and method should be initiated once network formation. Then it follows a two stage process i.e. measurement and localization algorithm. In measurement step we measure distance/angle between nodes.

Then an algorithm is applied to measure approximate location of each node [2] (Fig. 4).

2 Related Work

In the field of sound localization, researchers do lot of work and implemented different types of model such as Audio-Fingerprinting-based device Localization Method [3], Periodicity Analyzing Network Model [4], BiSoLaNN (Binaural Sound Source laterization Neural Network) [5], SNN (Spiking Neural Network) [6] etc.

Hon et al. [3] proposes an algorithm for audio fingerprinting. In this the distance between any two devices can be calculated using the extreme time differences of arrival of sounds. As sounds are produced from more than one source, a noisy environment is created. In their proposed method of localization the sound signals is first inputted for Audio fingerprint extraction which is then used for fingerprint matching and TDOA estimation through which the distance between the pair of devices is estimated.

Voutsas et al. [4, 5] presented a neural model that similar to human brain termed as periodicity analyzing network (PAN) for sound processing. In this model their present many complex model of neurons that are used in various audio signal processing applications and also used to understand the dynamics of neuron and neural network. They assume that a sound signal comes on the ear where the sound source is located is represented as $asin2\pi ft$, whereas this same sound signal is represented by $asin2\pi f(t + \Delta t)$ at the other ear, Here this represents extra time that is required by the sound signal to reach the other ear and this time difference is used in the process of localization of sound source.

Liu et al. [6] proposes a Spiking Neural Network (SNN) for binaural sound source localization that uses Interaural Time Difference and Level Difference both cues. They described that ITD cue is extracted by Medial Superior Olive (MSO) and ILD cue is extracted by Lateral Superior Olive (LSO). These both cues are then combined in Inferior Colliculus (IC) to achieve a sharp azimuthal localization of a sound source over a wide range of frequency.

Schaik et al. [7] studies the influence and the relationship for the synthesis and integration of five different acoustical cues i.e. interaural time difference (ITD), interaural level difference (ILD), band-edge spectral contrast, interaural and monaural spectrum to produce a coherent and robust percept of spatial location.

Raspaud et al. [8] provides a method for binaural source localization that is based on two cues i.e. interaural time differences (ITDs) and interaural level differences (ILDs). In this estimation of the azimuth of sources is computed using a two channel time-frequency representation that combined these given two cues. It provides a parametric model for ITD and ILD both, and validates the results of this approach by conducting several experiments on it.

Zoran et al. [9] studied the localization of electric sources within the neurons by using tetrodes as sensors which acquire the different action potential's measurement of neurons. Edmonds B. A. and Krumbholz K. [10] investigated that whether or not, at higher process levels, ITDs and ILDs stay independent or are integrated into a standard code of sound laterality.

Willert et al. [11] studies a sound localization system that is implemented technically and is inspired by biological sound localization functionality. This helps in estimating the azimuthal location of the sound source in the frontal half-plane. Valin et al. [12] proposes robust sound source localization and tracking method using an array of eight microphones. As artificial audition on mobile robots is a new area for researchers to explore, they do the most of the work in localization of sound sources and which is done using two microphones only.

Ben Reuven and Yoram [13] achieves azimuth prediction accuracy that exceeds extrema locating methods. Julie et al. [14] used only the interaural intensity (level) difference cue that is calculated by the lateral superior olive (LSO) nuclei of the brain for sound localization. In this a spiking neural network (SNN) is designed that is based on the architecture of LSO and uses the LIF excitatory and inhibitory spiking neurons.

3 Preliminaries

As in sound localization we detect the origin of sound generally into three distinct types of coordinate systems. This paper focuses on one coordinate system that is known as Azimuth. Azimuth is the horizontal coordinate in which the location of sound source is detected. To find out this Azimuth the ear uses the interaural differences which are defined as the disparities between the sounds reaching each ear. There are two disparities that help ears to determine the azimuth, ITD and ILD. ITD is the difference in amount of time taken by sound to reach each ear and ILD is defined as the difference between the levels of loudness of sound that reach to each ear.

3.1 Coincidence Detectors

The coincidence detectors are a type of neurons that takes several inputs within an assigned time interval from other neurons and produce single output. Coincidence detectors encode the input information that comes in the form of signals at the same time but are spatially distributed. Thus it reduces the temporal variations, minimizing spontaneous activity and forming associations between different neural inputs.

The functioning of these detectors termed as Coincidence detection depends on separate inputs from neurons that come together on a common target. For example if there are three excitatory neurons X, Y and Z that provides input for a neuron F, then neuron F may produce an action potential only when all the inputs on it are comes simultaneously and they move forward the membrane potential of that neuron above its threshold value. In case if each excitatory input neuron's potential is subthreshold for the neuron F, then F will not produce action potential and thus not fire unless all inputs comes together on F.

3.2 Kalman Filters

Kalman filter is a tool which is used to predict the next state of a system whose state is continuously changed and the reason of those changes is uncertain. These changes may be due to known influences and unknown external influences as well.

A state of a system is defined by its various parameters. These parameters are distributed randomly and Gaussian distributed as well, which means that each parameter has a mean value μ and a variance σ2. Here mean value and variance value represents the center value and uncertainty of random variable respectively that is randomly distributed. In case of navigation the state of an object can be defined using two parameters, one is its location (l) or position and other one is its moving velocity (v). It can be represented as

$$\vec{s} = \begin{bmatrix} l \\ v \end{bmatrix} \tag{1}$$

These variables of a state are related to each other and the relation or dependency between these variables is defined by a covariance matrix $\sum$. Since the exact state of an object is lying under a range. So we model the predicted state as a Gaussian blob using two parameters at time t. These parameters are best estimate state $\hat{s}_t$ of the object and its covariance matrix $\sum_t$.

$$\hat{s}_t = \begin{bmatrix} \text{location} \\ \text{velocity} \end{bmatrix} \tag{2}$$

$$\textstyle\sum_t = \begin{bmatrix} \sum_{ll} & \sum_{lv} \\ \sum_{vl} & \sum_{vv} \end{bmatrix} \tag{3}$$

There are two steps in kalman Filter. These are:

3.2.1 Prediction Step

In this step on the basis of object's current state (at time t−1), its next state is predicted (at time t).

In case of navigation the estimate of object's next state which is assumed best is predicted as

$$l_t = l_{t-1} + \Delta t v_{t-1} \tag{4}$$

$$v_t = v_{t-1} \tag{5}$$

In form of matrix it can be written as

$$\begin{aligned} \hat{S}_t &= \begin{bmatrix} 1 & \Delta t \\ 0 & 1 \end{bmatrix} \hat{s}_{t-1} \\ &= P_t \hat{s}_{t-1} \end{aligned} \tag{6}$$

Here P_t is known as prediction matrix. Similarly the covariance matrix $\sum_t$ is updated as $\sum_t = P_t \sum_t P_t^T$ (According to covariance matrix property). As their present some known and unknown external influences and uncertainties in the environment that cause to change the state of the system. So these entities must also be added in predicting the next state of the object. In navigation the known influence might be some acceleration a, that may be applied manually on the object. This known parameter change the state as

$$l_t = l_{t-1} + \Delta t v_{t-1} + \frac{1}{2} a \Delta t^2 \tag{7}$$

$$v_t = v_{t-1} + a \Delta t \tag{8}$$

In form of matrix it can be written as

$$\hat{s}_t = P_t \hat{s}_{t-1} + \begin{bmatrix} \frac{\Delta t^2}{2} \\ \Delta t \end{bmatrix} a \tag{9}$$

$$P_t \hat{s}_{t-1} + C_t \overrightarrow{u_t} \tag{10}$$

Here C_t is known as control matrix and $\overrightarrow{u_t}$ is known as control vector. The unknown parameters lead the estimated states move to a certain stage, so this uncertainty must be added in prediction in terms of covariance Ut as

$$\sum_t = P_t \sum_t P_t^T + U_t \tag{11}$$

3.2.2 Updation Step

There are multiple sensors that record the states of an object. These sensors are represented by a matrix Rt.

In order to find the expected state, different readings of sensors are distributed as,

$$\vec{\mu}_{expected} = R_t \hat{s}_t \quad (12)$$

$$\sum_{expected} = R_t \sum_t R_t^T \quad (13)$$

In order to get the best estimation the two Gaussian blobs having different mean and covariance values is multiplied, by which we get a new updated mean and covariance values. A 1 dimension Gaussian bell curve is given by the equation

$$N(s, \mu, \sigma) = \frac{1}{\sigma\sqrt{2\pi}} e^{-\frac{(s-\mu)^2}{2\sigma^2}} \quad (14)$$

Multiplication of two Gaussian blobs is given as

$$N(s, \mu_0, \sigma_0).N(s, \mu_1, \sigma_1) = N(s, \mu', \sigma') \quad (15)$$

From above two equations we get

$$\mu' = \mu_0 + \frac{\sigma_0^2(\mu_1 - \mu_0)}{\sigma_0^2 + \sigma_1^2} \quad (16)$$

$$\sigma'^2 = \sigma_0^2 - \frac{\sigma_0^4}{\sigma_0^2 + \sigma_1^2} \quad (17)$$

If

$$k = \frac{\sigma_0^2}{\sigma_0^2 + \sigma_1^2} \quad (18)$$

Then

$$\mu' = \mu_0 + k(\mu_1 - \mu_0) \quad (19)$$

$$\sigma'^2 = \sigma_0^2 - k\sigma_0^2 \quad (20)$$

In form of matrix it can be represented the mean and variance can be represented as

$$K = \sum_0 \left(\sum_0 + \sum_1\right)^{-1} \quad (21)$$

$$\vec{\mu}' = \vec{\mu}_0 + K(\vec{\mu}_1 - \vec{\mu}_0) \quad (22)$$

$$\sum' = \sum_0 - K\sum_0 \quad (23)$$

Here K is known Kalman Gain. Suppose the covariance and mean measured through the sensors is given by Vt and $\vec{m}$. Therefore the observed state of object is given as ($\mu 1$, $\sum 1$) = ($\vec{m}$, Vt), whereas the predicted state is given as

$$\left(\mu_0, \sum_0\right) = \left(R_t\hat{s}_t, R_t \sum_t R_t^T\right) \quad (24)$$

Put these in Eqs. 22 and 23, we get

$$R_t\hat{s}_t' = R_t\hat{s}_t + K(\vec{m} - R_t\hat{s}_t) \quad (25)$$

$$R_t \sum_t' R_t^T = R_t \sum_t R_t^T - K\ R_t \sum_t R_t^T \quad (26)$$

and also the Kalman Gain is now given as

$$K = R_t \sum_t R_t^T \left(R_t \sum_t R_t^T + V_t\right)^{-1} \quad (27)$$

from Eqs. 25, 26 and 27 we get

$$\hat{s}_t' = \hat{s}_t + K(\vec{m} - R_t\hat{s}_t) \quad (28)$$

$$\sum_t' = \sum_t - KR_t \sum_t \quad (29)$$

$$K' = \sum_t R_t^T \left(R_t \sum_t R_t^T + V_t\right)^{-1} \quad (30)$$

These are the final equations for the updation in state of an object to get the best estimation of the next state that is predicted.

4 Proposed System

The sound localization is an important functionality of human beings and animals. This gives an existence proof of the capabilities of binaural systems which helps them in perceiving the environment, react and take action on the basis of perceiving sound. This functionality is in-built in mammals which are performed by the neurons. Research on this subject provides that these principles of sound localization may be very useful to integrate in a machine implementation. If we are able to implement it technically then it can be useful in making smart robots and also help in treating the hearing loss problem in patients efficiently.

Here the proposed system used the Kalman filter that's trained to reduce the error using the multilayer preceptron neural network. The whole process will be understood by the proposed steps

- Detect N Sound Source and M Microphone (receivers) with (x_s, y_s) and (x_m, y_m).
- Calculate the interaural time difference (ITD) as

$$\mathrm{ITD} = \frac{\sqrt{(x_m - x_s)^2 + (y_m - y_s)^2}}{\text{sound speed}} \tag{31}$$

- Compute the delay for each Microphone where delay for *i*th microphone can be calculated as

$$\mathrm{delay}_i = \mathrm{ITD}/2 * \sin\left(\frac{i-1}{M-1} * \pi - \frac{\pi}{2}\right) \text{ [16]} \tag{32}$$

- Estimate the position (x_p, y_p) of microphone by using delay.
- Input (x_p, y_p) to the Kalman filter to get position say (x_k, y_k)
- Calculate Error

$$e = \sqrt{(x_k - x_m)^2 + (y_k - y_m)^2} \tag{33}$$

- Input e to MLP neural network.
- Output of neural network updates the Kalman.
- The process goes on until error gets minimized.

A priori estimation is the previously estimated value of the position by using the delay value as shown in the algorithm. This input is given to the Kalman filter. The procedure of Kalman filter will be explained as follow.

A linear dynamical system can be measured by:

$$S_{t+1} = Q_{t+1} S_t + w_t \tag{34}$$

where $Q_{t+1,t}$ is the transition matrix taking the state S_t from time t to time t + 1. W_t is the process noise which is taken as an additive, white, and Gaussian, with mean equal to zero and with covariance matrix. But now, we've got to estimate unknown state St. However we've got a previous estimate for the state say s_t. Now the new value of state can be calculated as

$$S_t = h_t s_t + H_t Y_t \tag{35}$$

where Y_t is the updated value at t, H_t and h_t is the matrix factors. The error in the measurement can be calculated as e = $S_t - s_t$. It can also be given as the

$$h_t s_t + H_t Y_t - s_t \tag{36}$$

This system estimates the value and calculates the error as shown above. This error is given for the updating and original position is given as the input. It trains the Kalman filter that updates the value and calculates a new estimation. The process is repeated 100 times to get the minimum error. The process is also shown using flowchart in Fig. 5.

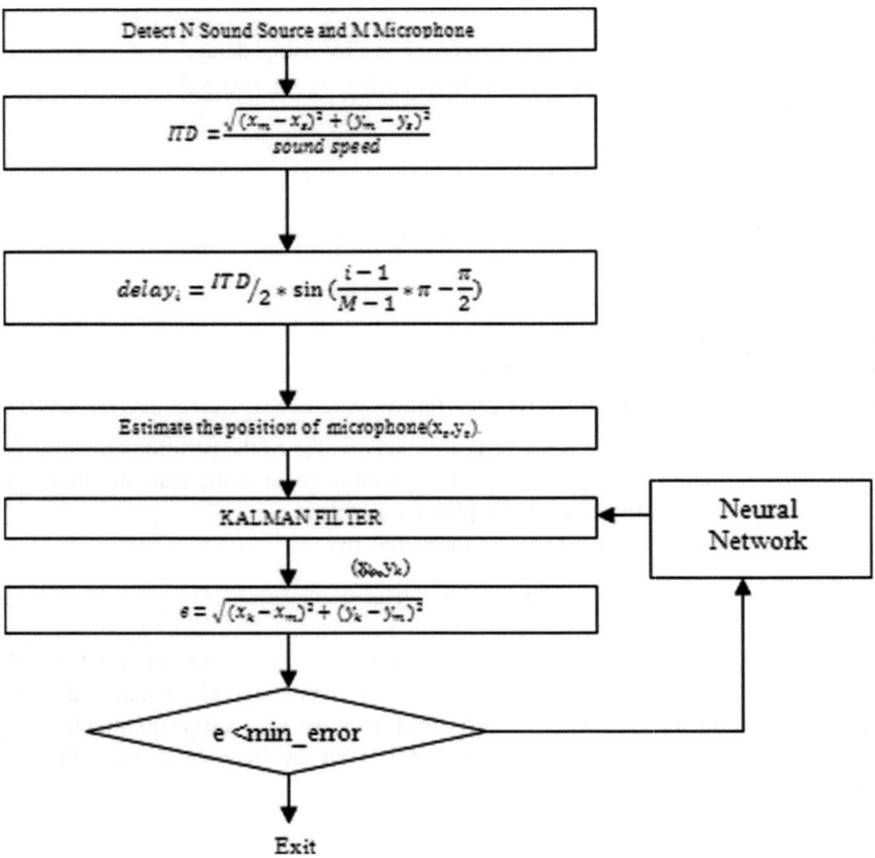

Fig. 5 Proposed flowchart

5 Results and Discussion

The implementation is done using the MATLAB. In this the sound that is given as an input is generated with the MATLAB that differs in left and right channel sound. The left channel and right channel sound is treated as the sound that comes on left side and right side ear respectively. In this the fast Fourier transform is also used. The training is provided to the Kalman filter using the MLP and this is applied to get the position of the microphone in 3D.

Currently the researchers that work on sound localization in 3D space uses the head related transfer functions (HRTFs) to localize the sound source [15] which is a Monaural cue. Here in order to making a system that behaves like a human auditory system and uses binaural cues instead of monaural cues we come up with the use of kalman filter and neural network. Here inspired by the human auditory system the delay in transmitting the signals by axon in different neurons is given by a delay function. In this the neurons are represented by coincidence detectors. The response of each neuron or coincidence detector is tested by the trained neural network for number of points or samples in a sound signal and the difference of this response with the actual response is treated as the error.

To show the accuracy of this proposed algorithm it is compared with the already existing sound localization in 2D space which used the same binaural cues as our algorithm in 3D uses i.e. ITD.

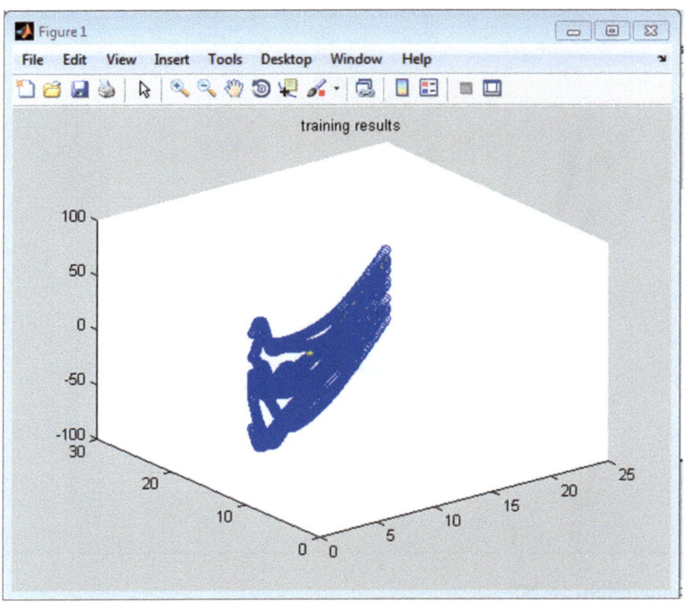

Fig. 6 Training results in 3D

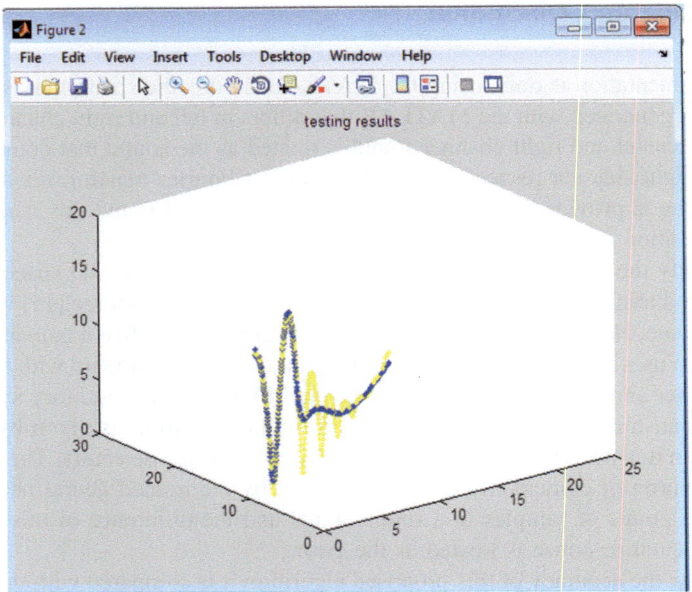

Fig. 7 Testing results in 3D

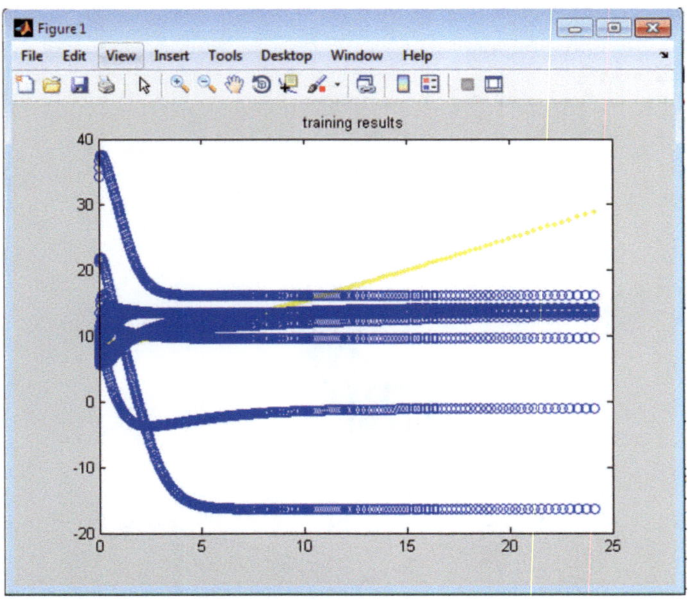

Fig. 8 Training results in 2D

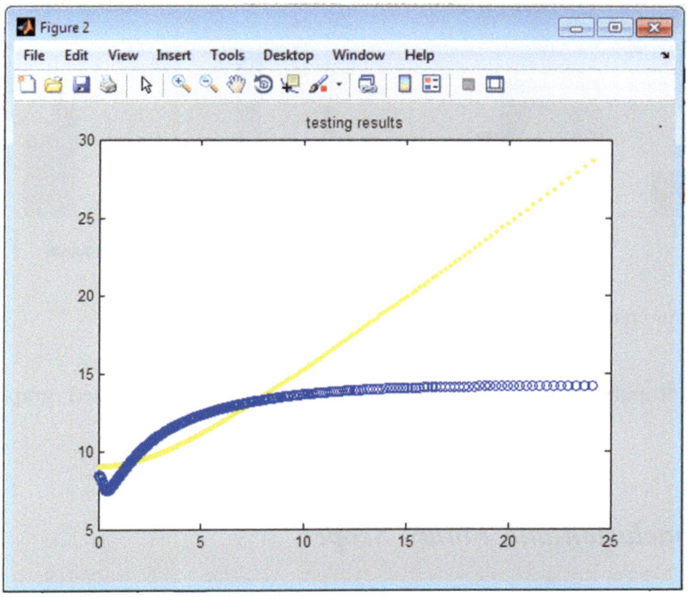

Fig. 9 Testing results in 2D

Table 1 Errors comparision in 2D (Existing) and 3D (Proposed)

Sound	Sum of square error		Average error	
	Existing	Proposed	Existing	Proposed
1	405.6067	113.6107	1.3796	0.3864
2	91.0284	85.5090	0.3096	0.2908
3	318.9075	200.9659	1.0847	0.6836
4	162.5236	87.4822	0.5528	0.2976

The comparison is of error obtained (difference between actual and desired values) between the 2D and 3D. Figures 6, 7 and Fig. 8, 9 shows the training and the testing results of the microphone location in 3D and 2D respectively. The Yellow dotes denotes the original position and the blue circle represents the estimated position. The calculated error is shown in the Table 1. For the purpose of comparing 2D and 3D we train and test the algorithms on same sounds.

Here in table we listed the results for 4 different frequency sounds i.e. 500, 900, 1000, 1500 Hz which is generated using a sin function. For making left and right channel sound different we use a constant delta value. To make a setup in which the sound source is on right side then the right channel sound signal is generated as sin (2 * pi * f * (t)) and left channel sound signal is generated as sin (2 * pi * f * (t + delta)).

The comparisons shows that the error is reduced i.e. the position is identified more accurately.

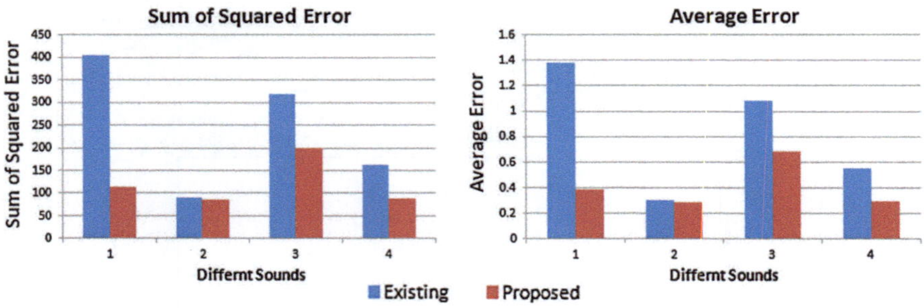

Fig. 10 Error comparisons

This comparison can be seen more clear when it is plotted graphically as (Fig. 10).

5.1 *Conclusion and Future Scope*

The advancement of the technology in intelligent robots needs the detection of sound source accurately. The work presented in this paper focuses on examining accurate position of the sound sources by using the Kalman filter and neural network. This work can be used to detect the location of the sound sources in three dimensions. This work gives the more accurate result even if two sources are placed at same position at different height. The proposed system used the Kalman filter that is trained to minimize the error using the Multilayer preceptor neural network. The implementation is done using the MATLAB and the sound signals are also generated through MATLAB. The results clearly signify the reduction in the error.

Present advanced robots are a type of intelligent machine those are supposed to behave like a human and perform complex actions. The implementation of proposed algorithm to make smart robots is the part of our future work. If robots are able to localize the sounds accurately then they will capable to understand the environment by sounds too and will be able to make the intelligent decisions as per the need. This algorithm is currently tested on sounds that are generated by simulation using MATLAB.

References

1. http://sorensen.myweb.usf.edu/Projects/Dev.%20of%20Tech%20based%20Inst/Webpages/Neuron%20Structure.html.
2. Hongyangchen et. al. "Accurate and efficient node localization in wireless sensor networks", February 2011.

3. Hon T., Wang L., Reiss J., Cavallaro A., "Audio Fingerprinting for Multi-Device Self-Localization" IEEE/ACM Transactions on Audio, Speech, and language processing, vol. 23, No. 10, October 2015.
4. Voutsas k., Langer G., Adamy J. and Ochse M, "A Brain-Like Neural Network for Periodicity Analysis" IEEE Transactions on Systems, Man, and Cybernetics, Vol. 35, No. 1, Feb 2005.
5. Voutsas K. and Adamy J., "A Biologically Inspired Spiking Neural Network for Sound Source Lateralization", IEEE transactions on Neural Networks, Vol. 18, No. 6, November 2007.
6. Liu J., Gonzalez D., Rees A., Erwin H., Wermter S., "A biologically inspired spiking neural network model of the auditory midbrain for sound source localization", Elsevier transaction on Neurocomputing 74(2010) 19–139.
7. Schaik A. V., Jin C., Carlile S., "Human Localisation of Band-pass Filtered Noise", International Journal of Neural Systems, Vol. 9. No. 5 (October, 1999) 441–446.
8. Raspaud M., Viste H., and Evangelista G., "Binaural Source Localization by Join Estimation of ILD and ITD", IEEE Transactions on Audio, Speech, and Language Processing, Vol. 18, No. 1, January 2010.
9. N.Zoran, D. Hieu, W.L. Chang, "An Efficient Algorithm for Current Source Localization with Tetrodes", 29th Annual International Conference of the IEEE EMBS France, August 2007.
10. Edmonds B. A. and Krumbholz K., "Are Interaural Time and Level Differences Represented by Independent or Integrated Codes in the Human AuditoryCortex?",Journal of the association for research in Otolaryngology (JARO) 15: 103–114 (2014), November 2014.
11. Willert V., Eggert J., Adamy J., Stahl R., and Körner R., "A Probabilistic Model for Binaural Sound Localization", IEEE Transactions on Systems, Man, and Cybernetics- Part B: Cybernetics, Vol. 36, No. 5, October 2006.
12. Valin J. M., Michaud F., Rouat J., "Robust localization and tracking of simultaneous moving sound sources using beamforming and particle filtering", Elsevier Robotics and Autonomous Systems 55 (2007) 216–228, October 2006.
13. Reuven Ehud Ben, Singer Yoram, "Discriminative Binaural Sound Localization"
14. Julie A. Wall, Liam J., Liam P. and Thomas M., "Spiking Neural Network Model of Sound Localization Using the Interaural Intensity Difference", IEEE Transactions on Neural Networks and Learning Systems, Vol. 23, No. 4, April 2012.
15. Fakheredine Keyrouz "Advanced Binaural Sound Localization in 3-D for Humanoid Robots", IEEE Transactions on Instrumentation and Measurement (Volume: 63, Issue: 9, Sept. 2014).
16. Roads, Curtis. The Computer Music Tutorial. Cambridge, MA: MIT, 2007. Print.
17. Calmes, L. (2009). Biologically Inspired Binaural Sound Source Localization and Tracking for Mobile Robots. PhD thesis, RWTH Aachen University (http://www.laurentcalmes.lu/index.html).

Analysis of Stock Prices and Its Associated Volume of Social Network Activity

Krishna Kinnal, K.V. Sanjeev and K. Chandrasekaran

Abstract Stock market prediction has been a convenient testing ground and a highly cited example for applying machine learning techniques to real-life scenarios. However, most of these problems using twitter feeds to analyze stock market prices make use of techniques such as sentimental analysis, mood scoring, financial behavioral analysis, and other such similar methods. In this paper, we propose to discover a correlation between the stock market prices and their associated twitter activity. It is always observed that whenever there are spikes in the stock market prices, there is a preceding twitter activity indicating the imminent spike in the aforementioned stock price. Our objective is to discover this existence of a correlation between the volumes of tweets observed when a market indices' stock price spikes and the amount by which the stock price changes, with the help of machine learning techniques. If this correlation does exist, then an attempt is made to figure out a mathematical relationship between the two factors.

Keywords Twitter · Social network analysis (SNA) · Sentimental analysis

1 Introduction

Social Network Analysis (SNA) is a tool used to investigate and analyze social structures and their associated networks through the use of graph theories. Such a network, analyzed using SNA methods, consist of a node, and a number of ties, or edges between these nodes to help connect these nodes. Some examples of social

K. Kinnal (✉) · K.V. Sanjeev · K. Chandrasekaran
Department of Computer Science and Engineering, National Institute of Technology Karnataka, Surathkal, Mangalore, India
e-mail: krishnakinnal@gmail.com

K.V. Sanjeev
e-mail: sanjeev.vadiraj@gmail.com

K. Chandrasekaran
e-mail: kchnitk@ieee.org

G.M. Perez et al. (eds.), *Networking Communication and Data Knowledge Engineering*, Lecture Notes on Data Engineering and Communications Technologies 3, https://doi.org/10.1007/978-981-10-4585-1_20

networks include friendship and acquaintance networks, disease transmission, kinship, and social media networks. A *sociogram* is used to visualize a social network in terms of these nodes and ties. SNA has emerged as a key technique in fields such as anthropology, communication studies, political science, sociolinguistics, and other such related fields. In the case of predicting the stock market, as investigated in this paper, social media networks, such as Twitter, are employed for analysis. Twitter is an online social networking platform for micro blogging, wherein it allows users to send and view short 140 character messages known as 'tweets'. Users with accounts can create and view tweets, but unregistered users may only view them. Today, tweets pertaining to the stock market are aplenty, and figuring out a technique to predict a sudden change, i.e., a spike in stock market prices by avoiding the trodden paths of sentimental analysis or behavioral analysis was a challenge, but the solution proposed in this paper aims to be a revolutionary and extraordinarily simple one, using only the volume of tweets on a trending topic as a parameter to judge the outcome of a stock price. The solution proposed is based by behavioral psychology. It is an empirical certainty that reaction to extreme news, whether good or bad, always carries more emotional weight, and hence, more tweets, than when compared to the reaction to more subtle news, which, on the contrary, is associated with a much more subdued social media activity, and hence, fewer tweets.

2 The Foundations of Social Network Analysis

A fundamental and comprehensive survey of network terminologies was studied [1] and its contents have been perused and explained in the sections below for the purpose of introduction.

2.1 Terminology

A network is a collection of nodes connected by a set of ties, or links. The nodes may be individuals, organisations, concepts, etc. In the case of SNA the nodes are individuals. If all the nodes are similar, the network is homogeneous, else, it is heterogeneous. In a social network, ties connect pairs of nodes. These ties may be directed or undirected, and in another type of classification, may be weighted, or dichotomous. It is important to note that as a matter of fact, all ties are weighted or have values. Even dichotomous relations have binary values (either the tie exists and is assigned a value of 1 or it doesn't and is assigned a value of 0). Relational data is the collected data on ties between a set of nodes. This data may be visualised in a graphic, or a matrix form. The following Table 1 summarizes the above.

Table 1 Social network terminologies and their corresponding definitions

Social network terminology	Definition
Node	The basic element of a network
Tie/Edge	A set of two nodes. Ties can be dichotomous (unweighted) or weighted/valued, directed or undirected
Directed tie	An ordered set of two nodes, i.e., with an initial/source and a terminal/destination node
Network	A set of nodes connected by a set of ties
Valued network	A network whose ties/edges are associated with a measure of magnitude or strength
Network size	The total number of nodes in a network
Relational data	The set of links in a network

2.2 *Structure of Social Networks*

There are two directions of thought when dealing with social network structures. In the first, only structure is focused on to derive behavior, and in the second, the focus is on both the diverse nature of the many different actors within the social groups and the structure to derive their respective behavior. However, these are only static descriptions of a social network, better known as 'frozen' networks. This, therefore, doesn't come to close considering the breaking and forming of new bonds between nodes, and the change in behavioral patterns of the nodes, etc. There is also no clear definition as to what defines a bond between nodes to be classified as 'weak' or 'strong'. Hence, it is only a rudimentary analytical tool, and not a useful, theoretical one.

2.3 *Network Evolution*

There are two types of dynamics that come into play inside a network—the dynamics *of* the network and dynamics *on* the network. The first type refers to the making and breaking of ties between the nodes. For the second type of dynamics, the nodes perform some actions, and these actions have specific outcomes that may or may not be influenced by their neighbors and by the structure of the network, Both of these aforementioned dynamics take place simultaneously within the network.

2.4 Network Density

Density is defined as the number of links in the network, which is denoted as a ratio of the maximum possible number of links in the network.

Networks also have the property of centrality. There are two types of centrality —local centrality and global centrality. Local centrality only considers the direct links between nodes, and global centrality is expressed in terms of the logical distances between the nodes. A geodesic is defined as the shortest path between any two nodes on the given social network.

Betweenness is a further manifestation of centrality, which measures the extent to which a particular node lies between the other nodes in a network.

Centralization is a combination of the aforementioned three factors of centrality, i.e., local, global and betweenness. The value of centralization of a network varies between zero and one. For example, a value of one is assigned to a star network, whereas a fully connected network would be assigned a value zero.

2.5 Network Performance

Network performance is measured using four important factors—robustness, efficiency, effectiveness, and diversity. The robustness of a network can be evaluated by analyzing the degree of fragmentation a network undergoes as its nodes are removed one by one. This measures the inherent inclination of nodes toward forming clusters or groups.

A network is said to be efficient if a few nodes can instantly reach out to a larger number of nodes with the help of a relatively lesser number of links. Hence, the lesser the number of non-redundant nodes, the more efficient is the network. Effectiveness of a network targets a local cluster of nodes that can be reached through a series of non-redundant nodes.

Also, it is critical for the network to consist of diverse nodes, i.e., the actors within a social network should be socially diverse.

3 Literature Survey

In the field of twitter analysis pertaining to stock market analysis, it is found that most of the literature available is in regards to sentimental analysis [2, 3, 4, 5], mood scoring [6, 7, 8], or financial behavior analysis. Most of the research on twitter analysis has mostly restricted itself to the aforementioned three domains, and there is a lot more scope.

3.1 *Correlation Between Public Sentiment and Market Sentiment*

In this method, sentiment analysis and machine learning principles to find some correlation between 'market sentiment' and 'public sentiment' [8]. Using twitter data, a prognosis of the public mood is calculated, and using the market index's previous day's price, the movements of the stock price can be calculated for the next day. This is implemented using a cross-validation method known as Self Organizing Fuzzy Neural Networks (SOFNN). These methods are used to collect stock price data over 6 months. The raw stock price values are fed into the pre-processor to obtain the processed values. Simultaneously, the tweets are fed into a sentimental analysis algorithm that calculates mood values for the four mood classes prevalent in each day. These moods and the aforementioned price values are then fed into a model framework which uses the SOFNN to learn a model to predict the next day's stock prices. The sentiment analysis portion of the project included the following methods

- Generating a list of words based on the profile of mood states (POMS) questionnaire
- Filtering of tweets, considering only those tweets that are more likely to express a certain emotion, or mood
- Computing a daily score for every POMS word found in each day by calculating a ratio the number of times the POMS word matches tweets in a day to the number of total matches of all words, and,
- Mapping the score of each word to six standard POMS states which are then mapped to one of four mood states.

A Granger causality analysis, which is based on linear regression, is performed over the tweets, measuring the ability one signal to generate predictive information over the other. Then an algorithm is run to create the SOFNN, from which neurons are generated, as and when new sample cases arrive from the live twitter feed. After the predicted stock value is found, it is checked if the price value is n standard deviations below the mean. If so, a buy decision is made. Else, if the predicted stock price value is m standard deviations more than the adjusted value during buying time, a selling decision is made. Hence, a causative relation between public mood, derived by live twitter feeds using simple natural language processing techniques, and the market mood is discovered, and a predictive binary buy or sell decision is made.

3.2 *Sentiment Analysis at the User Level*

It is seen that information about relationships between the nodes in a social network may be used to substantially improve user-level sentiment analysis [4]. All users are in some way or another connected, and all such connected users are likely to have similar opinions. Therefore, extracting a single node's viewpoints is an aggregation of all other neighboring nodes interactions. In the case of twitter, clusters within the social network can be determined either by the people that the user follows, or by the people the user references by using the '@' in his/her tweets. Hence a user relationship is more easily obtainable. A model is obtained based on this follower/followee network, indicating a cluster of people holding similar opinions as that of the user. A semi-supervised approach is followed, since the collection of all user-related data can consume a lot of time and prone to errors. Also, the sentiment is derived more from the user, and less from the tweets published by the user. An empirical confirmation is done to make sure that the cluster of users in fact shares the same opinion. Then using graphical models, and taking into effect social network information can substantially improve the user-level sentiment data, when compared with classical methods of sentimental analysis. Then the sentiment, either positive, indicating the user supports the topic, or negative, indicating the user doesn't support the topic. After graphical analysis, it is seen that the probability that two connected nodes sharing the same opinion of some particular topic is much higher than it might seem. Also, even the converse is true. i.e., two users are connected to each other whenever it is observed that they share a similar opinion on any particular topic. Hence, it is imperative to consider aspects of human behavioral psychology to perform effective social network analysis.

3.3 Logistical Regression Modeling

One can also study the relationship between the social media output and games of a certain sport to predict the result of said game, using datasets provided by twitter feeds [9]. Tweets pertaining to specific teams or matches of a particular sport and use them with statistical data to build models that predict the results of future games. A set of tweets of over 2 years is collected and tokenized with appropriate library functions, and related game data from those 2 years is also collected. This data is a comprehensive set of statistics. Timestamps of the tweets may be used to classify a certain tweet to a certain week of the sport's season. By using the hash tags in the tweets, each team can be mapped to a tweet based on the team the tweet refers to. However, tweets with hash tags of more than one team are discarded, so that the view of a particular game from the perspective of either one of the two teams is considered, and not just a general comment on the game in whole. Hence, for a correct prediction, a tweet is considered only if it is mapped to a certain team. A logistic regression classifier model is employed on these tweets. It is observed that an aggregate of fan behavior on twitter captures significant information relating to future games. It was also observed that initial experiments on feature sets with high predictive accuracy early on in the season may not be accurate as the season progresses, and hence, the feature set must be constantly updated so that it stays consistent and relevant. A range of feature sets are taken into consideration while classifying the tweets into either a 'win' or a 'loss' for the team under consideration. Hence, the result of a team's future game can be approximately predicted using sentimental analysis.

3.4 Disadvantage of Predictive Models

An interesting aspect of social media is its potential to accurately predict the outcome of an election by making use of a generated opinion, or buzz, just by analyzing the state's twitter activity [10]. Data sets from twitter are streamed live from the site regarding to the election under consideration using the twitter streaming API. The method used is similar to the method explained in the previous section, i.e., a tweet regarding to any particular candidate is only considered, and tweets referring to multiple candidates are discarded. Positive, negative and neutral words are filtered from the tweets. Then a predicted vote share is calculated for each candidate. Neutral tweets are not considered in the computation, because they are ambivalent and don't express any emotion toward a candidate. After comparing the predicted results with the actual ones, the margin of error was substantial, and it was hence discovered that in some cases, public opinion measured using twitter feeds aren't very reliable for prediction of election results, and when extrapolated, prediction of results pertaining to any particular domain in general. Therefore, it is observed that more research work is needed in the field of sentimental analysis when applied to a political context, and in general, any context. An in depth understanding of the dynamics involved in a quasi-shallow twitter universe, wherein millions of opinions and projections of users are constantly being shared for the world to see every day, is imperative to a better understanding of the matter at hand.

4 Volume of Twitter Activity

As pointed out in the literature survey, there are some key takeaway points,

- In most cases, public opinion projected on twitter, obtained via a thorough process of tweet analyzing is in fact an effective method to create predictive models.
- Most of the research done, and currently being done, on twitter analysis use machine learning techniques such as sentimental analysis, financial behavioral analysis, assigning mood scores, etc., albeit there is a lot more scope for differentiating these techniques.
- It is important to study not just the behavior of a single node in a social network, but also the behavior of all the connections and ties to other neighboring nodes to derive a wholesome understanding of the aforementioned node's behavior.

- It is important to discard tweets that may refer to more than one entity under consideration. For example, in reference to the stock market, it is imperative that only those tweets are used that contain '#sensex' (Sensex is a market weighted stock market index of 30 well established companies listed on the Bombay Stock exchange, similar to the Dow Jones Industrial Average (DJIA)) are considered in the analysis for better results, and not those tweets that may refer to multiple market averages.

Taking into consideration the aforementioned factors, the idea this paper wishes to propose is that whenever there is a sudden increase in twitter activity, it is indicative of an imminent increase or decrease, i.e. a spike in the Sensex's value. Although it cannot be stated for sure, it can be speculated that market index values such as that of the Sensex tend to either rise or fall drastically whenever financial news relating to a few significant companies (also known as blue chip stocks, whose price fluctuations can significantly affect the value of the Sensex, whose price fluctuations can have significant impacts on the Sensex) breaks out on social networking sites, and especially on sites such as twitter. This leads to re-tweets and generation of even more reactionary tweets by investors, leading to an increase in twitter activity. Building on this idea, it may be concluded that said financial news on social media subsequently leads to a spike in the stock market index value, i.e., a sudden rise, followed by a sudden fall in the value of the market index. Also, since there is no mention of this approach in literature, it is an exciting avenue to explore, making this a novel approach to stock market analysis using twitter, making it a novel approach to stock market analysis.

4.1 Procedure

After being provided with authorization keys from Twitter and using the *tweepy* library, a python script is run to extract the number of tweets per minute with the hash tag '#sensex' from Twitter and stored in a list. Simultaneously another python script is executed to extract Sensex point's information on a minute-wise basis from the *Yahoo Finance* website and is also stored into a list. Thus, there are now $n + 1$ Sensex point values for every n minute interval that the volume of tweets is recorded. Then plots of both the datasets contained in the lists are generated and the curves of the graphs are compared with each other to manually find out a correlation. This procedure was done for a one-month period from April 1st 2016 to May 1st 2016. Hence the dataset used in this experiment consists of the values of the Sensex index and the volume of tweets over the one-month period. As an example, the plot for the values on April 16th, 2016 is shown in Fig. 1. It is seen that whenever the Sensex index value spikes, there was a preceding spike in the volume of tweets. However, when the index values were steady and hovering around a particular value, the twitter activity recorded was placid, indicating little or almost no significant activity.

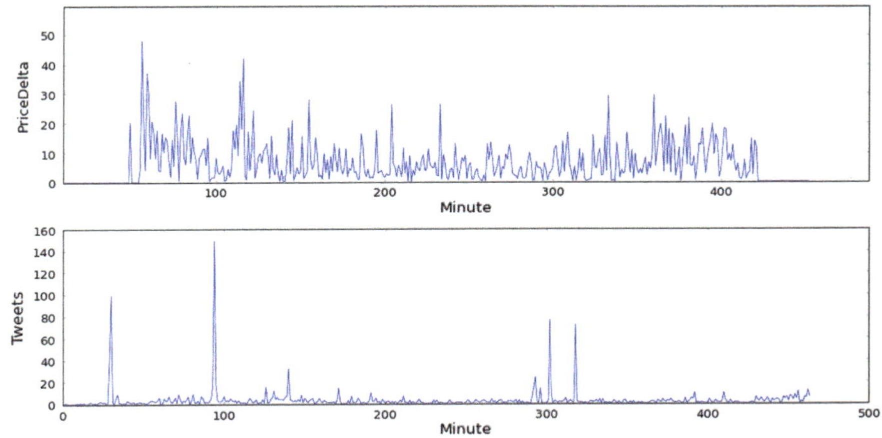

Fig. 1 Plot of the sample data collected on April 16th, 2016

5 Analysis

The python scripts were run over a period of one month and corresponding datasets for each day were collected ad collated. On all the days, the hypothesis tested true, and spikes in twitter volumes were indicative of imminent spikes in the value of the Sensex. The upper graph is a plot of change in Sensex value on the Y-Axis (the difference between the current value of the Sensex and its value at the start of the trading session) versus the minute interval, on the X-Axis, it was recorded in. The lower graph is a plot of the number of tweets, on the Y-Axis, versus the minute interval, on the X-Axis, it was recorded in. As an example, it is seen that just before the 100th minute, there is a sudden increase in the volume of twitter activity (shown in the lower graph), which then quickly dies down. Correspondingly, a few minutes after the 100th minute, the value of the Sensex records a similar change in its value (shown in the upper graph). As another example, at around the 300th minute, there are two spikes in the activity recorded on twitter (as shown in the lower graph) and there are corresponding spikes for the value of the Sensex a few minutes later (as shown in the upper graph. At other times, when there is no significant change in the value of the Sensex, the twitter activity remains relatively static.

6 Mathematical Analysis

In the example shown above for April 16th, 2016, and in most examples, it is seen that when the PriceDelta value is below Rs. 30, the preceding number of corresponding tweets with '#Sensex' was around thrice this number. As the value of

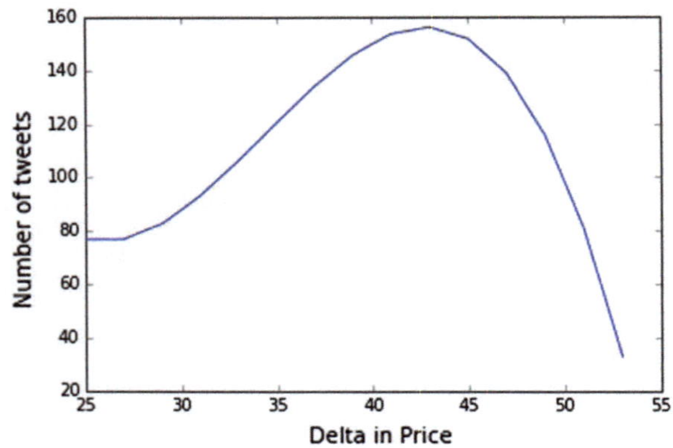

Fig. 2 Variation of the volume of tweets with respect to the change (*delta*) in price

PriceDelta increases to Rs. 40, the number of tweets quadruple, but then decrease back to double the number when the value of PriceDelta is Rs. 50. Hence, the value of Tweets as a function of PriceDelta increases at first, reaches a zenith, and then decreases. This can be seen from the graph below. Hence, there is a direct cause-effect relationship established between PriceDelta and Tweets. The explanation for this relationship can only be speculated, and has been stated in the previous section (Fig. 2).

7 Conclusion

The relevance of social media has never been greater. Social media platforms are increasingly becoming a quick, important and trustworthy source for daily news updates, and especially time-sensitive updates such as financial and business news that have direct implications on the stock market. It is important to unearth this correlation between the stock market and the volume of twitter activity for short-term investors and brokers who need to make investment decisions within minutes, and sometimes seconds. Taking a look at sudden increases in twitter activity will thus lead to optimism among investors, allowing them to safely invest in the stock market for immediate gains. This can have short-term as well as long-term financial gains, and can help increase investor awareness significantly. Thus, the existence of this correlation and its simple unearthing and discovery demonstrated in this research paper may have great implications in the world of investment and financial research.

References

1. Fabrice Coulon, PhD Candidate, "The use of Social Network Analysis in Innovation Research: A literature review," Division of Innovation - LTH Lund University, Sweden, January 17th, 2005
2. Ray Chen and Marius Lazer, "Sentiment Analysis of Twitter Feeds for the Prediction of Stock Market Movement."
3. Anshul Mittal and Arpit Goel, "Stock Prediction Using Twitter Sentiment Analysis," Stanford University
4. Chenhao Tan, Lillian Lee, Jie Tang, Long Jiang, Ming Zhou, Ping Li, "User-Level Sentiment Analysis Incorporating Social Networks,"
5. Serda Selin Ozturk and Kursad Ciftci, "A Sentiment Analysis of Twitter Content as a Predictor of Exchange Rate Movements," Istanbul Bilgi University, Review of Economic Analysis 6 (2014)
6. Michael Nofer and Oliver Hinz, "Using Twitter to Predict the Stock Market Where is the Mood Effect?", Springer Fachmedien Wiesbaden 2015
7. Ching-Yung Lin, Fellow IEEE, Lynn Wu, Zhen Wen, Senior Member IEEE, Hanghang Tong, Vicky Griffiths-Fisher, Lei Shi, Member IEEE, and David Lubensky, "Social Network Analysis in Enterprise."
8. Johan Bollen, Huina Mao, Xiao-Jun Zeng, "Twitter mood predicts the stock market.", School of Informatics and Computing, Indiana Unviersity Bloomington, United States, School of Computer Science, The University of Manchester, United Kingdom
9. Shiladitya Sinha, Chris Dyer, Kevin Gimpel, and Noah A. Smith, "Predicting the NFL Using Twitter," Carnegie Mellon University, Pittsburgh PA 15213, USA, Toyota Technological Institute at Chicago, Chicago IL 60637, USA
10. Daniel Gayo-Avello, Panagiotis T. Metaxas, Eni Mustafaraj, "Limits of Electoral Predictions Using Twitter," Departamento de Informatica Universidad de Oviedo, Department of Computer Science Wellesley College

On Solutions to Capacitated Vehicle Routing Problem Using an Enhanced Ant Colony Optimization Technique

Ashima Gupta and Sanjay Saini

Abstract This paper presents an enhanced ant colony optimization (ACO) algorithm for solving the capacitated vehicle routing problem (CVRP). CVRP is the most elementary version of VRP, but also a difficult combinatorial problem which contains both the TSP (routing) and BPP (packing) problems as special cases. In the CVRP a number of vehicles having uniform capacity starts and terminates at a common depot, services a set of customers with certain demands at minimum transit cost. In this paper, an enhanced version of ACO algorithm is implemented on Fisher and Christofides benchmark problems. Computational results compared with the performance of different algorithms and relaxations are presented. These results indicate that the proposed algorithm is a comparative approach to solve CVRP. The article is concluded by examining the possible future research directions in this field.

Keywords VRP · CVRP · Meta-heuristic · ACO · Optimal solution

1 Introduction

The Vehicle Routing Problem (VRP) is among some of the widely studied problems in the field of operation research and combinatorial optimization. It has several applications in important domains such as travel, transportation, logistics and distribution management [25] etc. Therefore, the importance of this field has motivated many researchers and mathematicians all over the world, to find all possible alternatives to solve the vehicle routing problem optimally.

A. Gupta (✉) · S. Saini
Department of Physics & Computer Science, Dayalbagh Educational Institute, Agra 282005, India
e-mail: ashimagupta17@gmail.com

S. Saini
e-mail: sanjay.s.saini@gmail.com

G.M. Perez et al. (eds.), *Networking Communication and Data Knowledge Engineering*, Lecture Notes on Data Engineering and Communications Technologies 3, https://doi.org/10.1007/978-981-10-4585-1_21

Reflecting the huge variety of conditions present in the system, VRP has dozens of variants that are introduced in the last few decades. There are variants that consider time windows (VRPTW) [10], pickups and deliveries (VRPPD), back-hauling (VRPB), multi depot (MDVRP), split delivery (SDVRP) [16] etc. In the vast landscape of VRP variants, CVRP occupies a central position.

Various exact and approximate approaches such as branch-and-bound, branch-and-cut [14], branch-and-price [23], have been applied to solve the capacitated VRP (see the surveys [16, 17]). As CVRP is NP-hard problem, only small size instances can be optimally solved using exact algorithms [3, 10, 25]. Therefore, researchers and practitioners moved to newly emerging soft computing techniques based on the intelligence present in nature, such as intelligence in birds, ants, fishes, bacteria, water drop etc. These techniques have the capability of addressing large and complex problems in tractable time frame.

During the past two-three decades an increasing number of meta-heuristics have been developed to solve the CVRP. The work can be categorized into simulated annealing [19], tabu search [7, 11, 25], large neighborhood search [9], variable neighborhood search [7], genetic algorithm [2, 4], evolutionary algorithms [20, 21, 24], particle swarm optimization [1, 15, 26] ant colony optimization [5, 8, 22, 27], artificial bee colony [18] etc. For extensive surveys on VRP meta-heuristics refer to papers [12, 13, 16].

Designing a better approach for the CVRP may reduce the effective cost of goods, travel, transportation, making remarkable impacts on our economy. In the pursuit to achieve better optimal solutions, the present attempt is aimed to solve CVRP using an enhanced ACO and then verifying it with experimental methods. The performance of the algorithm is evaluated on benchmark instances and results are compared with other heuristics results present in the literature.

2 CVRP Formulation

The terms and notations in CVRP can be formally defined as follows [18]:

Graph: an undirected graph $G = (V, E)$, $V = \{v_0, v_1, \ldots, v_n\}$ is the set of vertices and E is the set of all possible connections.

Depot: in the graph, vertex v_0 is the depot from where routes starts and terminates.

Customers: the problem is defined for n customers represented by vertices v_1, v_2, …, v_n. Each customer has a non-negative deterministic demand q.

Vehicles: vehicles that serve the customers have capacity Q. Vehicles can serve many customers provided the sum of demands for each customer should not exceed the vehicle capacity Q. Also, the vehicle must start and end at a common depot.

Travelling Cost: C_{ij} represents the travelling cost between customers i and j. It is generally calculated using Euclidian distance between the customers.

Route: is a sequence of nodes, starting and ending at the depot. The length of each route r depends upon the number of customers.

CVRP: each vehicle has limited capacity. It ensures that the sum of customers' demand q_i cannot exceed the vehicle capacity Q. Similarly, the total route distance d_{ij} of a vehicle can't exceed its route length limit. It also ensures that each customer can be served by only one vehicle.

- *Objectives*:
  - *Minimize the total cost of travelling.*
  - *Minimize the total number of vehicles.*
  - *Minimize the distance travelled by all the vehicles.*
- *Constraints*:
  - *Each vehicle must start and end at the depot.*
  - *Total demand of customers of any route does not exceed the vehicle capacity.*

3 Enhanced Ant Colony Optimization (ACO) Algorithm

Dorigo in 1992 proposed ACO, which is a probabilistic technique to find near optimal solutions to large and complex problems [8]. It is achieved by seeking the foraging behaviour of ants to find a shortest path from food to the nest, with the help of a trace called pheromone that permits them to communicate with each other.

In the proposed algorithm, solution is improved by considering various factors: (1) 1–1 swap heuristic is used to exchange two cities (customers) from different routes, i.e., a city c_1 from tour t_1 is exchanged with a city c_2 from tour t_2, if it can improve the solution. (2) After some iteration ants will not explore some edges due to less pheromone deposit, hence can be stuck in local minima. Therefore, to avoid being trapped in local minima pheromone will be reset for all edges and to achieve exploitation increase the pheromone of those edges, that belongs to best solution found so far by some factor. (3) Each ant has an associated memory to keep record of current iteration solution (which can be improved in next iteration) and a count value (no. of iterations for which solution is not improved). Hence, solution is improved in every iteration rather than building a new solution.

The pseudo code for the proposed ACO algorithm is given as follows:

(1) Solution Construction:
(i) for each ant K: if previous solution is null then build a new solution.
(ii) else improve the previous solution by choosing a new edge (C, X) (swap) and by building a new tour.
(2) Apply Local Search: 2-opt approach
(3) Update Memory:

(i) for each ant K: if new solution cost is less than the previous solution cost then reset count to 0.
(ii) else: increment the count till maximum count
(a) if count > maximum count then delete previous solution of ant K and reset count to zero
(4) Update Pheromone
(5) Reset Pheromone
(i) after every R iteration reset pheromone for each edge (i, j) to initial value T_0 as follow:
(a) if edge (i, j) belongs to best solution then pheromone (i, j) will be exploited by same value T_1
(b) otherwise, if edge (i, j) not belongs to best solution found so far, then Pheromone (i, j) = T0

The new tour is created using the equation:

$$p_{i,j} = \begin{cases} \frac{[T_{ij}]^{\alpha}[\eta_{ij}]^{\beta}}{\sum_{h \in \Omega}[T_{ih}]^{\alpha}[\eta_{ih}]^{\beta}} & if\ v_j \in \Omega \\ 0 & otherwise \end{cases}$$

Here, p_{ij} is probability of selecting edge (i, j), T_{ij} is pheromone present on edge (i, j), η_{ij} is the visibility of edge (i, j), f and g are parameters which scales visibility. The pheromones trails are updated by elitist ants only, by the rule given as:

$$T_{ij}^{new} = \rho T_{ij}^{old} + a \sum_{\mu=0}^{\sigma-1} \Delta T_{ij}^{\mu} + b\Delta T_{ij}^{*}$$

$$\Delta T_{ij}^{\mu} = \frac{(\sigma - \mu)}{L_{\mu}}$$

Here, ρ is the trail persistence with ($0 \leq \rho \leq 1$), Δ_{ij}^{μ} is the quantity by which each elitist ant μ increases pheromone and is scaled by a factor a.

$\Delta_{ij}^{*} = 1/L^{*}$ is the quantity by which b ants deposit pheromone and L^{*} is the cost of best solution found so far.

4 Computational Results

The proposed algorithm is evaluated on Fisher and Christofides benchmark instances. All the experiments were performed on 2.93 GHz computer and Matlab 2012 has been used for the implementation and simulation of the algorithm.

The number of ants *M* is employed as equal to the *N* – 1 customers in each set, and nearest neighborhood of each city is considered as *N*/4. Initial Pheromones, $T_0 = 0.9$ and $T_1 = 1.2$ and *alpha* and *beta* are set to *5*. *R* is taken as *20* and *Max_Count* is equal to 10. The number of elitist ant denoted by σ (sigma) is taken

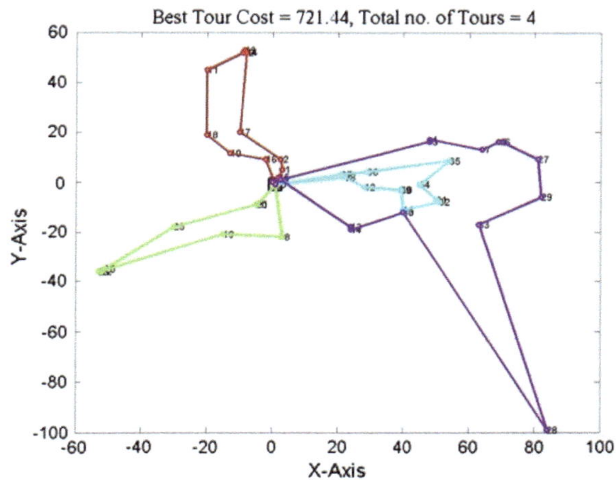

Fig. 1 VRP Instance: F-n45-k4, Customers: 44, Vehicles: 4, Capacity: 2010, Result: 721.44

as *6* and trail persistence ρ is equal to *0.98*. Other parameters are taken as: $a = 10$, $b = 10, f = 2$ and $g = 2$. To compare better, the number of iteration is taken as 350.

Figures 1 and 2 below shows the matlab plots for Fisher instances for customers 44 and 134 resp. and Figs. 3 and 4 shows the routes taken by the vehicles along with the sum of demands fulfilled for each route (Tables 1 and 2).

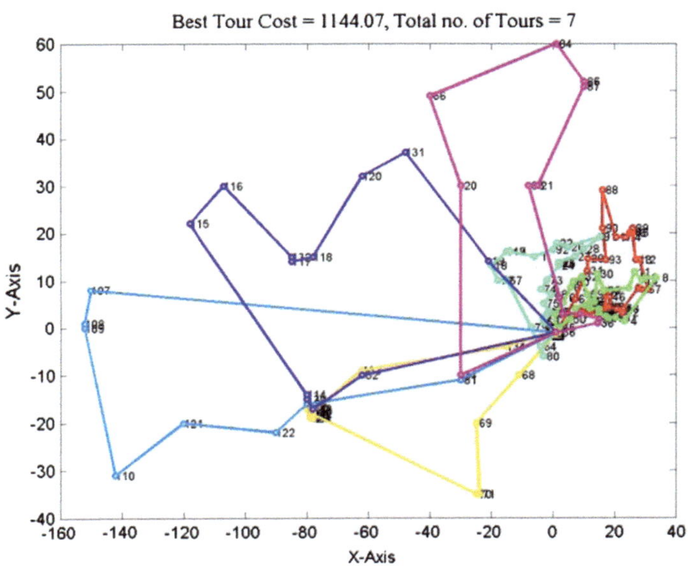

Fig. 2 VRP Instance: F-n135-k7, Customers: 134, Vehicles: 7, Capacity: 2210, Result: 1144.07

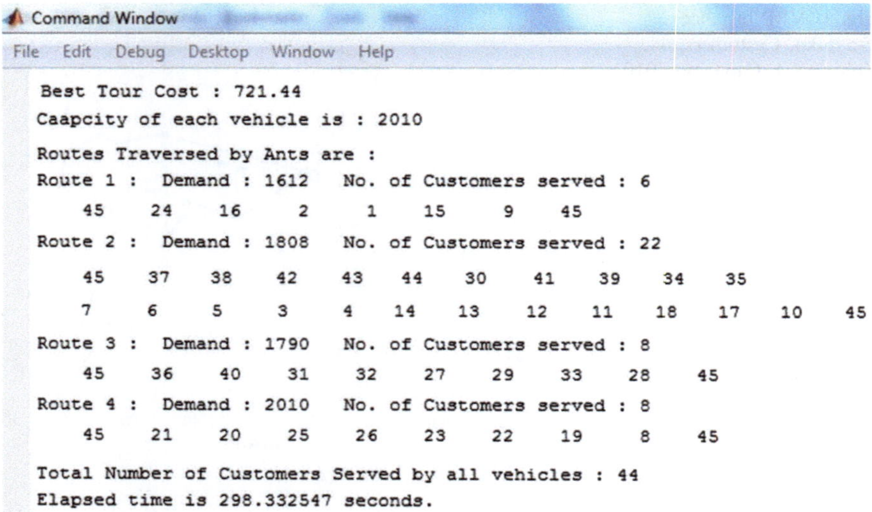

Fig. 3 Shows the routes traversed by all the 4 vehicles for Fisher F-n45-k4 instance

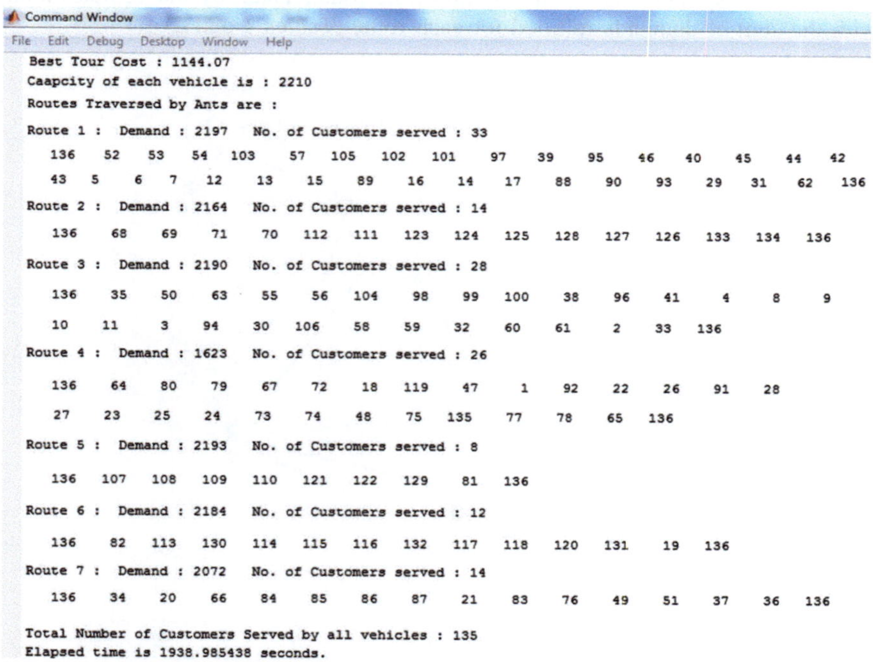

Fig. 4 No. of Iterations versus Cost (524.61) plot for VRPNC1

Table 1 Comparison of computation results of the present ACO and other existing heuristics for Christofides benchmark problem

Instance	Best known	1997 ACO[a]	1998 ACOb[b]	Chen-SR[c]	SS_ACO[d]	Present ACO[e]	ACO Avg.[f]
Vrpnc1	524.61	524.61	524.61	524.61	524.61	**524.61**	**524.61**
Vrpnc2	835.26	870.58	844.31	865.86	835.26	836.78	840.12
Vrpnc3	826.14	879.43	832.32	840.91	830.14	**826.16**	828.76
Vrpnc4	1028.42	1147.41	1061.55	1068.22	1038.20	1036.37	1038.34
Vrpnc5	1291.29	1331.4	1320.46	1365.15	1307.18	1306.3	1308.89
Vrpnc6	555.43	562.94	560.24	560.89	559.12	561.39	564.45
Vrpnc7	909.68	948.16	916.21	917.68	912.68	913.74	916.70
Vrpnc8	865.94	886.17	866.74	878.59	869.34	870.84	878.96
Vrpnc9	1162.55	1202.01	1195.99	1181.14	1079.4	1178.23	1182.02
Vrpnc10	1395.85	1504.79	1451.65	1428.46	1410.26	1414.91	1422.54
Vrpnc11	1042.11	1072.45	1065.21	1042.38	1044.12	1046.19	1054.38
Vrpnc12	819.56	819.96	819.56	820.62	824.31	**819.62**	825.84
Vrpnc13	1541.14	1590.52	1559.92	1569.14	1556.12	1556.52	1562.85
Vrpnc14	866.37	878.36	873.67	866.37	870.26	869.38	874.41

[a]Obtained from Bullnheimer (1997) ACO method
[b]Obtained from Bullnheimer (1998) ACO method
[c]Obtained from Chen et al. [6] hybrid method
[d]Obtained from Zang and Tang hybrid ACO (2006)
[e]Best solution obtained by ACO (present method)
[f]Average solution obtained by ACO (present method) in 10 runs

Table 2 Comparison of computation results of the present ACO and other existing heuristics for Fisher CVRP instances

Instance	Best known	DPSO SA[a]	CPSO-SA[b]	PACO[c]	SR 2[d] [1]	LB Tabu[e]	Present ACO[f]	ACO Avg.[g]
F_n45_k4	724	731	724	724	724	724	**721.44**	**722.62**
F_n72_k4	237	244	237	237	237	232.5	244	248.91
F_n135_k7	1162	1215	1200	1170	1162	1157.5	**1144.07**	1147.82

[a]Obtained from Chen et al. [6], hybrid PSO
[b]Obtained from Chen (2011) hybrid ACO
[c]Obtained from Yi-Ting (2012) hybrid ACO & PSO
[d]Obtained from Ai and Kachitvichyanukul (2009) [1]
[e]Obtained from Augerat et al. (1998), Tabu Search
[f]Best solution obtained by ACO (present method)
[g]Average solution obtained by ACO (present method) in 10 runs

Figures 5 and 6 above shows the plot for Christofides VRPNC1 instance of 50 customers and the routes traversed by all the vehicles with the sum of demands for each route. In Fig. 7 plot for cost versus number of iterations is given for VRPNC1 instance.

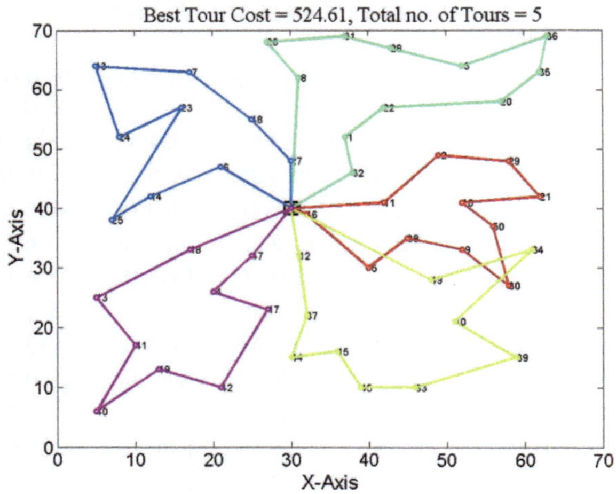

Fig. 5 VRP Instance: VRPNC1, Customers: 50, Vehicles: 5, Capacity: 160, Result: 524.61

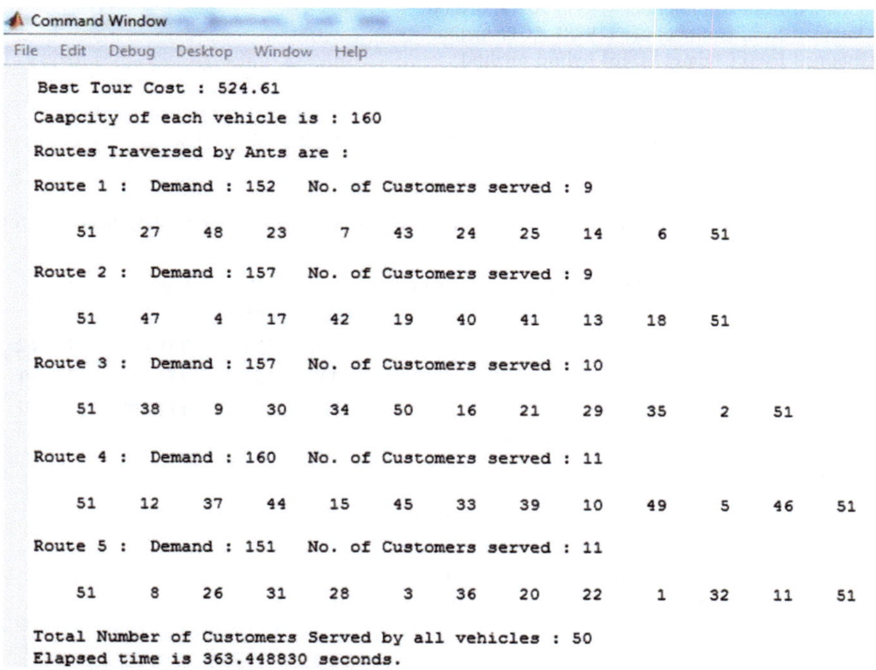


```
Best Tour Cost : 524.61
Caapcity of each vehicle is : 160
Routes Traversed by Ants are :
Route 1 :  Demand : 152   No. of Customers served : 9

   51   27   48   23    7   43   24   25   14    6   51

Route 2 :  Demand : 157   No. of Customers served : 9

   51   47    4   17   42   19   40   41   13   18   51

Route 3 :  Demand : 157   No. of Customers served : 10

   51   38    9   30   34   50   16   21   29   35    2   51

Route 4 :  Demand : 160   No. of Customers served : 11

   51   12   37   44   15   45   33   39   10   49    5   46   51

Route 5 :  Demand : 151   No. of Customers served : 11

   51    8   26   31   28    3   36   20   22    1   32   11   51

Total Number of Customers Served by all vehicles : 50
Elapsed time is 363.448830 seconds.
```

Fig. 6 Shows the routes traversed by all the 5 vehicles for Christofides VRPNC1 instance

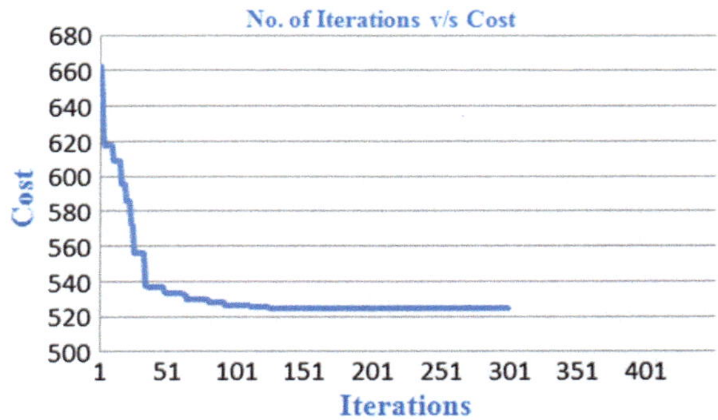

Fig. 7 No. of Iterations versus Cost (524.61) plot for VRPNC1

5 Conclusion

In this paper, we presented an enhanced ant colony optimization (ACO) meta-heuristic for the capacitated vehicle routing problem (CVRP). The proposed approach is implemented of Fisher and Christofides benchmark instances. The computational results show that the proposed algorithm is able to produce much better solutions as compare to other existing heuristics, but the best known solutions for all test problems could not be achieved. However, detailed parameter tuning can yield better results. From the results it is indicate that the present approach can be an alternative to solve the CVRP.

Hence, future research will focus on improving the scalability and performance of the algorithm by using parallel implementation along with hybridization and integration of ACO algorithm with other intelligent systems. Besides this, other variants of VRP such as VRPTW, VRPPD etc. can be solved using our ACO algorithm.

References

1. Ai, J., Kachitvichyanukul, V.: Particle swarm optimization solution representations for solving capacitated vehicle routing problem. Computers & Industrial Engineering 56 (2009) 380–387.
2. Alba, E., Dorronsoro, B.: The Exploration/Exploitation Tradeoff in Dynamic Cellular Genetic Algorithms. IEEE Transactions on Evolutionary Computation 9, 2 (2005) 126–142.
3. Baldacci, R., Toth, P., Vigo, D.: Exact algorithms for routing problems under vehicle capacity constraints. Annals of Operations Research 175 (2010) 213–245.
4. Berger, J., Barkoui, M.: A new hybrid genetic algorithm for the capacitated vehicle routing problem. Journal of the Operational Research Society 54(2003) 1254–1262.
5. Bullnheimer, B., Hartl, R.F., Strauss, C.: An improved ant system algorithm for the vehicle routing problem. Annals of Operations Research 89 (1999) 319–328.
6. Chen, A. L., Yang, G. K., Wu, Z. M.: Hybrid discrete particle swarm optimization algorithm for capacitated vehicle routing problem. Journal of Zhejiang University: Science, 7(4), (2006) 607–614.

7. Chen, P., Huang, K., Dong, Y.: Iterated variable neighborhood algorithm for capacitated vehicle routing problem. Expert Systems with Applications 37 (2010) 1620–1627.
8. Cordeau, J.F., Laporte, G., Mercier, A.: A unified tabu search heuristic for vehicle routing problems with time windows. Journal of Operational Research Society 52 (2001) 928–936.
9. Dorigo, M., Birattari, M., Stutzle, T.: Ant colony optimization. Computational Intelligence Magazine, IEEE, 1 (2006) 28–39.
10. Ergun, O., Orlin, J.B., Steele-Feldman, A.: Creating very large scale neighborhoods out of smaller ones by compounding moves. Journal of Heuristics 12 (2006) 115–140.
11. Gambardella, L. M., Taillard, E., Agazzi, G.: MACS-VRPTW: A Multiple Ant Colony System for Vehicle Routing Problems with Time Windows. [ed.] M. Dorigo and F. Glover D. Corne. New Ideas in Optimization. s.l.: McGraw-Hill, UK, (1999) 63–76.
12. Gendreau, M., Hertz, A., Laporte, G.: A tabu search heuristic for the vehicle routing problem. Management Science 40 (1994) 1276–1290.
13. Gendreau, M., Potvin, J.-Y., Braysy, O., Hasle, G., Løkketangen, A.: Metaheuristics for the vehicle routing and its extensions: A categorized bibliography. In: Golden, B., Raghavan, S., Wasil, E. (Eds.). The Vehicle Routing Problem: Latest Advances and New Challenges. Springer, (2008) 143–169.
14. Gupta, A., Saini, S.: On Solutions to Vehicle Routing Problems using Swarm Intelligence Techniques: A Review. In Proceedings of International Conference on Computer, Communication & Computational Sciences (IC4S'16), Ajmer (2016).
15. Gutierrez, J. G., Desaulniers, G., Laporte, G., Marianov, V.: A Branch-and-Price Algorithm for the Vehicle Routing Problem with Deliveries, Selective Pickups and Time Windows. European Journal of Operational Research 206 (12) (2010) 341–349. doi:10.1016/j.ejor.2010.02.037.
16. Kanthavel, K., Prasad, P.: Optimization of Capacitated Vehicle Routing Problem by Nested Particle Swarm Optimization. American Journal of Applied Sciences 8(2) (2011) 107–112.
17. Laporte, G., Osman, I. H.: Routing problems: A bibliography. Annals of Operations Research 61(1995) 227–262.
18. Laporte, G.: The Vehicle Routing Problem: An overview of exact and approximate algorithms. European Journal of Operational Research 59 (1992) 345–358.
19. Marinakis, Y., Marinaki, M., Dounias, G.: Honey bees mating optimization algorithm for large scale vehicle routing problems. Natural Computing 9 (2010) 5–27.
20. Osman, I. H.: Metastrategy simulated annealing and tabu search algorithms for the vehicle routing problem. Annals of Operations Research 41 (1993) 421–451.
21. Prins, C.: A simple and effective evolutionary algorithm for the vehicle routing problem. Computers & Operations Research 31 (2004) 1985–2002.
22. Prins, C.: A GRASP evolutionary local search hybrid for the vehicle routing problem. In: Pereira, F.B., Tavares, J. (Eds.), Bio-inspired Algorithms for the Vehicle Routing Problem. Springer-Verlag, Berlin, Heidelberg, (2009) pp. 35–53.
23. Reimann, M., Doerner, K., Hartl, R. F.: D-Ants: Savings based ants divide and conquer the vehicle routing problem. Computers & Operations Research 31 (2004) 563–591.
24. Ropke, S., Cordeau, J.-F., Iori, M. and Vigo, D.: Branch-and-Cut-and-Price for Capacitated Vehicle Routing Problem with Two-Dimensional Loading Constraints. Proceedings of ROUTE, Jekyll Island (2007).
25. Shen, Y., Murata, T.: Pick-up Scheduling of Two-Dimensional Loading in Vehicle Routing Problem using GA. Lecture Notes in Engineering & Computer Science 6 (2) (2012) 1532–1537.
26. Toth, P., Vigo, D.: The granular tabu search and its application to the vehicle routing problem. Journal on Computing 15 (2003) 333–346.
27. Yin, L., Liu, X.: A Single–depot Complex Vehicle Routing Problem and its PSO Solution. Proceedings of the Second Symposium International Computer Science and Computational Technology (ISCSCT '09) China (2009) 266–269.
28. Yu, B., Yang, Z.-Z., Yao, B.: An improved ant colony optimization for vehicle routing problem. European Journal of Operational Research 196 (2009) 171–176.

Music Recommendation System with User-Based and Item-Based Collaborative Filtering Technique

M. Sunitha and T. Adilakshmi

Abstract Internet and E-commerce are the generators of abundant of data, causing information Overloading. The problem of information overloading is addressed by Recommendation Systems (RS). RS can provide suggestions about a new product, movie or music etc. This paper is about Music Recommendation System, which will recommend songs to users based on their past history i.e. taste. In this paper we proposed a collaborative filtering technique based on users and items. First user-item rating matrix is used to form user clusters and item clusters. Next these clusters are used to find the most similar user cluster or most similar item cluster to a target user. Finally songs are recommended from the most similar user and item clusters. The proposed algorithm is implemented on the benchmark dataset Last.fm. Results show that the performance of proposed method is better than the most popular baseline method.

Keywords E-commerce · Music recommendation system · Collaborative filtering · Similarity measures

1 Introduction

With the start of E-commerce and Internet Era, information started flowing like a flood. Recommendation systems (RS) are the tools to address information overloading problem. Recommendation system consists of two types of entities users and items. Users can be customers in an online store and items can be products. There is wide variety of applications exists for RS such as movie recommendation system, books recommendation system, music recommendation system etc. This research work discuss about music recommendation system. In music recommendation system users are listeners and items are songs listened.

M. Sunitha (✉) · T. Adilakshmi
Department of CSE, Vasavi College of Engineering, Ibrahimbagh,
Hyderabad 500031, India
e-mail: sashu2006@gmail.com

G.M. Perez et al. (eds.), *Networking Communication and Data Knowledge Engineering*, Lecture Notes on Data Engineering and Communications Technologies 3, https://doi.org/10.1007/978-981-10-4585-1_22

Music is omnipresent. Millions of songs are present at a click away from everyone[1, 2]. With the number of songs, bands, and artists coming up, music listeners are overwhelmed by choices. Users will always try to find the music that will match their taste. This is the motivational fact for the field of music recommendations. In the recent years, there have been many services like Pandora, Spotify, and Last.fm [3, 4] that have come up in order to find a perfect solution, but haven't been completely successful. The Choice for music is influenced by taste, trust, and liking towards any particular artist. It is very difficult to quantify all these factors for a machine or software. Hence, it has been a very difficult experience for these service providers to find the music really interesting and satisfying the taste of a person. Every music recommendations system works on a given set of assumptions in order to provide effective recommendations.

Two standard recommendation systems exist in literature are Collaborative filtering (CF) and Content based. CF depends on the behavior of other [2].

Users' in order to provide recommendations whereas Content based systems uses content of the item for recommendations.

The rest of the paper is organized as follows. Section 2 explains Related work. Section 3 discusses about the proposed algorithm. Section 4 showcases the results obtained for the proposed algorithm and Sect. 5 describes conclusion and future directions of research.

2 Related Work

2.1 User-Item Rating Matrix

Collaborative Filtering technique is based on the data structure known as User-Item rating matrix. In this data structure each user's rating for each item is stored. If a user Ui listen an item Ij for Sij number of times then Sij is placed in the cell corresponding to user Ui and item Ij. The user-item matrix for m users and n items is shown in Fig. 1

Fig. 1 User-item matrix

Item /User	I_1	I_2	… …	I_n
U_1	S_{11}	S_{12}	… …	S_{1n}
Ur_2	S_{21}	S_{22}	… …	S_{2n}
… …	… …	… …	… …	… …
U_m	S_{m1}	S_{m2}	… …	S_{mn}

Fig. 2 User-based CF with nearest neighbors

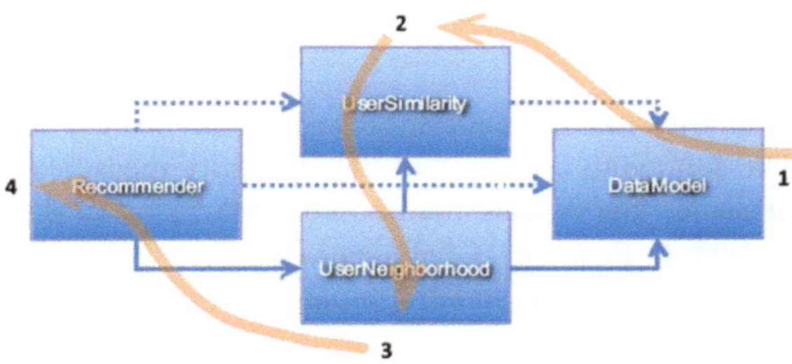

2.2 *Collaborative Filtering (CF) Technique*

CF is the most common method used in recommendation systems. Basic idea of CF [2, 5] is if users agree in the past i.e. if they have same taste in the past they will agree in future also. CF can be categorized into the following two types

2.2.1 User-Based CF

In the user-based collaborative filtering approach, we use user ratings for each item to infer interests and make recommendations. The bottom line of this approach is to find all the nearest neighbors of the current target user and try to recommend items that the target user would like as shown in the Fig. 2.

2.2.2 Item-Based CF

In the item based collaborative filtering approach, we construct item-profiles instead of user profiles and find similarities between any two given items using various measures like Euclidean distance, Tanimoto coefficient and Log likelihood similarity.

For any given item i, we compute its similarity with the item which is already present in the user profile to predict if the target item is worth recommending to the user or not. This type of approach is useful when new items are being added to the system too often.

3 Proposed Algorithm

This section describes normalization techniques, similarity measures and user-based and item-based methods to form user clusters and item clusters which will be used in recommendations.

3.1 Similarity Measures

Similarity measures are used to find the nearest neighbors of a given target user. The following are some of the similarity measures studied in literature. Similarity measures can be classified based on the kind of data.

Simple matching Coefficient (SMC) and Jaccard Coefficient (JC) are used for symmetric and asymmetric binary data respectively as defined below

$$\mathrm{SMC} = \frac{\mathrm{f00} + \mathrm{f11}}{\mathrm{f00} + \mathrm{f11} + \mathrm{f10} + \mathrm{f01}} \quad \mathrm{JC} = \frac{\mathrm{f11}}{\mathrm{f11} + \mathrm{f10} + \mathrm{f01}}$$

Euclidean Distance(ED) measure is the most common and shows how close the two users or items are to each other, which is given by the sum of square of the difference of the individual elements of the vectors representing the users or items.

The mathematical formula is given as shown below

The distance from p to q is given by

$$\begin{aligned} \mathrm{d}(\mathrm{p}, \mathrm{q}) = \mathrm{d}(\mathrm{q.p}) &= \sqrt{(\mathrm{q}_1 - \mathrm{p}_1)^2 + (\mathrm{q}_2 - \mathrm{p}_2)^2 + \cdots + (\mathrm{q}_n - \mathrm{p}_n)^2} \\ &= \sqrt{\sum_{i=1}^{n} (\mathrm{q}_i - \mathrm{p}_i)^2} \end{aligned}$$

where d(p, q) refers to the Euclidean distance between two users or items p, q represented as n dimensional vectors where n is the number of items.

3.2 Normalization Techniques

Normalization is a standardization technique generally performed in Data mining. It is used to transform data or normalize the data to give equal importance to all attributes. The most commonly used normalization technique is Length normalization (LN) and Root mean square normalization (RMSN).

Length normalized vectors for each user is obtained by dividing the rating of each user for each item with the sum of his/her ratings for all items. The mathematical formula to find length normalized vectors is given in Eq. 1. The purpose of length normalization is to bring all users on to the same scale irrespective of the number of items rated.

$$\mathrm{LN}(\mathrm{U}_i, \mathrm{I}_j) = \frac{\sum_{i=1}^{N} \sum_{j=1}^{M} R_{ij}}{N} \tag{1}$$

where R_{ij} is the rating of U_i for the item I_j, N is the number of items and M is the number of users.

RMSN is a generic normalization technique. The denominator is the square-root of the sums of the square of the all the elements of the vector, while the numerator has the frequency of the user 'i' for item 'j'. Mathematically RMSN is expressed as the equation given below.

$$\mathrm{RMSN}(t_i, d) = \frac{C_i}{\sqrt{\sum_{i=1}^{n} C_i^2}}$$

where RMSN is the Root Mean Square Normalization vector, I_j is the item and $user_i$ is a user, R_{ij} is the rating of $user_i$ for the I_j.

3.3 Proposed Approach

This paper proposes a method to perform music recommendation. The method used in this research work is model based collaborative filtering. A model is built by using user-based CF and item-based CF. The model is then used to provide recommendations for test user.

3.3.1 Pre-processing of Data

Initially pre-processing is performed on user-item rating matrix. First binarization is performed on user-item matrix as shown in Table 1.

$$B_{ij} = \{1 \text{ if } R_{ij} > 0 \\ 0 \text{ if } R_{ij} = 0\}$$

where R_{ij} is the rating of a user U_i for an item I_j

Table 1 Binary user-item rating matrix

	Item 1	Item 2	Item 3	Item 4
User 1	0	0	1	0
User 2	1	0	0	1
User 3	1	1	0	1
User 4	1	0	0	0
User 5	0	1	1	0
User 6	0	1	1	0
User 7	1	0	0	1
User 8	1	0	0	0
User 9	1	0	0	1
User 10	1	0	0	0

Table 2 Length normalized user-item rating matrix

	Item 1	Item 2	Item 3	Item 4
User 1	0	0	1	0
User 2	0.5	0	0	0.5
User 3	0.33	0.33	0	0.33
User 4	1	0	0	0
User 5	0	0.5	0.5	0
User 6	0	0.5	0.5	0
User 7	0.5	0	0	0.5
User 8	1	0	0	0
User 9	0.5	0	0	0.5
User 10	1	0	0	0

Table 3 RMSN normalized user-item rating matrix

	Item 1	Item 2	Item 3	Item 4
User 1	0	0	0	0
User 2	0.707	0	0	0.707
User 3	0.408	0.408	0	0.408
User 4	0	0	0	0
User 5	0	0.707	0.707	0
User 6	0	0.707	0.707	0
User 7	0.707	0	0	0.707
User 8	0	0	0	0
User 9	0.447	0.447	0.447	0.447
User 10	1	0	0	0

Second step in pre-processing is normalization.

Length normalization and Root Mean Square Normalization is used to normalize binary user-item rating matrix. The Table 2 shows the user-item matrix with length normalization and Table 3 shows user-item matrix with RMSN normalization.

3.3.2 Model Building

The data taken as input to build the model is user-item rating matrix. Each row is considered as a user vector and each column is considered as an item vector. Hierarchical threshold based clustering algorithm is used to build user or item model [6]. User model is formed by using the pseudocode Fig. 3 and item model is formed [5, 7] by using the pseudocode Fig. 4.

Test data is given as input to the user model and item model to provide recommendations based on user clusters and item clusters respectively. The algorithm used for recommendation is shown in Fig. 5.

Fig. 3 Pseudocode for user-based model

```
Algorithm Threshold_clusters()
Begin
Initialize the threshold value to th_cutoff
For each user in u_1, u_2 ...............u_n
Assign u_1 to cluster C_1
For each user u_i in u_2 ...............u_n
  begin
  Find the similarity of each user u_i with C_1
  Assign u_i to C_1 if the sim(u_i , C_1 ) ≤ th_cutoff
    Otherwise create a new cluster C_2
    end
Return the clusters C_1, C_2 ........... C_k
End
```

Fig. 4 Pseudocode for item-based model

```
Algorithm Threshold_Itemclusters()
Begin
Initialize the threshold value to th_cutoff
For each item in i_1, i_2 ...............i_n
Assign i_1 to cluster C_1
For each user i_i in i_2 ...............i_n
  begin
  Find the similarity of each item i_i with C_1
  Assign i_1 to C_1 if the sim(i_i , C_1 ) ≤ th_cutoff
    Otherwise create a new cluster C_2
    end
Return the clusters C_1, C_2 ........... C_k
End
```

Fig. 5 Pseudocode for recommendation from user and item model

```
Algorithm Recommendation_itemclusters()
Begin
For each test_user in u_1 , u_2,....u_k
    Find the similarity with each item-cluster
    Find the cluster with highest similarity
    Recommend the items of the mapped item-cluster
End
```

4 Results

Experimental analysis is performed on benchmark data set obtained from Last.fm[1]. Last.fm is a website which allows users to listen music online. The data considered for this work is the listening history of 50 Last, fm users from 2006 to 2009. For conducting experiment we considered 1 year data (2008). The number of records are 2,20,338. Sample record is shown in Fig. 6.

Each record consists of the following fields.

User_000001	2009-04-09T12:49:50	Z078a9376-3c04-4280-b7d7-b20e158f345d	A Perfect Circle
5ca13249-26da-47bd-bba7-80c2efebe9cd	People Are People		
User_000002	2009-04-09T12:20:50	Z078a9376-3c04-4280-b7d7-b20e158f345d	A Perfect Circle
5ca13249-26da-47bd-bba7-80c2efebe9cd			
People Are People			

Fig. 6 Sample log records of Last.fm users

User Id (User_000004)—Since the data is captured anonymously, we assigned each

user, a user-id of the format user_000004.

Date–Time (2009-04-09T12:49:50Z)—Time of activity is recorded

Album Id (078a9376-3c04-4280-b7d720e158f345d)—A unique identifier is Attributed to each Album.

Album name (Frightened Rabbit)—An album to which that song belongs to.

Track Id (5ca13249-26da-47bd-bba7-80c2efebe9cd)—A unique identifier is attributed to each track/song.

Track name (Old Old Fashioned)—The songs which the user listened to.

Evaluation Measures

Precision, Recall and F-measure are used to evaluate the performance of the proposed method. The measures are defined as shown below

$$Precision = \frac{\text{TP}}{\text{TP} + \text{FP}} \quad F - measure = 2 * \frac{(\text{Precision} * \text{Recall})}{(\text{Precision} + \text{Recall})}$$

$$Recall = \frac{\text{TP}}{\text{TP} + \text{FN}}$$

4.1 Experimental Results

The experiment is repeated for threshold values varying from 0.05 to 0.5 for different normalization techniques (Length normalization, Root Mean Square normalization) and similarity measures (Euclidean, Cosine). The results are shown below. From the experiments we can show that the proposed music recommendation system works well compared to the baseline method. The results are shown for various threshold values with only user clusters, only item clusters and both with user and item clusters (Figs. 7, 8, 9, 10, 11 and 12, Tables 4, 5, 6, 7, 8 and 9).

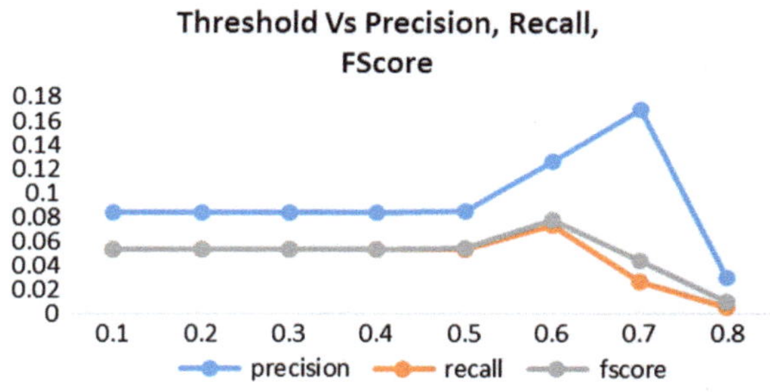

Fig. 7 Items clusters with cosine similarity

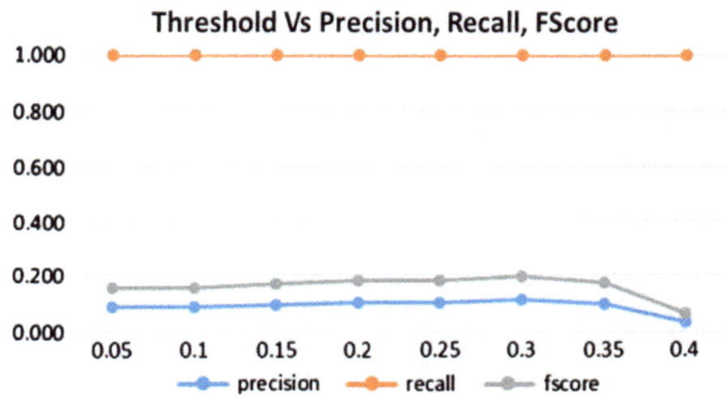

Fig. 8 User clusters with cosine similarity item

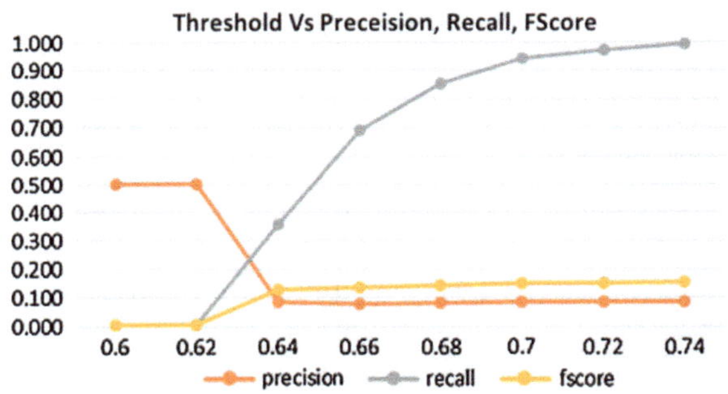

Fig. 9 Item clusters with Euclidean distance

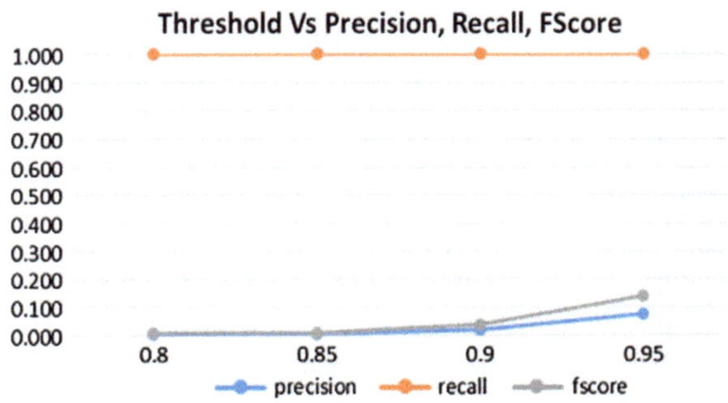

Fig. 10 User clusters with Euclidean distance

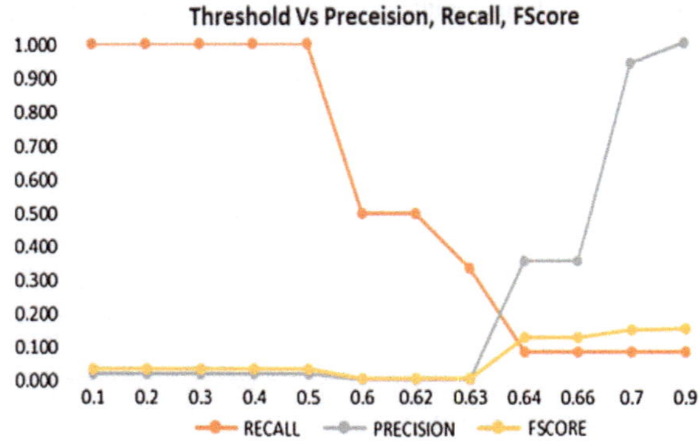

Fig. 11 User and item clusters with euclidean distance

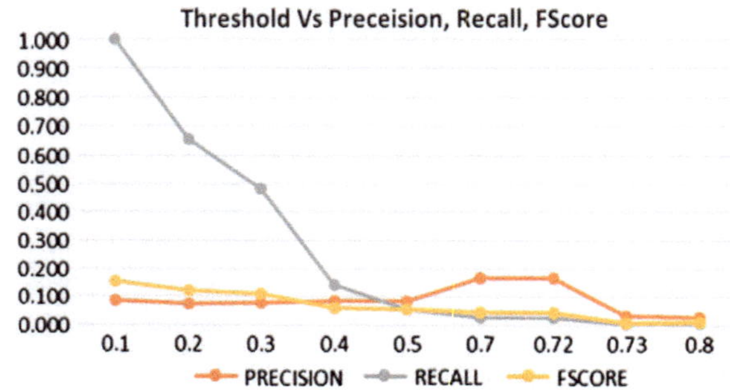

Fig. 12 User and item clusters with cosine similarity

Table 4 Items clusters with cosine similarity

Threshold	Precision	Recall	F-score
0.1	0.084	0.053	0.054
0.2	0.084	0.053	0.054
0.3	0.084	0.053	0.054
0.4	0.084	0.053	0.054
0.5	0.085	0.054	0.054
0.6	0.126	0.073	0.078
0.7	0.169	0.027	0.044
0.8	0.029	0.006	0.010
0.9	1.000	0.168	0.286

5 Conclusion and Future Scope

This paper addresses the problem of music recommendation by using CF model based approach with users and items. We conducted experiments by using user based and item based CF model with user clusters and item clusters on last.fm bench mark data set. The results of proposed method is compared with baseline

Table 5 User clusters with cosine similarity

Threshold	Precision	Recall	F-score
0.05	0.086	1.000	0.154
0.1	0.086	1.000	0.154
0.15	0.095	1.000	0.170
0.2	0.101	1.000	0.180
0.25	0.101	1.000	0.179
0.3	0.112	1.000	0.196
0.35	0.097	1.000	0.171
0.4	0.034	1.000	0.065

Table 6 Item clusters with Euclidean distance

Threshold	Precision	Recall	F-score
0.6	0.500	0.003	0.006
0.62	0.500	0.003	0.006
0.64	0.086	0.357	0.128
0.66	0.079	0.690	0.136
0.68	0.082	0.855	0.145
0.7	0.084	0.942	0.150
0.72	0.085	0.973	0.151
0.74	0.085	0.995	0.153

Table 7 User clusters with Euclidean distance

Threshold	Precision	Recall	F-score
0.8	0.003	1.000	0.007
0.85	0.003	1.000	0.007
0.9	0.019	1.000	0.036
0.95	0.076	1.000	0.138

Table 8 User and item clusters with Euclidean distance

Threshold	Recall	Precision	F-score
0.1	1.000	0.019	0.035
0.2	1.000	0.019	0.035
0.3	1.000	0.019	0.035
0.4	1.000	0.019	0.035
0.5	1.000	0.019	0.035
0.6	0.500	0.003	0.006
0.62	0.500	0.003	0.006
0.63	0.333	0.003	0.005
0.64	0.086	0.357	0.128
0.66	0.086	0.357	0.128
0.7	0.084	0.942	0.150
0.9	0.086	1.000	0.154

Table 9 User and item clusters with cosine similarity

Threshold	Precision	Recall	F-score
0.1	0.086	1.000	0.154
0.2	0.074	0.655	0.122
0.3	0.078	0.484	0.111
0.4	0.086	0.143	0.059
0.5	0.085	0.054	0.054
0.7	0.169	0.027	0.044
0.72	0.169	0.027	0.044
0.73	0.032	0.004	0.007
0.8	0.029	0.006	0.010

(popular) method. Results obtained shows that the performance of proposed method for RS improves over baseline method. This work can be enhanced to address various issues of CF such as Sparsity, the long tail problem and cold-start problem.

References

1. NetflixPrize, http://www.netflixprize.com/, 2012.
2. Adomavicius, G., Tuzhilin, A. 2005. Toward the Next Generation of Recommender Systems: A Survey of the State-of-the-Art and Possible Extensions. IEEE Transactions on Knowledge and Data Engineering 17, 734–749.
3. Last. FM—A popular music web portal http://www.last.fm
4. The Million Song Dataset Challenge, Brian McFee, Thierry Bertin-Mahieux, Daniel P.W. Ellis, Gert R.G. Lanckriet, WWW 2012 Companion, April 16–20, 2012, Lyon, France
5. A Collaborative Filtering Recommendation Algorithm Based on User Clustering and Item Clustering, Song-Jie Gong, Journal of Software, Vol. 5, No. 7, July 2010
6. M. Sunitha Reddy, Dr. T. Adilakshmi, User Based Collaborative Filtering For Music Recommendation System, International Journal of Innovative Research and Development, Dec 2013, Volume 2, Issue 12 pg no 185–190
7. Context-aware item-to-item recommendation within the factorization framework, Balázs Hidasi, Domonkos Tikk, CaRR'13, February 5, 2013, Rome, Italy
8. Introduction to Recommender Systems, Markus Zanker, Dietmar Jannach, Tutorial at ACM Symposium on Applied Computing 2010, Sierre, Switzerland, 22 March 2010

GLCM Feature Extraction for Insect Bites Pattern Recognition

Abdul Rehman Khan, Nitin Rakesh, Rakesh Matam and Shailesh Tiwari

Abstract This paper describes the elements that are vital for feature extraction process from a Grey Level Co-occurrence Matrix. Every pattern recognition model consists of a primary phase where Feature Extraction is implemented, that focuses on determining distinct parameters from a given data. With respect to data set of images, which are a complex form of data, it is very difficult to analyze it's features due to its nature. Image processing community is inundated with research on classification processes, none has been done on classification of insect bites ever before. This paper will propose a model to extract features from images of insect bites which can further be used so as to classify insect bites based on their vectors. Computer aided diagnosis can be achieved with successful detection of insect bites, that can aid at remote locations, such as Forests. Textural analysis of insect bite can help in classification of insect bites. The search for image point correspondences involves finding the interest points, neighborhood of those interest points are represented using vectors and finally the vectors are matched with the targeted image.

Keywords Pattern recognition · Insect bites · Textural analysis
Grey level co-occurrence matrix · Feature extraction · Statistical analysis
Syntactical analysis

A.R. Khan (✉) · N. Rakesh
Department of Computer Science and Engineering, Amity University, Noida, Uttar Pradesh, India
e-mail: arkspeed@gmail.com

N. Rakesh
e-mail: nitnrakesh@gmail.com

R. Matam
Department of Computer Science and Engineering, Indian Institute of Information Technology, Guwahati, India
e-mail: matamrakesh@gmail.com

S. Tiwari
Department of Computer Science and Engineering, A.B.E.S Engineering College, Ghaziabad, India
e-mail: shail.tiwari@gmail.com

G.M. Perez et al. (eds.), *Networking Communication and Data Knowledge Engineering*, Lecture Notes on Data Engineering and Communications Technologies 3, https://doi.org/10.1007/978-981-10-4585-1_23

1 Introduction

Feature extraction helps in determining the vital interest points from a large dataset, which can lead to complex analysis due to large number of variables involved. Such complex analysis can not only increase the computation time, but also use up large amount of memory, both of which can be a determining factor when it comes to creating an effective pattern recognition. Over-fitting is another indication of a poor feature space, which can cause any pattern recognition model more complex not to achieve accuracy in reality but rather unseen samples can't be generalized with better accuracy. During feature extraction we not only aim at extracting features but to be able to generalize those feature so that each feature can be uniquely understood by the trained model. Eventually, with the data being described with certain accuracy, classification model can construct combinations of variables to avoid any problem that can cause the model to be incompetent or inaccurate. Texture analysis helps in ubiquitous construction of relations among the textural characteristics of a sample image, which can be utilized for vigorous, precise characterization and categorization of items. To be able to make consciousness and true classification of banal images of insect bites, would require tremendous load on many classification model. Many opting for SIFT, SURF or more complex algorithms on tasks such as bone age assessment [1], classification of animals [2], 3-D Object classification [3], or classification of hep-2 cell [4] can't be utilized for this specific problem. Simply because of the lack of distinctive features present on a bite marks—one can distinguish between a snake bite or a dog bite due to their teeth marks—which led us to rely on textural analysis. Section 2 will further discuss the novelty of this approach with volubility. In this paper, feature vectors are calculated from a Gray level co-occurrence matrix, involving inverse difference moment, entropy to angular second moment, fourteen textural features can be extracted from the matrix. Section 3 provides a contemplation on GLCM while implementation of GLCM is shown in Sect. 4 concerning with task to extract features of five classes namely bees, mosquitos, tick, spiders and healthy looking skin. Finally, providing results of this approach which has never been attempted on insect bites classification in Sect. 5, concluding the paper in Sect. 6, discussing the future scope of the work that has been presented here.

2 Texture Analysis

When one talks about textures in Computer Vision, we entail to the physical variations present in an image, which can be any regular recurrence of an element that can give it distinctiveness of a pattern in an image. Properties like brightness, shape or color, can help in determining the texture by describing a shape or structure of any image. Continuous texture can be seen, if the characteristics are constant, which helps in grouping of a similar kind of images. Considering its utility—when it comes to insect bites—it can provide us with wide spectrum of distinctiveness which can help in further deducing the insect bites. Many medical personnel can utilize the ability to identify the insect bites using the structure of the bites, since the venom of each insect is different, obviously the infection of each venom will produce different reaction with the skin, producing distinct pattern associated with each insect bites. The structure of each bites differ greatly for every insect making it easier for medical personnel to identify the insect bites just by making an analysis of its structure. That is to say, if a pattern is generalized among a group of insects we can utilize this information to train a classifier.

Further, this paper will discuss extraction of textural features (coarseness, brightness, smoothness, etc.) from any insect bite image, which can be used to train a classifier. Textural pattern recognition can assist in computer aided diagnosis of insect bites by helping in either replacing or improving anomalies detection and evaluation of what is observed by human eyes.

3 GLCM

Gray Level Co-occurrence Matrix [5] aims at evaluating a function which shows the frequency of a gray level occurrence at a pixel whose position is located with respect to each other pixel. It is a square matrix with size corresponding to the threshold levels of pairs of pixels separated by a distance. Distance among a neighboring pixel is taken as one, while relative distances among other pixels will be different hence creating a different matrix every time. To determine co-occurrence of two pixels (i,j) of a grayscale image I of size $S \times S$, and construct a matrix P_d, we use following equation.

$$P_d(i,j) = |\{(r,s),(t,v): I(r,s) = i, I(t,v) = j\}| \tag{1}$$

where $(r,s) \in i$ and $(t,v) \in j$ and displacement vector $d(d_x, d_y) = (t - r,\ v - s)$. Consider an entry in matrix, (1, 2), for a right neighbor would be interpreted as the frequency of finding a gray level 2 immediately to the right of pixel with gray level

Fig. 1 Illustration of a co-occurrence matrix

0	0	2	1
1	0	2	1
1	0	0	1
0	0	1	0

1. For example, in Fig. 1, a co-occurrence matrix is constructed of an image with threshold of 4. If we are looking at 2 at second row and third column, it would indicate frequency of having a gray level pixel of level 3 to the right of pixel with gray level 2, which is twice in this case.

4 Extraction of Features from GLCM

Texture Analysis can be accomplished with the data provided by GLCM, as we can utilize simple scalar measures to determine texture values [6, 7]. Smoothness for example, can easily be comprehended if most of the non-zero entries in the matrix are gradually near the main diagonal. Since the values of neighboring pixels will be closer and hence we could estimate the smoothness in an image with the help of entries in matrix along the diagonal. In next section, we would be discussing some of the features that we extracted from images of insect bites, namely of classes, Bees, Mosquitos, Spider, Tick Bites and Healthy Skin. Variety of statistics can be derived with the help of grey level co-occurrence matrix, as stated by Harlick et al. fourteen measurements were recognized, we will be discussing seven of those measurements.

Using the formula of *Angular Second Moment*, we can measure the uniformity of textures in an image, making it an ideal measure for analyzing disorders in any image. *Mean* of GLCM matrix can help us to analyze the gray level tone of each image differently, thereby creating a distinctive feature to distinguish a feature. With the mean of grey level pixels, relative measurement of changes can further be quantified by their frequency, which can help in analyzing the textures of various images. When it comes to measurement of changes in any image, *Correlation* can be of great assistance. It is frequently used to quantify distortion, changes, strain as well as optical continuity, yet it is generally connected in numerous regions of Science and Engineering. Among its many application, "Optical Mouse" functions based on correlation. Fundamentally, correlation helps by measuring the relation among grey histograms of neighboring pixels, with this principle, tracking as well as object registration techniques for accurate measurement of variations in images can be achieved.

Further, to be able to measure disorder in an image is of high value with respect to textural analysis. *Entropy* helps us to understand the disorder in an image. Elements in a GLCM having small values can be an indicator of high entropy, which is why entropy is considered to be inversely related to energy and angular second moment. If all the values in a matrix are equal, maximum entropy is reached. *Kurtosis* helps in deducing a concise information with regards to shape of data distribution. The peakedness and tail weight of the shape, plays an important role in inferring statistical information about the data, as well as robustness of the

Table 1 Seven texture parameters determined through GLCM

S.No	Parameters	Formula
1	Angular second moment	$\sum_i \sum_j \{p(i,j)\}^2$
2	Mean	$\sum_i \sum_j p(i,j)$
3	Correlation	$\frac{\sum_i \sum_j (i,j)p(i,j) - \mu_x \mu_y}{\sigma_x \sigma_y}$
4	Entropy	$-\sum_i \sum_j p(i,j) \log p(i,j)$
5	Kurtosis	$\frac{\sum_i \sum_j (p(i,j) - \mu)^4}{\mu_x \mu_y \sigma^4}$
6	Contrast	$\sum_i \sum_j (i-j)^2 p(i,j)$
7	Skewness	$\sqrt{\frac{\sum_i \sum_j (p(i,j) - \mu)^2}{\mu_x \mu_y}}$

data provided. *Contrast* helps in measuring variation in the pixel density of an image. It can help in understanding the density of details in an image, which can be enhanced so as to differentiate immersed objects with clarity. Another measurement we will be discussing is *Skewness*. Which is a measure of symmetry of the textures. Zero Skewness refers to symmetric data, while a negative can relate to data distribution to the left and positive Skewness if data distribution is to right. Table 1 will provide with all the formulas for calculating these features.

Once these parameters are deduced from GLCM, we can analyze textures of various images. These results will help in training a classification model to understand the characteristics of each image for categorization. An example of training a hierarchical classifier is performed by [8]. Any query image having their feature vector closest to the feature vector of a certain class will be categorized to that class.

5 Result

Implementation of GLCM was done using MATLAB on images consisting of four insect bites and one healthy skin. Since changing the resolution of images would give different GLCM, the Images of bites we analyzed consisted of image scaled to a uniform scale of 256 by 256. The GLCM matrices are calculated from a single image from each class. Table 1 provides the data we collected. The Images that were used for the implementation are provided in Fig. 2. Normalization is done for each image with the help contrast enhancement and final results of all features are provided in Table 2.

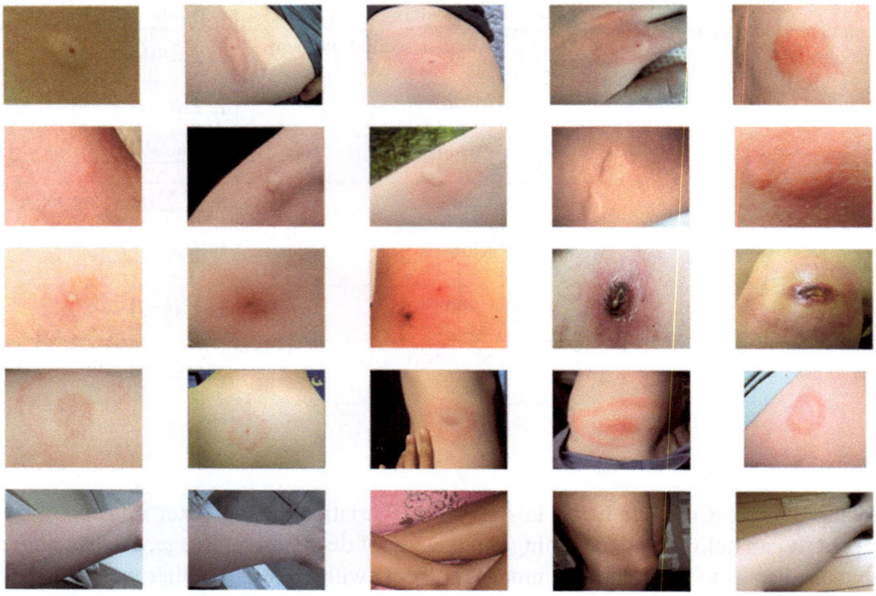

Fig. 2 (Row-wise) Bee Sting [9], Mosquito Bite [10], Spider Bite [11], Tick Bite [12], Healthy Skin [13]

Table 2 Results of textural parameters

Bite classes	ASM	MEAN	COR	ENT	KUR	COS	SKW
Bee	0.809729	44.4861	0.621109	3.03437	2.9454	2.18134	1.12846
Mosquito	0.949358	55.9546	0.923591	4.18	2.12588	0.57380	0.77782
Spider	0.956662	70.0399	0.979303	4.00263	1.80222	0.29038	0.726875
Tick	0.973594	47.8116	0.956065	3.08761	3.05248	0.46849	1.31346
Healthy	0.958101	108.236	0.948762	4.24879	1.07205	1.18261	0.195636

6 Conclusion and Future Work

This paper presented a novel approach for determining features from images of insect bites. Insect bites due to its ambiguousness can cause inaccuracy to many classification methods, using the knowledge of each insect bite, we can comprehend the uniqueness among them. Every insect has distinctive venom reaction to the skin it bites, causing a pattern to form on the skin. Evaluating the textural description of each insect, a classifier can be trained which can overcome this obscurity to insect's minuscule puncture wound. Many image pattern recognition tool have been proposed in past which doesn't possess the correct model to deduce features of an insect bite, this paper provides a solution from which many features can be utilized to classify numerous insect bites among images. A successful detection and recognition of insect bites can help at providing computer aided diagnosis at remote locations, image processing soft wares can be used to implement GLCM for determining seven Statistical Texture Parameters, i.e., Angular Second Moment, Mean, Correlation, Entropy, Kurtosis, Contrast and Skewness. This paper provides the use of texture analysis as compared to traditional methods of detectors and descriptors and shows that it provides effective way to extract interest points from an image.

The future of this paper involves utilizing the model proposed in this paper, and work upon it to create a pattern recognition model. Additionally, emphasis to improve the classification accuracy which can led to application of dimensionality reduction algorithms on the feature vectors extracted in this paper. Consequently, leading to a well formed pattern recognition model with better accuracy and less obscurity.

References

1. Kashif, Muhammad, et al. "Feature description with SIFT, SURF, BRIEF, BRISK, or FREAK? A general question answered for bone age assessment." *Computers in biology and medicine* 68 (2016): 67–75.
2. Matuska, S., et al. "Classification of animals to determine the migration potential at the construction of new infrastructure." *Archives of Transport System Telematics* 6 (2013).
3. Socher, Richard, et al. "Convolutional-recursive deep learning for 3d object classification." *Advances in Neural Information Processing Systems*. 2012.
4. Ensafi, Shahab, et al. "A bag of words based approach for classification of hep-2 cell images." *Pattern Recognition Techniques for Indirect Immunofluorescence Images (I3A), 2014 1st Workshop on*. IEEE, 2014.
5. Gebejes, A., & Huertas, R. (2013, March). Texture characterization based on grey-level co-occurrence matrix. In *Proceedings in Conference of Informatics and Management Sciences* (No. 1).
6. Ou, X., Pan, W., & Xiao, P. (2014). In vivo skin capacitive imaging analysis by using grey level co-occurrence matrix (GLCM). *International journal of pharmaceutics*, *460*(1), 28–32.
7. Dutta, S., et al. "Detection of tool condition from the turned surface images using an accurate grey level co-occurrence technique." *Precision Engineering* 36.3 (2012): 458–466.
8. Heisele, Bernd, et al. "Hierarchical classification and feature reduction for fast face detection with support vector machines." *Pattern Recognition* 36.9 (2003): 2007–2017.
9. Sograzdan, B., Sograzdan, B., & profile, V. (2015). *Sweating: Do Sweat Bees Sting*. *Sweatingstatel.blogspot.in*. Retrieved 14 June 2016, from http://sweatingstatel.blogspot.in/2015/09/do-sweat-bees-sting.html
10. *Five Home Remedies to Soothe Mosquito Bite Itch*. (2016). *95.7 KEZJ - Southern Idaho's New Country*. Retrieved 14 June 2016, from http://kezj.com/five-ways-to-soothe-mosquite-bite-itch/
11. *Australian Spider Bites - Here's an Australian spider bite on…* (2016). *Insectbitespictures.-com*. Retrieved 14 June 2016, from http://insectbitespictures.com/p/35/australian-spider-bites/picture-35

12. *Lyme Disease Rashes and Look-alikes | Lyme Disease | CDC*. (2016). *Cdc.gov*. Retrieved 14 June 2016, from http://www.cdc.gov/lyme/signs_symptoms/rashes.html
13. Expert, A., Questions, M., & Dermatologist, A. (2012). *I found this helpful answer from a dermatologist on JustAnswer.com. JustAnswer*. Retrieved 14 June 2016, from http://www.justanswer.com/dermatology/7ea3r-please-help-skin-symptoms-are-dry-skin-burning-itching.html

A Novel Hybrid PSO-DA Algorithm for Global Numerical Optimization

Indrajit N. Trivedi, Pradeep Jangir, Arvind Kumar, Narottam Jangir, R.H. Bhesdadiya and Rahul Totlani

Abstract Recent trend of research is to hybridize two and more algorithms to obtain superior solution in the field of optimization problems. In this context, a new technique hybrid Particle Swarm Optimization (PSO)—Dragonfly Algorithm (DA) is exercised on some unconstraint benchmark test functions and overcurrent relay co-ordination optimization problems in contrast to test results on constrained/complex design problem. Hybrid PSO-DA is combination of PSO used for exploitation phase and DA for exploration phase in uncertain environment. Position and Velocity of particle is updated according to position of dragonflies in each iteration. Analysis of competitive results obtained from PSO-DA validates its effectiveness compare to standard PSO and DA algorithm separately.

Keywords Meta-heuristic · Dragonfly Algorithm · Particle Swarm Optimization · HPSO-DA · Overcurrent Relay

---

I.N. Trivedi (✉)
Electrical Engineering Department, G.E. College, Gandhinagar, Gujarat, India
e-mail: forumtrivedi@gmail.com

P. Jangir
Electrical Engineering Department, LEC, Morbi, Gujarat, India
e-mail: pkjmtech@gmail.com

A. Kumar
Electrical Engineering Department, GCET, Bikaner, Rajasthan, India
e-mail: akbharia8@gmail.com

N. Jangir
Electrical Engineering Department, ECB, Bikaner, Rajasthan, India
e-mail: nkjmtech@gmail.com

R.H. Bhesdadiya
LEC, Morbi, Gujarat, India
e-mail: rhbhesdadiya@gmail.com

R. Totlani
JECRC, Jaipur, Rajasthan, India
e-mail: rhl.totlani@gmail.com

G.M. Perez et al. (eds.), *Networking Communication and Data Knowledge Engineering*, Lecture Notes on Data Engineering and Communications Technologies 3, https://doi.org/10.1007/978-981-10-4585-1_24

1 Introduction

HPSO-DA comprises of best characteristic of both Particle Swarm Optimization [1] and Dragonfly algorithm [2]. HPSO-DA result expresses that it has ability to converse faster with comparatively optimum solution for both unconstrained and constrained function.

Population based algorithms based on randomization consists of two main phases for obtaining better results that are exploration (unknown search space) and exploitation (best solution).

In this HPSO-DA, DA is applied for exploration as it uses static swarming behavior done by Dragonflies in search of food so covers large uncertain search space with less computational time to explore possible solution or to converse particle towards optimum value. Most popular PSO algorithms have ability to attain near optimal solution avoiding local solution.

Contemporary works in hybridization are: HGAPSO [3], PSACO [4], HSABA [5], PBILKH [6], KH-QPSO [7], IFA-HS [8], HS/FA [9], CKH [10], HS/BA [11], HPSACO [12], CSKH [13], HS-CSS [14], PSOHS [15], DEKH [16], HS/CS [17], HSBBO [18], CSS-PSO [19] etc.

Recently trend of optimization is to improve performance of meta-heuristic algorithms [20] by integrating with chaos theory, Levy flights strategy, Adaptive randomization technique, Evolutionary boundary handling scheme, and genetic operators like as crossover and mutation. Popular genetic operators used in KH [21] that can accelerate its global convergence speed. Evolutionary constraint handling scheme is used in Interior Search Algorithm (ISA) [22] that avoid upper and lower limits of variables.

The structure of the paper can be given as follows: Sect. 1 consists of Introduction; Sect. 2 includes description of participated algorithms; Sect. 3 consists of competitive results analysis of unconstraint test benchmark problem and Overcurrent Relay Co-ordination with common configuration in Power System, finally acknowledgement and conclusion based on results is drawn.

2 Particle Swarm Optimization

The particle swarm optimization algorithm (PSO) was discovered by James Kennedy and Russell C. Eberhart in 1995 [1]. This algorithm is inspired by simulation of social psychological expression of birds and fishes. PSO includes two terms P best and G best. Position and velocity are updated over the course of iteration from these mathematical equations:

$$v_{ij}^{t+1} = wv_{ij}^{t} + c_1 R_1 (Pbest^t - X^t) + c_2 R_2 (Gbest^t - X^t) \tag{1}$$

$$X^{t+1} = X^{t} + v^{t+1}, (\mathrm{i}=1,2,\ldots,\mathrm{NP})\ and\ (\mathrm{j}=1,2,\ldots,\mathrm{NG}) \quad (2)$$

where,

$$w = w^{\max} - \frac{(w^{\max} - w^{\min}) * iteration}{\max\ iteration} \quad (3)$$

$w^{min} = 0.4$ and $w^{max} = 0.9$.

v_{ij}^{t}, v_{ij}^{t+1} is the velocity of *j*th member of *i*th particle at iteration number (t) and (t + 1). (Usually $C_1 = C_2 = 2$), r_1 and r_2 Random number (0, 1).

3 Dragonfly Algorithm

Dragonfly Algorithm [2] proposed by Seyedali Mirjalili, inspired from natural swarming behavior of Dragonflies which makes small group and fly different direction covers vast area in search of food technically term in optimization is exploration phase and makes large group to shift other place termed similar exploitation phase, called static and dynamic swarm. Figures 1 and 2 show the Dragonfly Algorithm principle. Figure 3 show the flowchart of Dragonfly Algorithm.

Each portion of Dragonfly Algorithm is formulated by algebraic equations are:

For Separation part formulating equation:

$$\mathrm{Si} = -\sum_{j=1}^{N} X - Xj \quad (4)$$

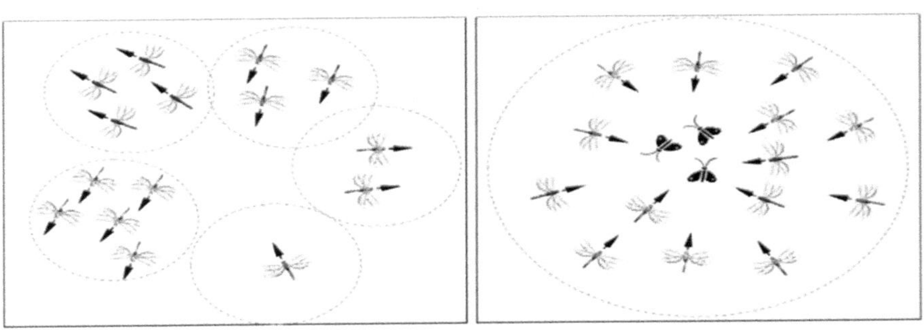

Fig. 1 Dynamic versus static dragonfly swarms

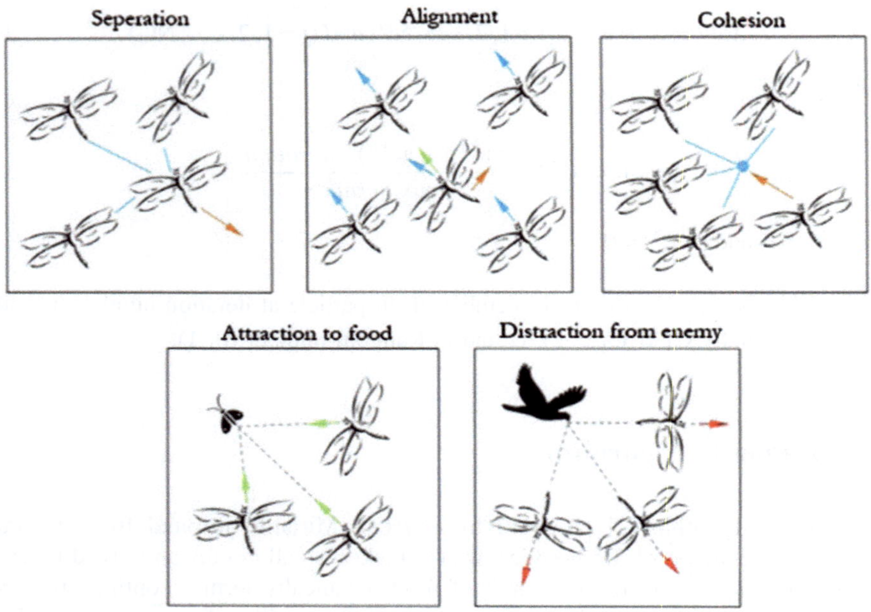

Fig. 2 Dragonfly Algorithm principle

Fig. 3 Flow chart for Dragonfly Algorithm

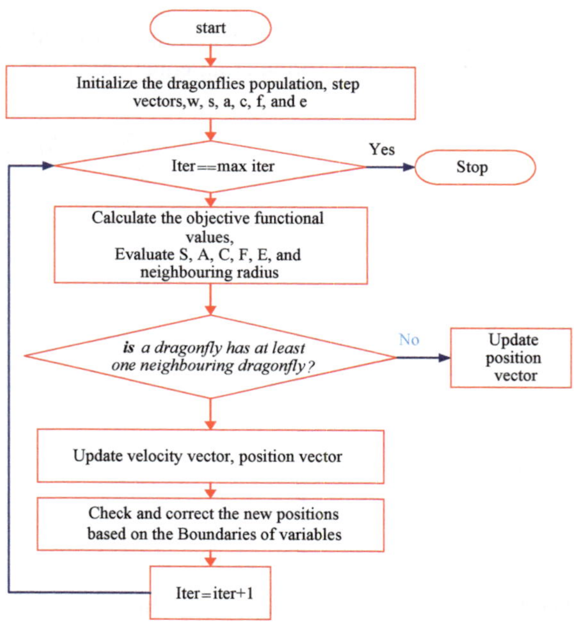

For Alignment part formulating equation:

$$\mathrm{Ai} = \frac{\sum_{j=1}^{N} Xj}{N} \tag{5}$$

For cohesion part formulating equation:

$$\mathrm{Ci} = \frac{\sum_{j=1}^{N} Xj}{N} - X \tag{6}$$

For Attraction towards a food source part formulating equation:

$$\mathrm{F_i} = \mathrm{X}^{+} - \mathrm{X} \tag{7}$$

For Attraction towards a food source part formulating equation:

$$\mathrm{E_i} = \mathrm{X}^{-} + \mathrm{X} \tag{8}$$

Step vector is formulating equation:

$$\Delta \mathrm{X}_{t+1} = (\mathrm{sS_i} + \mathrm{aA_i} + \mathrm{cC_i} + \mathrm{fF_i} + \mathrm{eE_i}) + \mathrm{w}\Delta \mathrm{X_t} \tag{9}$$

Position vector is calculated using equation

$$\mathrm{X}_{t+1} = \mathrm{X_t} + \Delta \mathrm{X}_{t+1} \tag{10}$$

Position of dragonfly updated using equation

$$\mathrm{X}_{t+1} = \mathrm{X_t} + \mathrm{Levy(X)} * \mathrm{X_t} \tag{11}$$

where, X = Location of the current individuals, N = Neighboring individuals, X^{+} = positions of food source, X^{-} = positions of enemy, s = separation weight, a = alignment weight, c = cohesion weight, f = food weight, e = enemy weight, w = inertia weight, t = iteration counter and d = dimension of position vectors that Levy flight step calculated.

4 The Hybrid PSO-DA Algorithm

A set of Hybrid PSO-WOA is combination of different PSO and WOA. Similarity between these two algorithm is that both the algorithms are swarming based. The most required condition for proposing new hybridization technique that the number of algorithms participating in hybridization must have at least one similar functionality. The drawback of PSO is the limitation to cover small search space while solving higher order or complex design problem due to limited range of inertia

Table 1 Internal parameters

Parameter name	Search agents no.	Max. iteration no.	No. of evolution
F_1, F_2, F_3	30	500	20–30
Over current relay co-ordination problem			
Case 1	30	100	20
Case 2	50	100	50

weight. This problem can be easily tackled by Hybrid PSO-WOA as it extracts the quality characteristics of both PSO and WOA. Whale Optimizer algorithm is used for exploration phase as it uses logarithmic spiral function so it covers broader area in uncertain search space and all the parameters involved in this technique having range in between (0, 1) that extensively enhance the range of solution in uncertain search space. Because both of the algorithms are randomization techniques so we use term uncertain search space during the computation over the course of iteration from starting to maximum iteration limit. Exploration phase means capability of algorithm to try out large number of possible solutions. Position of particle that is responsible for finding the optimum solution of the complex non-linear problem is replaced with the position of whale that is equivalent to position of particle but highly efficient to move solution towards optimal one. WOA directs the particles faster towards optimal value, reduces computational time. As we know that PSO is a well-known algorithm that exploits the best possible solution from its unknown search space. So combination of best characteristic (exploration with WOA and exploitation with PSO) guarantees to obtain best possible optimal solution of the problem that also avoids local stagnation or local optima of problem. Hybrid PSO-WOA merges the best strength of both PSO in exploitation and WOA in exploration phase towards the targeted optimum solution. Internal parameters are given in Table 1.

$$v_{ij}^{t+1} = wv_{ij}^{t} + c_1 R_1 (Dragonfly\_Position^t - X^t) + c_2 R_2 (Gbest^t - X^t) \quad (12)$$

5 Simulation Results for Unconstraint Test Benchmark Function

Unconstraint benchmark test functions are solved using HPSO-DA algorithm. 8 benchmark test functions, Uni-modal (F_1), Multi-modal (F_2) and Fixed dimension multi-modal (F_3) are performing to verify the HPSO-DA algorithm in terms of exploration and exploitation. These test functions are shown in Table 2. Results are shown in Table 3, HPSO-DA algorithm able to given more competitive results compare to standard PSO and DA algorithm. The convergence characteristics of

Table 2 Unconstraint benchmark test functions

Function	Dim	Range	F_{min}
$f_1(x) = \sum_{k=1}^{n} (x_k)^2 * R(x)$	30	[−100, 100]	0
$F_5(x) = \frac{1}{4000}\sum_{i=1}^{n} x_i^2 - \prod_{i=1}^{n} \cos\left(\frac{x_i}{\sqrt{i}}\right) + 1 * R(x)$	30	[−600, 600]	0
$f_6(x) = \sum_{i=1}^{11} a_i - \left[\frac{x_i\left(b_i^2 + b_i x_2\right)}{b_i^2 + b_i x_3 + x_4}\right]^2$	4	[−5, 5]	0.00030

Table 3 Result of benchmark functions

Fun	Std. PSO			Std. DA			HPSO-DA		
	Ave	Best	S.D.	Ave	Best	S.D.	Ave	Best	S.D.
F_1	7.6E-11	7.5E-11	1.8E-12	5.234	0.4465	6.7739	3.4E-23	**8.3E-29**	4.8E-23
F_2	0.1970	0.1847	0.0174	0.365	0.2452	0.1699	1.3E-9	**0**	1.9E-09
F_3	0.0124	0.0023	0.0144	0.001	6.7E-4	7E-04	6.4E-4	**4.5E-04**	2.6E-04

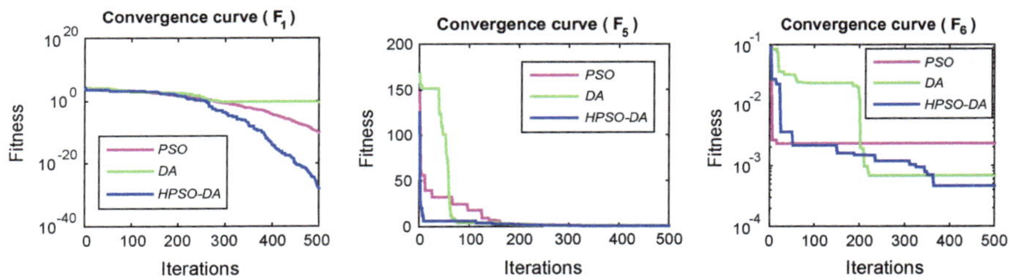

Fig. 4 Convergence characteristics of benchmark test functions

HPSO-DA is shown in Fig. 4. Search agent no. is 30 and maximum iteration no. is 500 used for all Unconstraint benchmark test functions.

6 Overcurrent Relay Co-ordination with Common Configuration in Power System

Over current relay used for primary and backup protection in distribution power systems. To minimize the total operating time of relays should be coordinated and set at the optimum values [23, 24]. A parallel feeder system for case 1 and multi loop distribution system for case 2 shown in Fig. 5.

All relays used in this paper are identical and they show the normal IDMT (Inverse Definite Minimum Time) characteristics represented in terms of equations are as follows:

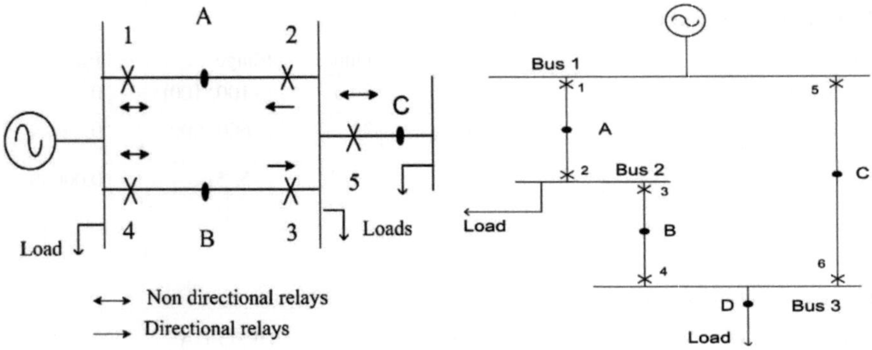

Fig. 5 A parallel feeder system and a multi-loop distribution system

$$t = \frac{0.14 * (TMS)}{PSM^{(0.02)} - 1} \quad (13)$$

where: t is the operating time of relay, PSM is plug setting multiplier and TMS represents time multiplier setting.

$$PSM = \frac{I_{relay}}{PS} \quad (14)$$

For linear problem PSM is constant, so t decreases to

$$t = \alpha_p * (TMS) \quad (15)$$

$$\alpha_p = \frac{0.14}{PSM^{(0.02)} - 1} \quad (16)$$

The target is to minimize the objective function given by:

$$F_{\min} = \sum_{p=1}^{n} \alpha_p * (TMS)_p \quad (17)$$

Case 1: Parallel Feeders (Fed from a Single End) [23]

The optimal results are given in Table 4. Figure 6 show the Convergence Characteristics of Overcurrent Relay Coordination for Parallel Feeder, fed from a single end.

Minimize

$$Z = \left\{ \begin{array}{c} 3.106X_1 + 6.265X_2 + 3.106X_3 \\ + 6.265X_4 + 2.004X_5 \end{array} \right\} \quad (18)$$

Table 4 Values of TMS for parallel feeder system, fed from a single end

Relay	TMS	DA	PSO	HPSO-DA
R_1	TMS_1	0.071297	0.11078	0.07806
R_2	TMS_2	0.025	0.025	0.025
R_3	TMS_3	0.078016	0.033037	0.045144
R_4	TMS_4	0.069706	0.096536	0.074826
R_5	TMS_5	0.051191	0.054176	0.052613
Total operating time		0.29521	0.319529	**0.275643**

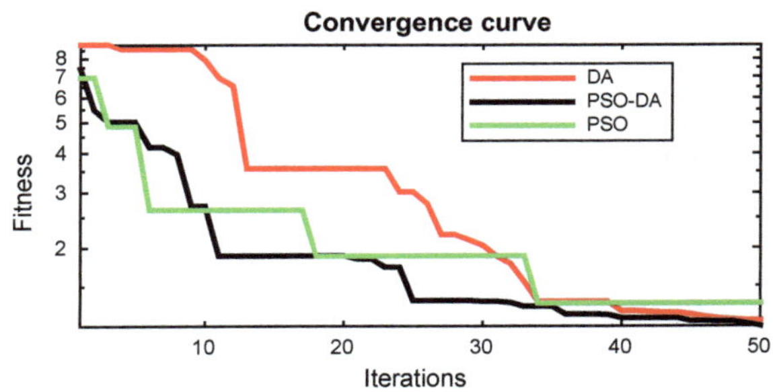

Fig. 6 Convergence characteristics of overcurrent relay coordination for parallel feeder, fed from a single end

Table 5 Values of TMS for single end fed distribution system (Six Over-Current Relays)

Relay	TMS	DA	PSO	HPSO-DA
R_1	TMS_1	0.63852	1.1944	0.025
R_2	TMS_2	0.82503	0.025	0.025
R_3	TMS_3	0.64416	1.2	0.025
R_4	TMS_4	0.48974	0.02999	0.031804
R_5	TMS_5	0.82453	0.068174	0.066269
R_6	TMS_6	1.2	0.025	0.025
Total operating time (s)		4.62198	2.542564	**0.198073**

$$\begin{aligned} &6.265X_4 - 3.106X_2 \geq 0.2, \quad 6.265X_1 - 3.106X_3 \geq 0.2, \\ &4.341X_1 - 2.004X_5 \geq 0.2, \quad 4.341X_4 - 2.004X_5 \geq 0.2, \end{aligned} \tag{19}$$

The constraints used are,

$$0.1 \leq 3.106X_1, 0.1 \leq 6.2655(X_2), 0.1 \leq 3.106(X_3), 0.1 \leq 6.2655(X_4), 0.1 \leq 2.004(X_5) \tag{20}$$

Case 2: Single End Fed Distribution System (Six Over-Current Relays)
The optimal values of TMS for single end fed distribution system (Six Over-Current Relays) given in Table 5. Convergence characteristics of overcurrent

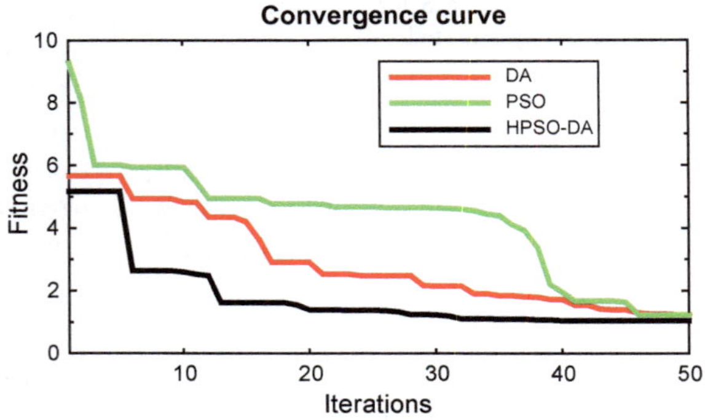

Fig. 7 Convergence characteristics of overcurrent relay coordination for single end fed distribution system (Six Over-Current Relays)

relay coordination for single end fed distribution system (Six Over-Current Relays) is shown in Fig. 7.

Minimize

$$Z = \left\{ \begin{array}{c} 102.4X_1 + 6.06X_2 + 98.75X_3 \\ + 24.4X_4 + 35.31X_5 + 11.53X_6 \end{array} \right\} \tag{19}$$

Constraints are used,

$$3.646X_1 \geq 0.1, 6.055X_2 \geq 0.1, 8.844X_3 \geq 0.1, 8.844X_4 \geq 0.1, 4.044X_5 \geq 0.1, 11.539X_6 \geq 0.1, \tag{20}$$

7 Conclusions

The drawback of PSO is the limitation to cover small search space while solving higher order or complex design problem due to constant inertia weight. This problem can be tackled with H PSO-DA as it extracts the quality characteristics of both PSO and DA. DA is used for exploration phase as it uses static swarming nature of Dragonflies so it covers broader area in uncertain search space. So DA directs the particles faster towards optimal value, reduces computational time. HPSO-DA is tested on twenty-three unconstrained and overcurrent relay as constrained problems. HPSO-DA gives better results in most of the cases and in some cases results are inferior that demonstrate the enhanced performance with respect to original PSO and DA.

Acknowledgements The authors would also like to thank Prof. Seyedali Mirjalili for his valuable comments and support. http://www.alimirjalili.com/DA.html.

References

1. J. Kennedy, R. Eberhart, Particle swarm optimization, in: Proceedings of the IEEE International Conference on Neural Networks, Perth, Australia, 1995, pp. 1942–1948.
2. Seyedali Mirjalili, "Dragonfly algorithm: a new meta-heuristic optimization technique for solving single-objective, discrete, and multi-objective problems", "The Natural Computing Applications Forum 2015", 30 April 2015. http://dx.doi.org/10.1007/s00521-015-1920-1.
3. A. Kaveh, S. Malakouti Rad "Hybrid Genetic Algorithm and Particle Swarm Optimization for the Force Method-Based Simultaneous Analysis and Design" Iranian Journal of Science & Technology, Transaction B: Engineering, Vol. 34, No. B1, pp. 15–34.
4. A. Kaveh and S. Talatahari, A Hybrid Particle Swarm and Ant Colony Optimization for Design of Truss Structures, Asian Journal of Civil Engineering (Building And Housing) Vol. 9, No. 4 (2008) Pages 329–348.
5. Iztok Fister Jr., Simon Fong, Janez Brest, and Iztok Fister, A Novel Hybrid Self-Adaptive Bat Algorithm, Hindawi Publishing Corporation the Scientific World Journal Volume 2014, Article ID 709738, 12 pages http://dx.doi.org/10.1155/2014/709738.
6. Gai-Ge Wang, Amir H. Gandomi, Amir H. Alavi, Suash Deb, A hybrid PBIL-based Krill Herd Algorithm, December 2015.
7. Gai-Ge Wang, Amir H. Gandomi, Amir H. Alavi, Suash Deb, A hybrid method based on krill herd and quantum-behaved particle swarm optimization, Neural Computing and Applications, 2015, doi: 10.1007/s00521-015-1914-z.
8. A. Tahershamsi, A. Kaveh, R. Sheikholeslami and S. Kazemzadeh Azad, An improved_rey algorithm with harmony search scheme for optimization of water distribution systems, Scientia Iranica A (2014) 21(5), 1591–1607.
9. Lihong Guo, Gai-Ge Wang, Heqi Wang, and Dinan Wang, An Effective Hybrid Firefly Algorithm with Harmony Search for Global Numerical Optimization, Hindawi Publishing Corporation The Scientific World Journal Volume 2013, Article ID 125625, 9 pages http://dx.doi.org/10.1155/2013/125625.
10. Gai-Ge Wang, Lihong Guo, Amir Hossein Gandomi, Guo-Sheng Hao, Heqi Wang. Chaotic krill herd algorithm. Information Sciences, Vol. 274, pp. 17–34, 2014.
11. GaigeWang and Lihong Guo, A Novel Hybrid Bat Algorithm with Harmony Search for Global Numerical Optimization, Hindawi Publishing Corporation Journal of Applied Mathematics Volume 2013, Article ID 696491, 21 pages http://dx.doi.org/10.1155/2013/696491.
12. A. Kaveh and S. Talatahari "Hybrid Algorithm of Harmony Search, Particle Swarm and Ant Colony for Structural Design Optimization" Z.W. Geem (Ed.): Harmony Search Algo. For Structural Design Optimization, SCI 239, pp. 159–198.
13. Gai-Ge Wang, Amir H. Gandomi, Xin-She Yang, Amir H. Alavi, A new hybrid method based on krill herd and cuckoo search for global optimization tasks. Int J of Bio-Inspired Computation, 2012, in press.
14. Ali Kaveh/Omid Khadem Hosseini, A hybrid HS-CSS algorithm for simultaneous analysis, design and optimization of trusses via force method, Civil Engineering 56/2 (2012) 197–212 doi: 10.3311/pp.ci.2012-2.06 web: http://www.pp.bme.hu/ ci Periodica Polytechnica 2012.
15. A. Kaveh, and A. Nasrollahi, Engineering Design Optimization Using A Hybrid PSO And HS Algorithm, Asian Journal Of Civil Engineering (Bhrc) Vol. 14, No. 2 (2013) Pages 201–223.
16. Gai-Ge Wang, Amir Hossein Gandomi, Amir Hossein Alavi, Guo-Sheng Hao. Hybrid krill herd algorithm with differential evolution for global numerical optimization. Neural Computing & Applications, Vol. 25, No. 2, pp. 297–308, 2014.
17. Gai-Ge Wang, Amir Hossein Gandomi, Xiangjun Zhao, HaiCheng Eric Chu. Hybridizing harmony search algorithm with cuckoo search for global numerical optimization. Soft Computing, 2014. doi: 10.1007/s00500-014-1502-7.

18. Gaige Wang, Lihong Guo, Hong Duan, Heqi Wang, Luo Liu, and Mingzhen Shao, Hybridizing Harmony Search with Biogeography Based Optimization for Global Numerical Optimization, Journal of Computational and Theoretical Nanoscience Vol. 10, 2312–2322, 2013.
19. S. Talatahari, R. Sheikholeslami, B. Farahmand Azar, and H. Daneshpajouh, Optimal Parameter Estimation for Muskingum Model Using a CSS-PSO Method, Hindawi Publishing Corporation Advances in Mechanical Engineering Volume 2013, Article ID 480954, 6 pages doi.org/10.1155/2013/480954.
20. A.H. Gandomi, X.S. Yang, S. Talatahari, A.H. Alavi, Metaheuristic Applications in Structures and Infrastructures, Elsevier, 2013.
21. A.H. Gandomi, A.H. Alavi, Krill Herd: a new bio-inspired optimization algorithm, Common Nonlinear Sci. Numer. Simul. 17 (12) (2012) 4831–4845.
22. Gandomi A.H. "Interior Search Algorithm (ISA): A Novel Approach for Global Optimization." ISA Transactions, Elsevier, 53(4), 1168–1183, 2014.
23. S. S. Gokhle, Dr. V. S. Kale, "Application of the Firefly Algorithm to Optimal Overcurrent Relay Coordination", IEEE Conference on Optimization of Electrical and Electronic equipment, Bran, 2014.
24. I.N. Trivedi, S.V. Purani, Pradeep Jangir, "Optimized over-current relay coordination using Flower Pollination Algorithm", "Advance Computing Conference (IACC), 2015 IEEE International", pages 72–77.

Short-Term Electricity Price Forecasting Using Hybrid SARIMA and GJR-GARCH Model

Vipin Kumar, Nitin Singh, Deepak Kumar Singh and S.R. Mohanty

Abstract The liberalization of the power markets gained a remarkable momentum in the context of trading electricity as a commodity. With the upsurge in restructuring of the power markets, electricity price plays a dominant role in the current deregulated market scenario which is majorly influenced by the economics being governed. Electricity price has got great affect on the market and is used as a basic information device to evaluate the future markets. However, highly volatile nature of the electricity price makes it even more difficult to forecast. In order to achieve better forecast from any model, the volatility of the electricity price need to be considered. This paper proposes a price forecasting approach combining wavelet, SARIMA and GJR-GARCH models. The input price series is transformed using wavelet transform and the obtained approximate and detail components are predicted separately using SARIMA and GJR-GARCH model respectively. The case study of New South Wales electricity market is considered to check the performance of the proposed model.

Keywords Time series model · Electricity price forecasting · SARIMA ARIMA · Nonlinear model · Volatility · GARCH model GJR-GARCH model

V. Kumar · N. Singh (✉) · D.K. Singh · S.R. Mohanty
Department of Electrical Engineering, MNNIT, Allahabad, India
e-mail: nitins@mnnit.ac.in

V. Kumar
e-mail: rathorvipin@gmail.com

D.K. Singh
e-mail: yadav.k.deepak@gmail.com

S.R. Mohanty
e-mail: soumya@mnnit.ac.in

G.M. Perez et al. (eds.), *Networking Communication and Data Knowledge Engineering*, Lecture Notes on Data Engineering and Communications Technologies 3, https://doi.org/10.1007/978-981-10-4585-1_25

1 Introduction

In the past few decades the vertically integrated electric utility structure has been disintegrated and replaced by deregulated structure which provides more choices to the customers through competition. In deregulated market structure the profit of the generating companies depends upon competitive bidding and decisions on investment made by the generating companies. In order to develop the bidding strategies, making investment decision and hedging against risk the accurate electricity price forecast is important tool used by the various market players.

The market operator receives bids from the generating companies and consumers containing information regarding electricity price and demand, after receiving the bids the market operator (MO) clears the market according to supply demand curve based on the bids. The market is cleared at a price known as market clearing price (MCP). In the presence of transmission congestion, the price of providing electricity in congested zone is called zonal market clearing price (ZMCP) or locational marginal price (LMP) [1].

The bids of the generating companies should be close to the MCP in order to make profit. The bid price higher than MCP will not be selected by any consumer and bid price lower than MCP will result in loss. This makes the accurate electricity price forecast an important tool for the generating companies to maximize their profits. Comprehensive literature does exist in the context of electricity price forecasting. Among the various existing models/techniques for EPF, the time series and soft computing based models have gained wide popularity among the researchers. However, one of the major drawbacks of the existing literature is to not include the volatility and the considered statistics usually takes mean and variance as constant value. On the contrary different statistical parameters are the function of environmental parameters which needs to be adopted in order to make price forecasting more accurate and efficient.

In order to tackle this problem the artificial neural network (ANN) is considered as a strong candidate of pattern classifier for estimating the parameters accurately without bothering the actual model of the parameters. However, ANN suffers to certain limitation with the probability of being trapped in local minima and large convergence error due to the consideration of large data with under fitting and over fitting of the weights. A three layered artificial neural network for predicting electricity price is shown in [2]. A neural network based approach for predicting system marginal prices is shown in [3]. The influence of different input parameters on the architecture selection of the ANN is presented in [4]. Some important ANN based models with preprocessing of the input data have been presented in [5–7].

Models based on time series analysis models such as autoregressive integrated moving average (ARIMA) [8, 9], multivariate transfer function (TF) models and dynamic regression (DR) models [10], multivariate time series model based on wavelet transform [11], generalized autoregressive conditional heteroskedasticity (GARCH) models [12], have also been proposed. The major shortcoming with these models is that they require lot of information about the system before predicting the future values which make them computationally inefficient.

The linear time series models are not capable of capturing the volatility present in the electricity price which makes them insufficient for predicting the future electricity prices [8]. However, the non-linear time series models are good in capturing the volatility present in the electricity price series but they are highly complex. A new hybrid model with combination of linear time series model, i.e., SARIMA and nonlinear time series model, i.e., GJR-GARCH is used for forecasting the NSW electricity price. The proposed model is capable of capturing the volatility present in the time series and at the same time has lesser complexity due to use of linear time series models.

Wavelet transform have been utilized to transform the input price series for providing robustness to the forecasting models [11–13]. The ill behaved price series is decomposed into a set of constitutive series by using wavelet transform, individually approximate (A) and detailed series (D) are then transformed back to the time domain and then each constitutive series based on their feature is predicted using an appropriate time series model. The high frequency components, i.e., the detail series which essentially contains the information regarding the price spike are predicted using GJR-GARCH model and the low frequency component i.e., approximate series which essentially contains the information regarding electricity price is predicted using SARIMA model.

2 The Proposed Approach

2.1 *Wavelet Transform*

The wavelet transform have gained lot of attention in the recent past in the field of electricity price forecasting due to their ability to detect the spikes present in the price series. The time series is usually not the deterministic series, but most of the researchers have considered it as a stationary while others have modeled it as deterministic function with white noise. The de-noising capability of WT is exploited for improving the characteristics of ill-behaved price series for obtaining better forecasting accuracy.

Wavelet transform decomposes the original signal into flexible function called wavelets which are localized in time and frequency [14]. Wavelet transform is a mathematical tool that decomposes the time series into a set of constitutive series called approximate and detail series. The wavelet transforms are very useful for the non-stationary data, i.e., data with non-constant mean and autocorrelation.

The mother wavelet is adopted as wavelet prototype and the temporal analysis is done with a high frequency version of the mother wavelet whereas the frequency analysis is done with the low frequency version of the mother wavelet. The wavelet transform for decomposition of the continuous signal is called continuous wavelet

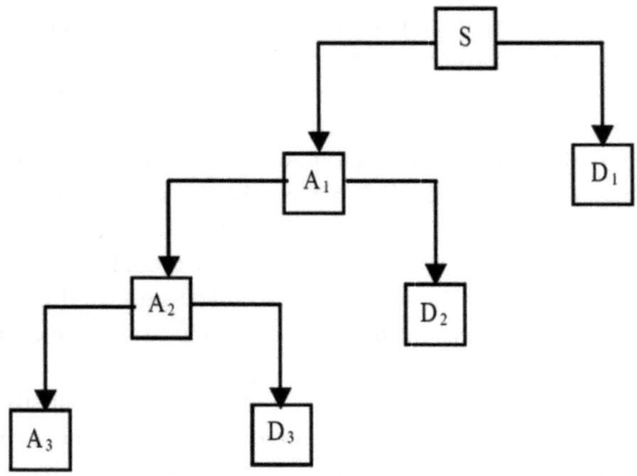

Fig. 1 Three level wavelet decomposition ($S = A_3 + D_3 + D_2 + D_1$)

transform (CWT) whereas for the discrete time signal is called discrete wavelet transforms (DWT). The time signal is decomposed into several other signals with multiple resolution, this is called as multi-resolution signal decomposition and it is proposed by Mallat [15]. The original signal can be recovered without loss of information by taking the inverse wavelet transform [16].

The wavelet transform is implemented using a multi-resolution pyramidal decomposition technique, i.e., a recorded digitized time signal S(n) can be analyzed into its detailed $cD_1(n)$ and approximated $cA_1(n)$ signals by using high pass and low pass filters respectively. Consequently, the detailed signal components are the high frequency component of the signal whereas the approximated signal is the low frequency component of the original signal. There exist various mother wavelets in the literature but among them Haar wavelet, Meyer wavelet, Coiflet wavelet, Daubechies wavelet and Morlet wavelet are some of the popular mother wavelet's used by the researchers. The level 3 wavelet decomposition is shown in the Fig. 1.

DWT is more efficient than and just as accurate as the CWT. The DWT for any discrete signal can be defined as (1).

$$W(m,n) = 2^{-(m/2)} \sum_{t=0}^{T-1} f(t)\phi\left(\frac{t-n2^m}{2^m}\right) \tag{1}$$

where, T is the length of the signal $f(t)$, the integer m and n represent the scaling and translation parameters, ($a = 2^m$, $b = n2^m$) and t is the discrete time index.

The proposed work utilizes the discrete wavelet transform (DWT) which is most suited for time series analysis. The multi-resolution algorithm proposed by Mallat is used for decomposing the signal into approximate and detailed series. The Daubechies (db) mother wavelet function of order 2 and 3 level decomposition is used for decomposing the original price series into a set of approximate (A_3) and three detailed series (D_1, D_2 and D_3).

2.2 SARIMA Model

The Seasonal Auto Regressive Integrated Moving Averge (SARIMA) model is the extension of popular Auto Regressive Integrated Moving Average (ARIMA) model it takes the seasonality present in the time series into consideration. The SARIMA model is widely used for forecasting the seasonal non-stationary time series. Often, time series possess a seasonal component that repeats after fixed number of observations. The SARIMA model is the seasonal version of the ARIMA model which is incapable of handling the seasonality present in the time series. SARIMA model is formed by including additional seasonal terms in the ARIMA models as given in (2):

$$SARIMA \quad \underbrace{(p, d, q)}_{\substack{\uparrow \\ \text{(Non seasonal part)}}} \quad \underbrace{(P, D, Q)_m}_{\substack{\uparrow \\ \text{(seasonal part)}}} \tag{2}$$

where m represents the number of periods per season. The uppercase notations are used for the seasonal parts of the model, and lowercase notation for the non-seasonal parts of the model.

2.3 GJR-GARCH Model

The Generalized Auto Regressive Conditional Heteroskedasticity (GARCH) models were successfully applied to model the relation between conditional variance and risk involved in bidding due to variance, (i.e. volatility) but these models have got certain limitations like:

- It might not give desired results when the return volatility respond asymmetrically with respect to the negative or positive shocks.
- GARCH model impose parameter restrictions that are often violated by estimated coefficients that may unduly restrict the dynamics of the conditional variance process.
- Interpretation of shocks to conditional variance "persist" or not is cumbersome in these models since, the usual norms for measuring persistence do not agree [17].

The volatility models suggested for coping up with the asymmetric effects are the exponential GARCH (EGARCH) [17] and the Glosten, Jagannathan and Runkle GARCH (GJR-GARCH) Model [18]. Later the performance of the GJR-GARCH model was found out to be better than the EGARCH model by the Marcucci [19].

Considering a return series as $r_t = \mu + \varepsilon_t$, where μ is the expected return and ε_t is the zero-mean white noise process, the series ε_t can be assumed to present conditional heteroskedasticity, the GJR-GARCH model assumes a specific parametric form for this conditional heteroskedasticity. The specific GJR GARCH (p, q) model can be written as (3) and (4), it can be generalized to incorporate more lags in the conditional variance, the generalize GJR-GARCH (p, q) model can be written as (5).

$$\sigma_t^2 = \omega + (\alpha + \gamma I_{t-1})\varepsilon_{t-1}^2 + \beta\sigma_{t-1}^2 \tag{3}$$

where,

$$I_{t-1} = \begin{Bmatrix} 0 & if & r_{t-1} \geq \mu \\ 1 & if & \mathrm{r}_{t-1} < \mu \end{Bmatrix} \tag{4}$$

$$\sigma^2 = \omega + \sum_{i=1}^{p} (\alpha_i + \gamma_i I_{t-1})\varepsilon_{t-1}^2 + \sum_{j=1}^{q} \beta_j \sigma_{t-j}^2 \tag{5}$$

The parameters $(\mu, \omega, \alpha, \gamma, \beta)$ are calculated by maximizing the log likelihood. The value of the lags (i.e. p and q) can be selected using the Bayesian Information Criterion (BIC) or Akaike Information Criterion (AIC), for the proposed model the value of p and q is chosen as 1 (i.e., p = 1 and q = 1) since, this option best fits the electricity price series of NSW electricity market.

2.4 Hybrid SARIMA-GJR GARCH Model

The price series is transformed using wavelet transform into a better behaved constitutive series. The constitutive series obtained is transformed into time domain by taking inverse wavelet transform. Thus, the original input price signal is transformed and reconstructed into approximation and detail components. Daubechies (db) wavelet of order 2 is used as the mother wavelet in the proposed work, as it provides an appropriate tradeoff between wave-length and smoothness. The decomposition level is chosen as three, it is based on the skewness and kurtosis analysis of the original price signal and moreover, it represents the price series in a more thorough and meaningful way than others.

An appropriate time series model, i.e., SARIMA or GJR-GARCH is chosen for predicting the obtained constitutive series based on their features. The low frequency component, i.e., approximation series (A3) which essentially follows the trend of the signal and contains the information regarding electricity price is therefore predicted using SARIMA model.

The high frequency component, i.e., detail series, contains information regarding noise content of the original signal and contains information regarding spike and change in price due to the bidding strategies used by the operator and variation in other influential factors. It contains the information regarding volatility of the price signal and therefore it best suited to be predicted by the GJR-GARCH model.

This hybrid model is an effective way to forecast electricity price as it can capture all the characteristics present in the price series in a true sense. The complete procedure is shown using Fig. 2, and a detailed explanation of above procedure is as follows:

(1) The original price signal Pt is decomposed using Daubechies (db2) wavelet transform into approximation series a3 and detail series d1, d2, d3, respectively.
(2) The transformed version of the original series is obtained by reconstructing the approximate series and detailed series using inverse wavelet transform series $a3_t$, $d1_t$, $d2_t$, $d3_t$ are reconstructed individually and denoted as $A3_t$, $D1_t$, $D2_t$, $D3_t$, the original series can be defined as (6)

$$p_t = A3_t + D3_t + D2_t + D1_t \tag{6}$$

(3) The hybrid SARIMA+GJR-GARCH model is used to predict the future electricity price of New South Wales electricity market (NSW). The future values of $A3_t$ are predicted using SARIMA model and GJR-GARCH model is used for predicting the future values of detail series (including $D1_t$, $D2_t$ and $D3_t$).
(4) The final predicted price is obtained by taking summation of the prediction of series $A3_t$, $D1_t$, $D2_t$, $D3_t$ respectively.

3 Results and Discussion

The forecasting ability of the proposed method is verified for the different seasons of the New South Wales (NSW) electricity market [20]. For the performance evaluation of the proposed method, mean absolute percentage error (MAPE) is chosen as the performance indices as shown in (7)

$$MAPE = \frac{1}{n} \sum_{t=1}^{n} \left| \frac{e_t}{y_t} \right| \times 100 \tag{7}$$

where $e_t = y_t - x_t$, is the forecast error and n is the size of test set, and y_t is the actual value of the price and x_t is the forecasted value of price. The performance of the proposed method is compared with the performance of the other existing methods. The results obtained are shown using Tables 1 and 2, Figs. 3, 4, 5 and 6.

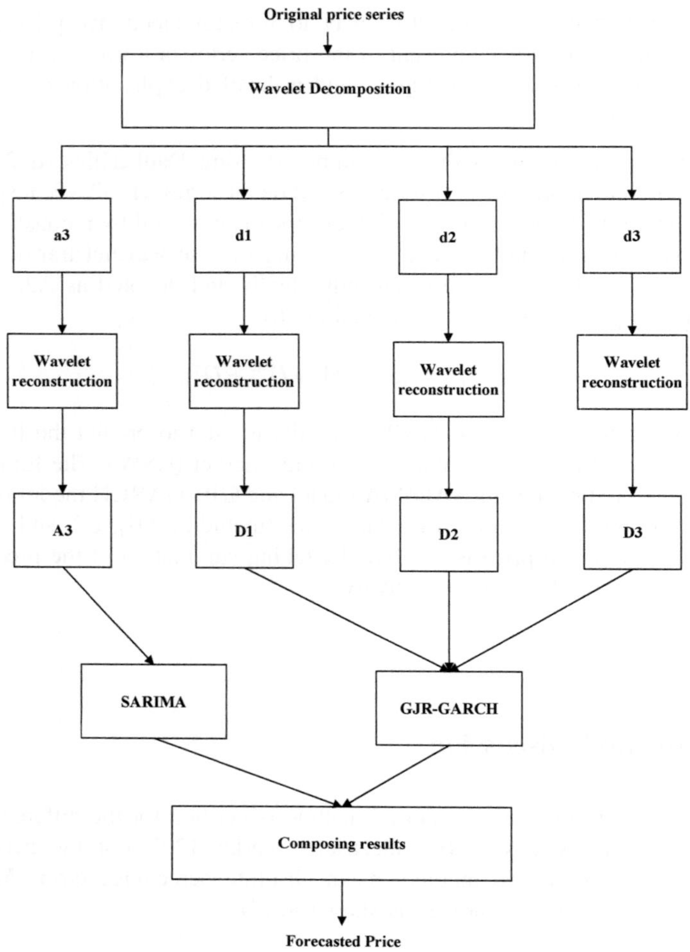

Fig. 2 Block diagram of the proposed model

Table 1 MAPE (%) for the four test week of NSW electricity market in year 2014

Season	ARIMA	SARIMA	GJR-ARCH	Proposed method
Winter	7.62	6.21	6.52	1.18
Spring	8.30	6.76	7.39	1.37
Summer	10.84	8.89	8.90	2.53
Fall	11.32	9.15	10.66	2.96
Average	9.52	7.75	8.36	2.55

Table 2 MAPE (%) for 12 weeks of NSW market in 2014

Month	SARIMA	SARIMA-GJR GARCH	WT-SARIMA	Proposed method
January	7.25	6.56	4.32	1.96
February	5.72	5.12	3.38	1.06
March	6.25	5.86	4.75	1.32
April	9.35	8.77	5.98	2.15
May	7.13	6.46	4.16	1.89
June	6.88	6.14	4.82	1.42
July	7.10	6.87	5.13	1.75
August	8.32	7.98	6.64	1.92
September	9.27	6.11	7.21	2.06
October	7.39	5.27	6.88	1.84
November	9.83	8.46	5.69	2.13
December	11.75	10.68	7.84	2.55
Average	8.02	7.02	5.56	1.83

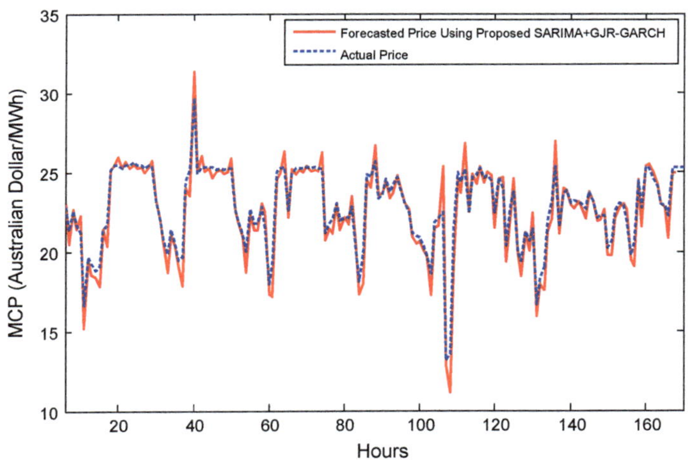

Fig. 3 Results of the price forecast for the Winter season of NSW market

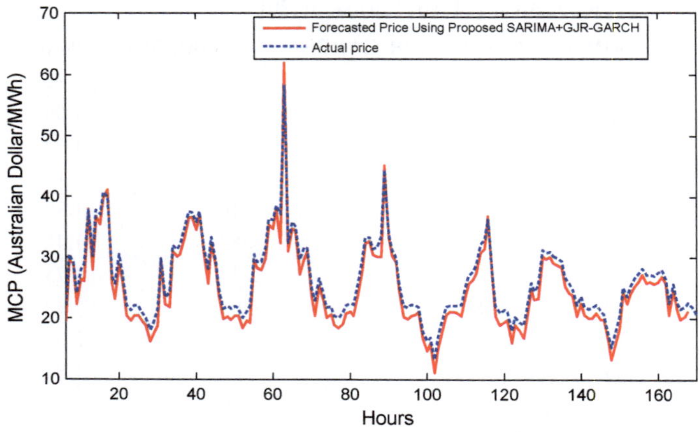

Fig. 4 Results of the price forecast for the Spring season of NSW market

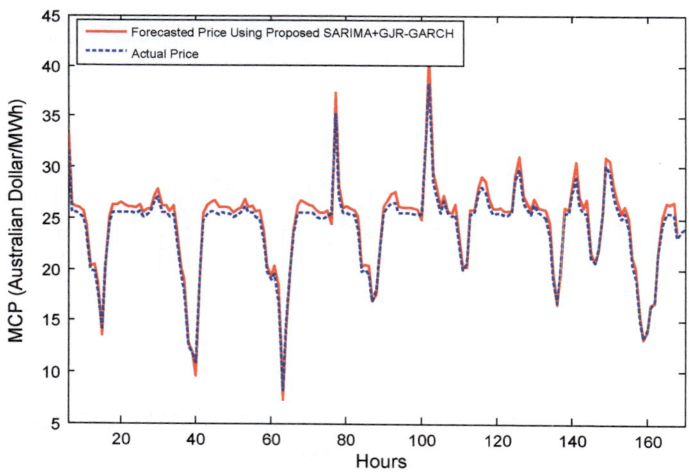

Fig. 5 Results of the price forecast for the Summer season of NSW market

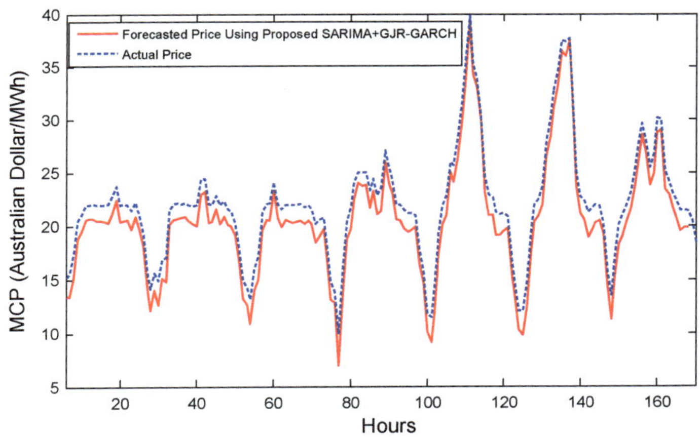

Fig. 6 Results of the price forecast for the Fall season of NSW market

4 Conclusion

A hybrid forecasting model combining a linear, i.e., SARIMA and nonlinear, i.e., GJR-GARCH model is proposed in the presented work. The wavelet transform in used for converting the input price series into constitutive series with better characteristics to improve the forecasting ability of the model. The transformed price series is predicted using SARIMA and GJR-GARCH model individually. The proposed model is capable of considering the volatility present in the price series in order to predict the future electricity price. The proposed hybrid model is used for forecasting the MCP of the NSW electricity market. The results of the proposed model are compared with some of the existing price forecasting models based on the value of MAPE and the performance of the proposed hybrid model is found better than the other existing models.

References

1. V. P. Gountis and A. G. Bakirtzis, "Bidding Strategies for Electricity Producers in a Competitive Electricity Marketplace," *IEEE Transactions on Power Systems*, vol. 19, no. 1, pp. 356–365, Feb. 2004.
2. B. R. Szkuta, L. A. Sanabria, and T. S. Dillon, "Electricity price short-term forecasting using artificial neural networks," *Power Systems, IEEE Transactions on*, vol. 14, no. 3, pp. 851–857, 1999.
3. D. Singhal and K. S. Swarup, "Electricity price forecasting using artificial neural networks," *International Journal of Electrical Power & Energy Systems*, vol. 33, no. 3, pp. 550–555, Mar. 2011.
4. Q. Tang and D. Gu, "Day-Ahead Electricity Prices Forecasting Using Artificial Neural Networks," 2009, pp. 511–514.

5. A. Azadeh, S. F. Ghadrei, and B. P. Nokhandan, "One day-ahead price forecasting for electricity market of Iran using combined time series and neural network model," 2009, pp. 44–47.
6. S. Anbazhagan and N. Kumarappan, "Day-Ahead Deregulated Electricity Market Price Forecasting Using Recurrent Neural Network," *IEEE Systems Journal*, vol. 7, no. 4, pp. 866–872, Dec. 2013.
7. A. M. González, A. M. S. Roque, and J. García-González, "Modeling and forecasting electricity prices with input/output hidden Markov models," *Power Systems, IEEE Transactions on*, vol. 20, no. 1, pp. 13–24, 2005.
8. J. Contreras, R. Espinola, F. J. Nogales, and A. J. Conejo, "ARIMA models to predict next-day electricity prices," *IEEE Transactions on Power Systems*, vol. 18, no. 3, pp. 1014–1020, Aug. 2003.
9. F. J. Nogales, J. Contreras, A. J. Conejo, and R. Espinola, "Forecasting next-day electricity prices by time series models," *IEEE Transactions on Power Systems*, vol. 17, no. 2, pp. 342–348, May 2002.
10. F. J. Nogales and A. J. Conejo, "Electricity price forecasting through transfer function models," *Journal of the Operational Research Society*, pp. 350–356, 2006.
11. H. Xu and T. Niimura, "Short-term electricity price modeling and forecasting using wavelets and multivariate time series," presented at the Power Systems Conference and Exposition, 2004. IEEE PES, 2004, pp. 208–212 vol. 1.
12. R. C. Garcia, J. Contreras, M. van Akkeren, and J. B. C. Garcia, "A GARCH Forecasting Model to Predict Day-Ahead Electricity Prices," *IEEE Transactions on Power Systems*, vol. 20, no. 2, pp. 867–874, May 2005.
13. A. J. Conejo, M. A. Plazas, R. Espinola, and A. B. Molina, "Day-Ahead Electricity Price Forecasting Using the Wavelet Transform and ARIMA Models," *IEEE Transactions on Power Systems*, vol. 20, no. 2, pp. 1035–1042, May 2005.
14. M. T. I. S. Al Wadia and M. Tahir Ismail, "Selecting wavelet transforms model in forecasting financial time series data based on ARIMA model," *Applied Mathematical Sciences*, vol. 5, no. 7, pp. 315–326, 2011.
15. S. G. Mallat, "A theory for multiresolution signal decomposition: the wavelet representation," *IEEE Transactions on Pattern Analysis and Machine Intelligence*, vol. 11, no. 7, pp. 674–693, Jul. 1989.
16. I.-K. Yu, Chang-Il Kim, Y. H. Song, "A Novel Short-Term Load Forecasting Technique Using Wavelet Transform Analysis," *Electric Machines & Power Systems*, vol. 28, no. 6, pp. 537–549, Jun. 2000.
17. D. B. Nelson, "Conditional Heteroskedasticity in Asset Returns: A New Approach," *Econometrica*, vol. 59, no. 2, p. 347, Mar. 1991.
18. L. R. Glosten, R. Jagannathan, and D. E. Runkle, "On the Relation between the Expected Value and the Volatility of the Nominal Excess Return on Stocks," *The Journal of Finance*, vol. 48, no. 5, pp. 1779–1801, Dec. 1993.
19. J. Marcucci, "Forecasting stock market volatility with regime-switching GARCH models," *Studies in Nonlinear Dynamics & Econometrics*, vol. 9, no. 4, 2005.
20. "AEMO Home." [Online]. Available: http://www.aemo.com.au/. [Accessed: 18-Aug-2015].

Erratum to: Design of Low Cost Blood Glucose Sensing System Using Diffused Reflectance Near-Infrared Light

Jyoti Yadav, Asha Rani, Vijander Singh and Bhaskar Mohan Murari

Erratum to: Chapter "Design of Low Cost Blood Glucose Sensing System Using Diffused Reflectance Near-Infrared Light" in: G.M. Perez et al. (eds.), *Networking Communication and Data Knowledge Engineering*, Lecture Notes on Data Engineering and Communications Technologies 3, https://doi.org/10.1007/978-981-10-4585-1_17

In the original version of the book, Reference 17 "G. Guevara, Joint optical–electrical technique for non-invasive glucose monitoring, *Mex. J. Phys.* 56(5): 430–434, 2010. doi:http://www.scielo.org.mx/pdfrmf/v56n5/v56n5a12" has been now changed as "E. Guevara, F.J. González, Joint optical–electrical technique for non-invasive glucose monitoring, *Mex. J. Phys.* 56(5): 430–434, 2010. doi:http://www.scielo.org.mx/scielo.php?script=sci_arttext&pid=S0035-001X201000050001" in chapter "Design of Low Cost Blood Glucose Sensing System Using Diffused Reflectance Near-Infrared Light", which is a belated correction.

The updated online version of this chapter can be found at https://doi.org/10.1007/978-981-10-4585-1_17

G.M. Perez et al. (eds.), *Networking Communication and Data Knowledge Engineering*, Lecture Notes on Data Engineering and Communications Technologies 3, https://doi.org/10.1007/978-981-10-4585-1_26

Author Index

G.M. Perez et al. (eds.), *Networking Communication and Data Knowledge Engineering*, Lecture Notes on Data Engineering and Communications Technologies 3, https://doi.org/10.1007/978-981-10-4585-1

MIX
Papier aus verantwortungsvollen Quellen
Paper from responsible sources
FSC® C105338

If you have any concerns about our products,
you can contact us on
ProductSafety@springernature.com

In case Publisher is established outside the EU,
the EU authorized representative is:
Springer Nature Customer Service Center GmbH
Europaplatz 3, 69115 Heidelberg, Germany

Printed by Libri Plureos GmbH
in Hamburg, Germany